Entomology in Focus

Volume 7

Insects are fundamentally important in the ecology of terrestrial habitats. What is more, they affect diverse human activities, notably agriculture, as well as human health and wellbeing. Meanwhile, much of modern biology has been developed using insects as subjects of study.

To reflect this, our aim with *Entomology in Focus* is to offer a range of titles that either capture different aspects of the diverse biology of insects or their management, or that offer updates and reviews of particular species or taxonomic groups that are important for agriculture, the environment or public health.

The series results from an agreement between Springer and the Entomological Society of Brazil (SEB) and as such may lean towards tropical entomology. The aim throughout is to provide reference texts that are simple in their conception and organization but that offer up-to-date syntheses of the respective areas, offer suggestions of future directions for research (and for management where relevant) and that don't shy away from offering considered opinions.

Editorial Committee

Series Editor **Sam Elliot** is Associate Professor in Entomology at the Universidade Federal de Viçosa (Brazil), also coordinates the Postgraduate Programme in Entomology currently rated maximally by the relevant authority in Brazil (CAPES) and is Associate Editor at *Ecology and Evolution*. He works on diverse aspects of insect-microbe interactions, with emphases on leafcutter ants, noctuid caterpillars, triatomine bugs, entomopathogenic fungi and microbial control of pests.

Adam Hart is Professor of Science Communication at the University of Gloucestershire (UK). His particular interest is in social insects but he has written and broadcasted on a broad range of biological subjects. He presents documentaries for BBC Radio 4, BBC4 and BBC2, as well as the weekly BBC radio programme Science in Action.

Eugenio Oliveira is Assistant Professor in Entomology at the Universidade Federal de Viçosa (Brazil), and scholar researcher of the Brazilian National Council of Scientific and Technologic Development (CNPq). He has also working as Associate Editor at the *journals Neotropical Entomology* and *Invertebrate Neuroscience*. He works principally on insect neurophysiology, applying this in particular to entomological/agricultural questions.

Ken Wilson is Professor of Evolutionary Ecology at Lancaster University (UK), and is Executive Editor of the *Journal of Animal Ecology*. He is particularly interested in host-parasite interactions and investigates these in invertebrate and vertebrate hosts. Noctuid caterpillars, especially armyworms, have been one of his main model systems and he is currently working on their ecology and biocontrol in Africa.

Walter R. Terra • Clelia Ferreira • Carlos P. Silva

Molecular Physiology and Evolution of Insect Digestive Systems

Springer

Walter R. Terra
Instituto de Química
Universidade de São Paulo
São Paulo, Brazil

Clelia Ferreira
Instituto de Química
Universidade de São Paulo
São Paulo, Brazil

Carlos P. Silva
Departamento de Bioquímica
Universidade Federal de Santa Catarina
Florianópolis - SC, Brazil

ISSN 2405-853X ISSN 2405-8548 (electronic)
Entomology in Focus
ISBN 978-3-031-39232-0 ISBN 978-3-031-39233-7 (eBook)
https://doi.org/10.1007/978-3-031-39233-7

This Springer imprint is published by the registered company Springer Nature Switzerland AG
The registered company address is: Gewerbestrasse 11, 6330 Cham, Switzerland

Paper in this product is recyclable.

Dedication: To our families

Preface

The whole field of insect midgut studies was covered in a book entitled *Biology of the Insect Midgut* edited by M.J. Lehane and P.F. Billingsley in 1996 (Chapman & Hall, London). Since then, highly efficient techniques of protein separation were developed and a great amount of molecular manipulation procedures were introduced or improved, like gene cloning and production of recombinant proteins, new-generation sequencing methods, and techniques of suppression of protein expression. The impact of the use of these methods on midgut studies was impressive, particularly regarding the molecular mechanisms underlying digestion and absorption and their evolution. These advances have been reviewed several times, but never in a comprehensive book form.

The aim of this book is to provide a balanced blend of introductory and specialized aspects of the molecular mechanisms underlying insect midgut buffering, digestion, nutrient absorption, and their changes along the evolution, ending with a discussion on new technologies of insect control based on the information gathered in the previous chapters.

The first part of the book (Chapters 1, 2, 3, and 4) comprises four major introductory subjects: the omics-based patterns of insect evolution, to be used to organize the findings in an evolutionary perspective; overview of the methods and results regarding the spatial organization of digestion and absorption; description of the types and chemistry of insect diets; ordinary digestive enzymes that degrade dietary molecules.

The second part of the book (Chaps. 5, 6, 7, and 8) describes the organization of midgut cells, the function of their microvillar membranes and extracellular layers, and, finally, their role in secretory processes. With this background, the function of membrane transporters in the absorption of nutrients and maintenance of midgut pH and water fluxes are detailed.

The third part of the book (Chaps. 8, 9, 10, 11, and 12) deals with special topics like endocrine regulation of midgut function; adaptations to overcome the action of digestive enzyme inhibitors; specializations to deal with plant, bacterial, and fungi cell walls; and, finally, the role of microorganisms in nutrition.

The last part of the book (Chaps. 13, 14, and 15) reviews the molecular mechanisms underlying digestion and nutrient absorption in the major insect orders and postulates how these processes changed along the evolution. The book ends with an overview of the techniques of insect control that employs the knowledge gathered along the midgut studies.

This book was written to be suitable for students and advanced scientists of entomology, biology, biochemistry, genetics, pest control specialists, and for those working with degradation of plant cell walls by insect or midgut microbial enzymes. Work done in the laboratories of the authors was financed by the São Paulo Research Foundation (FAPESP, Grant "Temático" no. 2017/08103-4), National Council of Research (CNPq), and National Institute of Science and Technology-Molecular Entomology (INCT-Entomologia Molecular).

São Paulo, Brazil
June 2023

Walter R. Terra
Clelia Ferreira
Carlos P. Silva

Contents

Chapter 1
Patterns of Insect Evolution

Abstract Insects are the most diverse living beings and their ancestors moved from the sea and colonized the land long before the chordates. Phylogenetic and fossil data were combined to detail the patterns of insect evolution. Insects able to flex their wings over the back (Neoptera) correspond to most of the insects and evolved along the major lineages: Polyneoptera, Condylognatha, and Holometabola. Polyneoptera includes Dictyoptera – cockroaches and termites that are omnivorous or wood feeders and the carnivorous mantids – and Orthoptera – the omnivorous crickets and grass-feeding grasshoppers. Condylognatha includes Hemiptera which are the only insects able to live entirely on plant sap such as aphids, cicadas, and spittlebugs; and others like bugs adapted to different diets. Holometabola is the most successful lineage with 86% of the insect species, have complete metamorphosis (larva, pupa, and adult) and comprises the major insect orders: Coleoptera (beetles); Hymenoptera (wasps, ants, and bees); Diptera (mosquitoes and flies); and Lepidoptera (butterflies and moths). Insects of these orders explore the most diverse food sources like other insects, stems, leaves and wood, seeds, keratin (like woolen carpets), pollen, nectar, fungi, and vertebrate blood. The major selective pressures affecting insect guts identified were: (a) adaptations to deal with large amounts of dilute fluid food, (b) adaptations to digesting plant and fungal cells as a result of horizontal transfer of genes from microorganisms and recruitment of lysosomal proteins as digestive enzymes, (c) adaptations to avoiding plant inhibitors by gene expansion and new functionalization, (d) and adaptations to avoiding prolonged exposure to natural enemies and to living in short-lived media by reduction of life span permitted by more efficient midguts.

1.1 Initial Considerations

This chapter overviews the patterns of insect evolution, highlighting the insect taxa and specific insects that are more important for midgut studies.

W. R. Terra et al., *Molecular Physiology and Evolution of Insect Digestive Systems*, Entomology in Focus 7, https://doi.org/10.1007/978-3-031-39233-7_1

Insects are the most diverse living beings and are found in practically all land surfaces, except in the extreme polar regions and in the highest mountain peaks (Daly et al. 1998). The success of the insects is measured by their large number of species amounting to about half of the known living species and to almost 75% of the animals. This success is supposed to rely on several of the insect's characteristics, such as its exoskeleton, flight capacity, small size, and ability to use diverse nutritional sources.

The insect exoskeleton made of repetitive segments and appendages pre-adapted them to walking on land, thus favoring their exit from the sea and the colonization of terrestrial environments much earlier than the chordates. Flight permitted the insects to escape from enemies and find mates, food, and sites to lay eggs. A small size allowed the insects to live in small places and have short generation time because less time is necessary to attain adulthood. Finally, the insect's capacity to rely on the most varied nutritional sources favors its adaptation to almost any environment (Daly et al. 1998).

A list of the major insect orders with common names of the most known insects is shown in Table 1.1.1. This table is expected to make it easier to follow the description of insect evolution. Non-winged insect ancestors, together with a few other arthropods, were the first animals to invade the land, approximately at the same time as plants in the Early Ordovician (circa 420 Ma) (Grimaldi and Engel 2005; Misof et al. 2014). The most primitive insects were wingless (Infraclass Apterygotha) and will not be considered here. The first winged insects (Infraclass Pterygotha) were characterized by wings that cannot be flexed over the back at rest (Paleoptera) and include Ephemeroptera (the mayflies) and the predatory Odonata (dragonflies and

Table 1.1 The major insect orders

Infraclass	Superorder	Order	Examples
Paleoptera	–	Odonata	Damselflies, dragonflies
		Ephemeroptera	Mayflies
Neoptera	Polyneoptera	Dictyoptera	Cockroaches, termites, praying mantids
		Orthoptera	Grasshoppers, crickets
		Phasmatodea	Stick and leaf insects
	Condylognatha	Hemiptera	Bugs, cicadas, aphids
		Thysanoptera	Thrips
	Psocodea	Phthiraptera	Lice
		Psocoptera	Booklice
	Neuropteroidea	Megaloptera	Alderflies, dobsonflies
		Coleoptera	Beetles
	Hymenopteroidea	Hymenoptera	Wasps, bees, ants
	Panorpoidea	Siphonaptera	Fleas
		Diptera	Mosquitoes, flies
		Lepidoptera	Moths, butterflies
		Trichoptera	Caddisflies

Paleoptera, Polyneoptera, and Condylognatha undergo incomplete metamorphosis, whereas Neuropteroidea and Panorpoidea have complete metamorphosis damselflies). Neoptera corresponds to most insects, and they have wing flexing that permits them to enter underneath environments like liter and bark.

Neoptera evolved along the major lineages Polyneoptera (Plecoptera, Orthoptera, Phasmatodea, and Dictyoptera); Condylognatha (Thysanoptera and Hemiptera); and a branch corresponding to Psocodea (Phthyraptera and Psocoptera) plus Holometabola (Fig. 1.1). The once proposed clade Paraneoptera, comprising Condylognatha and Psocodea, is paraphyletic because, as shown above, Condylognatha is a sister group of Psocodea+Holometabola. Holometabola includes Hymenoptera and a large grouping divided into two branches: Neuroptera+Coleoptera and Panorpoidea. Panorpoidea is divided into two sister clades: Trichoptera+Lepidoptera and Siphonaptera+Diptera (Fig. 1.1). The expansion of Hymenoptera, Diptera, and Lepidoptera occurred in parallel with the radiation of angiosperms in Early cretaceous (circa 100 Ma) (Misof et al. 2014).

Holometabola comprises 86%, Polyneoptera (mainly Orthoptera and Dyctioptera) 3.2%, and Condylognatha (mainly Hemiptera) 7.1% of the insect species. The evolutionary success of Holometabola is supposed to result from the fact that their young forms (larva) are adapted to ecological niches different from those of adults, thus avoiding their competition for the same food, as occurs with insects without complete metamorphosis (Polyneoptera, Condylognatha, and Psocodea). After this overview of insect evolution, the major insect orders will be reviewed, calling

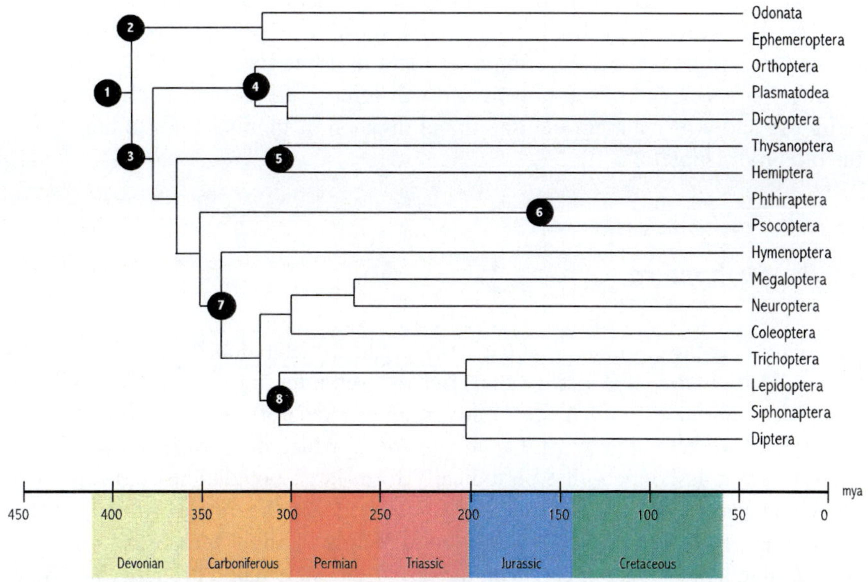

Fig. 1.1 Simplified phylogenetic tree of the relationships of the major insect orders. Number in nodes mean: 1, Pterygotha; 2, Paleoptera; 3, Neoptera; 4, Polyneoptera; 5, Condylognatha; 6, Psocodea; 7, Holometabola; 8, Panorpoidea. Geological periods: Devonian, Carboniferous, Permian, Triassic, Jurassic, Cretaceous. (Details in text)

attention to some remarkable species that are subject of studies of their digestive systems and, finally, their internal phylogenetics will be discussed.

1.2 Polyneoptera

1.2.1 Dictyoptera

Dictyoptera comprises Blattodea (cockroaches and termites) and Mantodea (mantis). Cockroaches are believed to be ancient insects arising in the Carboniferous. However, though modern families evolved since the Cretaceous and are usually omnivorous, some species only feed on dead wood (wood roaches) (Grimaldi and Engel 2005). Mantids are usually carnivorous with some species feeding on pollen as a complement (Beckman and Hurd 2003). Blattodea includes Blattidae, the most ancient group of modern cockroaches, exemplified by *Periplaneta americana* (American cockroach) and *Blata orientalis* (Oriental cockroach), and two other major branches: the branch Cryptocercidae-Termitidae, where one finds the termites that are no longer considered to be a separate order (Isoptera) (Inward et al. 2007), and the branch Ectobiidae (early Blatellidae)-Blaberidae. An example of Ectobiidae is *Blatella germanica* and of Blaberidae are the wood roaches *Panesthia cribata* and *Nauphoeta cineri* (Grimaldi and Engel 2005).

Lo et al. (2007) agree with Inward et al. (2007) that the order Isoptera should be dismissed. However, they think that the name Termitidae should not be employed to refer to all termites, because Termitidae is still in use to describe the more derived termites. We accepted this argument and will refer to all termites as the "taxon of termites" and will maintain the traditional division of families and subfamilies of the old order Isoptera.

1.2.2 Orthoptera

Orthoptera are the dominant group of chewing hemimetabolous insects. The ancestors of *Orthoptera* gave rise to the crickets (suborder Ensifera, exemplified by *Gryllodes sigillatus*), which are omnivorous or predatory, and the grasshoppers (suborder Caelifera like *Abracris flavolineata*), which feed mainly on grasses. Locusts are grasshoppers that periodically form large populations, attacking one and then swarming to other places. *Locusta migratoria*, the migratory locust, is the most widespread locust species being found mainly in all of Africa, Australia, and New Zealand. *Schistocerca gregaria*, the desert locust, is found primarily in Africa, Arabia, and West Asia. The phylogeny of Orthoptera is discussed by Zhang et al. (2013).

1.3 Condylognatha

1.3.1 *Thysanoptera*

Thysanoptera (thrips) is sister of the order Hemiptera and together forms the taxon Condylognatha. Thysanoptera have piercing-sucking mouthparts like Hemiptera, but less specialized for sucking fluids. The mouthparts are usually too short to tap into the vascular system of the plants. Feeding thrips perforate the surface of the plant tissues (their most common food) and insert their mouthparts in them. Then, in back and forth movements, fluid, particles of plants, or prey (in predatory thrips) are ingested (Grimaldi and Engel 2005).

1.3.2 *Hemiptera*

Hemiptera corresponds to about 7.1% of the insect species and are surpassed in number of species only by the holometabolous order of Coleoptera, Hymenoptera, Lepidoptera, and Diptera. The internal phylogenetic relationships of Hemiptera are summarized in Fig. 1.2 and discussed in detail by Johnson et al. (2018). Only

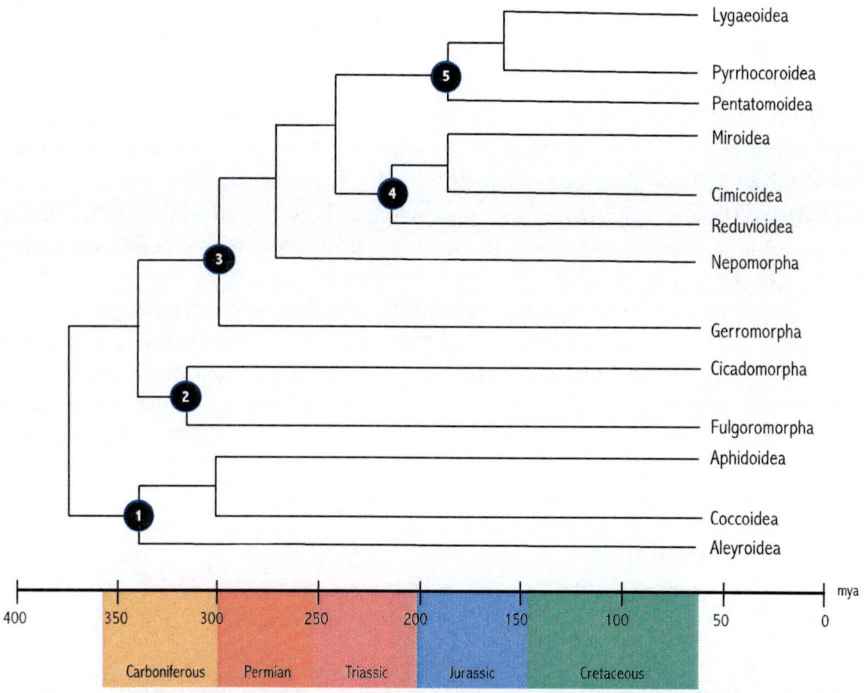

Fig. 1.2 Simplified phylogenetic tree of the relationships of the major Hemiptera taxa. Numbers in nodes: 1, Sternorrhyncha; 2, Auchenorrhyncha; 3, Heteroptera; 4, Cimicomorpha; 5, Pentatomomorpha. (Details in text)

Hemiptera developed a sap-sucking habit, though insects of many other orders have suitable piercing-sucking mouthparts. Thus, it is likely that the sap-sucking habit depends on modifications in the midguts that enable hemipterans to deal rapidly with large amounts of dilute fluid food.

Hemiptera ancestors on evolution gave origin to Sternorrhyncha (aphids and white flies), exemplified by *Acyrthosiphon pisum*, which together with Auchenorrhyncha (cicadas, spittlebugs, leafhoppers, and planthoppers), like *Homalodisca vitripennis and Nilapavarta lugens*, usually feed on plant sap, either from the phloem or xylem.

Phloem fluid sap is transported downward through the outer layers of the plant stems. It is usually rich in sugars, with significant amounts of free amino acids, potassium, and some organic acids. Some phloem fluids are rich in proteins, functioning in wound signaling and plugging the sieve elements to avoid nutrient loss. Other proteins are involved in defense against herbivorous and phloem feeders, exemplified by protease inhibitors and lectins (Kehr 2006; Zhang et al. 2012). Xylem fluid is transported upward through the inner layers of the plant stems, containing large amounts of potassium and trace amounts of amino acids and sugars (Lopez-Millan et al. 2000). There are also small quantities of proteins active in repair and defense that are taken up or produced by the roots (Buhtz et al. 2004).

Sternorrhyncha sucks phloem sap. A shift to xylem feeding, which is more internal than phloem, coincided with the origin of Cicadomorpha among Auchenorrhyncha (Johnson et al. (2018). Both Sternorrhyncha and Auchenorrhyncha excrete honeydew (excess sugar ingested by the insects), but Sternorrhyncha excrete more, often several times the body weight of the insect per day (Grimaldi and Engel 2005). The honeydew is collected by ants.

Heteroptera are adapted to different diets and have two major infraorders Cimicomorpha and Pentatomomorpha. Among the Cimicomorpha, the largest family are the Miridae (plant bugs), exemplified by *Lygus* sp. and the bed bug *Cimex lectularis.* Another large family of Cimicomorpha is Reduviidae, thought to be the basal group of the infraorder and its most studied representative is *Rhodnius prolixus.* The largest families of Pentatomomorpha are the Pentatomidae that have scent glands (e.g., *Nezara viridula*) and virtually all feed on plant sap or fruit, rather than on vegetative tissues, whereas the Lygaeoidea (e.g., *Oncopeltus fasciatus*, the common milkweed bug, and *Halyomorpha halys*, the brown marmorated stink bug) and Pyrrhochoridae (e.g., *Dysdercus peruvianus*) are seed-sucker bugs (Weirauch et al. 2019).

1.4 Holometabola

1.4.1 Coleoptera

Coleoptera is the most successful order of Holometabola, corresponding to about 35% of the insect species. This success probably is a consequence of Coleoptera being the first land insects with complete metamorphosis to massively occupy hidden niches on the ground (Grimaldi and Engel 2005). Furthermore, there was an adaptive radiation of specialized herbivorous beetles (Phytophaga), following the horizontal transfer of microbial genes encoding plant cell wall–digesting enzymes (see Chap. 10). Those enzymes made possible beetle leaf and seed mining and stem and wood boring.

The internal phylogenetic relationships of Coleoptera are summarized in Fig. 1.3 and discussed in detail by Zhang et al. (2018) and McKenna et al. (2019). The beetle ancestor gave rise to the major suborders Adephaga and Polyphaga with the infraorders Elateriformia, Staphyliniformia, Scarabaeiformia, Bostrichiformia, and Cucujiformia (Zhang et al. 2018; McKenna et al. 2019). Adephaga are the largely predaceous ground and water beetles comprising approximately 10% of all beetles, most of which are from the family Carabidae (e.g., *Pheropsophus aequinoctialis*).

Among Polyphaga, Elateriformia is sister of a branch comprising two taxons: Sta phyliniformia+Scarabaeiformia and Bostrichiformia+Cucujiformia. The major superfamilies of Elateriformia are Buprestoidea, with a single family of metallic wood-boring beetles (Buprestidae, jewel beetles), and Elateroidea, exemplified by the luminescent beetles of the family Elateridae (click beetles as *Pyrophorus divergens* and *Pyrearinus termitilluminans*) and Lampyridae (fireflies exemplified by

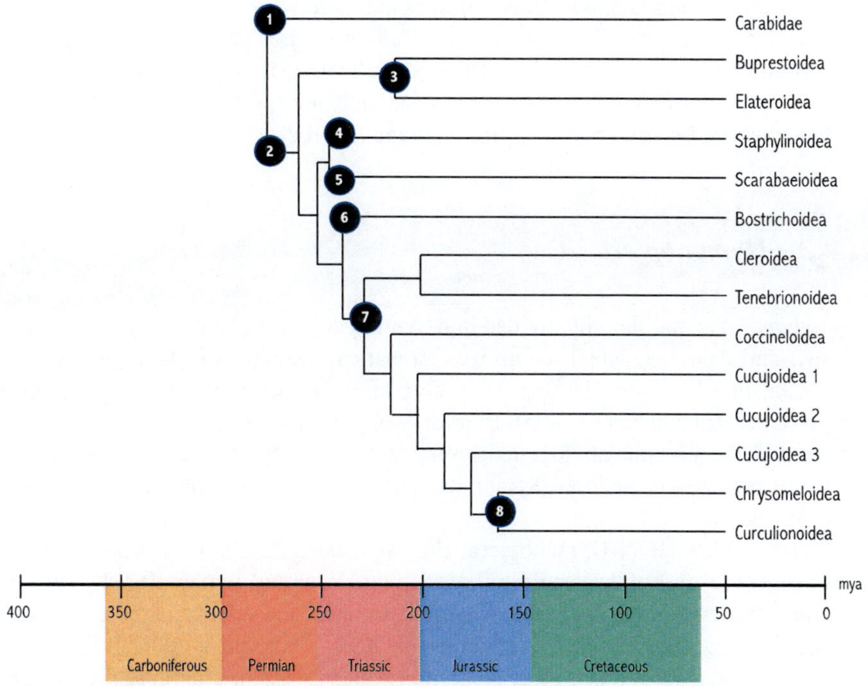

Fig. 1.3 Simplified phylogenetic tree of relationships of the major Coleoptera taxa. Numbers in nodes: 1, Adephaga; 2, Polyphaga; 3, Elateriformia; 4, Staphyliniformia; 5, Scarabaeiformia; 6, Bostrichiformia; 7, Cucujiformia; *, Phytophaga. (Details in text)

Aspisoma lineatum) that feed using extraoral digestion. For this, they regurgitate onto their preys and suck the partly digested material (Grimaldi and Engel 2005).

Scarabaeidae in the superfamily Scarabaeoidea is the largest family of the group, including the well-known dung-rolling beetles (dung beetles) that occur throughout the world, mainly in tropical grasslands. Females periodically lay eggs on the dung balls, which serve as food for the larvae (Grimaldi and Engel 2005). The most studied scarabs are *Oryctes nasicornis*, *Costelytra zealandica*, and *Pachnoda ephippiata*.

Bostrichiformia, a sister taxon of Cucujiformia, includes the family Dermestidae, the carpet beetles that feed on keratin (like woolen carpets) and the very dry proteinaceous remains of carcasses, exemplified by *Dermestes maculatus*. Among Cucujiformia, Coccineloidea is sister of the remaining superfamilies of the infraorder, which grouped into Cucujoidea (which now includes Phytophaga) and Cleroidea+Tenebrionoidea. Old lineage Cucujoidea is paraphyletic, as Phytophaga was recovered by McKenna et al. (2019) within Cucujoidea. Phytophaga comprises the superfamilies Curculionoidea (weevils, e.g., *Sphenophorus levis*), from which the main family is Curculionidae and Chrysomeloidea, with major families Chrysomelidae and Cerambycidae. Chrysomelidae includes a subfamily Bruchinae that borrow seeds which contain toxins. Examples are the chewing seed beetles *Callosobruchus maculatus* and *Zabrotes subfasciatus*. Cerambycidae (long-horned beetles, e.g., *Migdolus fryanus*) feed as adults on leaves bark and sometimes pollen, whereas the larvae mine the phloem of trees and bore wood.

The largest family of Tenebrionoidea is Tenebrionidae. They are generally scavengers on dried plant remains, but also feed on lichens, fungi, and decaying wood. Important pests of stored grain and insect models are the flour beetles *Tribolium castaneum* and *Tenebrio molitor* (Grimaldi and Engel 2005).

1.4.2 Hymenoptera

Hymenoptera is traditionally divided into Symphyta (sawflies and horntails or wood wasps) and Apocrita, which comprises Parasitica, parasites of other insects, and Aculeata, in which the piercing ovipositor of Parasitica evolved into a stinging organ (Grimaldi and Engel 2005). Extensive phylogenetic studies detailed the evolution and the internal phylogenetic relationships of Hymenoptera (Peters et al. 2017). It is now accepted that Symphyta is paraphyletic and some old clades were reorganized.

The basal lineage of Hymenoptera (Eusymphita) is formed by sawflies that are external leaf feeders of the superfamilies Pamphilioidea and Tenthredinoidea (which resemble caterpillars, e.g., *Themos malaisei*) or are wood wasps that bore wood as larvae. Among the endophytic sawflies lineages, Orussoidea (parasitoid wood wasps) are the closest relatives of Apocrita (waisted wasps). The rapid diversification of Apocrita is thought to be helped by the evolution of the wasp waist, a constriction between abdominal segments that favor movements of the abdomen, including the ovipositor. Apocrita evolved along two major lineages: Parasitoida (parasitoid wasps, e.g., *Bracon hebetor*), characterized by endoparasitism and miniaturization, and Aculeata (stinging wasps). Aculeata evolved from lineages close to Parasitoida, whose females used their ovipositor to sting and immobilize the host larvae. Wasp females then lay eggs on the host and their larvae develop inside them. Parasitoida includes Ichneumonoidea, Chalcidoidea, and Cynipoidea (respectively ichneumon, chalcid, and gall wasps).

Aculeata gave origin to Vespoidea and Formicidae-Apoidea. Vespoidea comprises the potter, honey, and social wasps. Potter or mason wasps are so named because of their mud nests; honey wasps because they produce honey, and social wasps, which are also known as paper wasps, because they build and live in communal nests of a paper-like material made by mixing wood fibers with saliva. Formicidae are the ants and Apoidea comprises digger wasps (Crabonidae) and the bees (Anthophila). Ants are among the most ubiquitous insects amounting in number of individuals to about 1% of all insects. All ants are social insects and may be herbivorous, scavengers, or predators, and some collect seeds or pieces of leaves and flowers on which they cultivate the fungus they eat (leaf-cutting ants of the tribe Attine). Bees (e.g., *Apis mellifera* and *Scaptotrigona bipunctata*) originate in evolution within the apoid wasp family Crabonidae with a change from a predatory to a herbivorous (pollen gathering) lifestyle associated with the diversification of angiosperms, which was followed by a great expansion of bees. A table detailing the evolution of food habits of Hymenoptera is found in Daly et al. (1998).

1.4.3 Diptera

Diptera account for about 15% of insect species and are the most ecologically variable group of insects, repeatedly changing between habits and habitats along the evolution. Most Diptera are saprophytes (consumers of vegetal or animal remains) and their larvae are predominantly found in wet media. There were several independent origins of predation, phytophagy (plant feeding), mycophagy (fungus feeding), hematophagy (vertebrate blood feeding), myiasis (internal feeding in vertebrates), besides the basal habit of saprophagy and dung feeding (coprophagy) (Wigmann and Yeats 2017). A list of habitats and feeding food habits of Diptera is found in Daly et al. (1998). As a consequence of the remarkable diversification of Diptera, the study of their internal phylogenetic relationship is difficult and is still underway (Yeats et al. 2007; Wigmann and Yeats 2017).

Traditionally Diptera is divided into two suborders: Nematocera and Brachycera. Nematocera include the most structurally primitive Diptera and according to modern studies employing a phylogenetic approach it is paraphyletic, whereas Brachycera is monophyletic. There is a trend now to recognize four infraorders corresponding to the old Nematocera clade: Tipulomorpha, Culicomorpha, Psychomorpha, and Bibionomorpha (Wigmann and Yeats 2017).

Culicomorpha include Chironomidae (midges, exemplified by *Chironomus thummi)*, Culicidae (mosquitoes as *Aedes aegypti* and *Anopheles gambia*), and Simuliidae (blackflies, usually of the genus *Simulium*). Culicomorpha larvae are plankton feeders living in standing water (Culicidae) or feeders of organic matter at the bottom of bodies of water (Chironomidae). The adult forms do not feed (Chironomidae), and others feed on nectar (Simuliidae and Culicidae males) or on vertebrate blood (Simuliidae and Culicidae females) (Daly et al. 1998).

Psycomorpha includes the psychodidae blood-feeders and disease vector *Phlebotomus.* Bibionomorpha comprises 17 families, exemplified by Mycetophilidae (fungus gnats) and Sciaridae (black fungus gnats, e.g., *Rhynchosciara americana*), which usually are litter decomposers in forests.

Brachicera is divided into a branch with three infraorders, one of which is Tabanomorpha, plus Muscomorpha, which includes two superfamilies and Eremoneura. Tabanomorpha includes Tabanidae (horse flies), which cut the skin of the host with their blade-shaped mandibles and then lap up the flowing blood. Eremoneura comprises two branches: one containing a single superorder (Empidoidea) and Cyclorrhapha, a large grouping with seven superorders, from which the more important are Muscoidea, Oestroidea, and Acalyptrata. Well-known Muscoidea insects are the housefly (*Musca domestica*, Muscidae), the tsetse fly (*Glossina palpalis*, Glossinidae), and the stable fly (*Stomoxys calcitrans,* Muscidae).

Among the Oestroidea, there are the grey flesh fly *Sarcophaga bullata* and the blow fly *Calliphora erytrocephala,* which are important in the field of forensic entomology, because of their value in post-mortem interval estimation. Other blow flies like *Lucilia cuprina* lay eggs in wounds of living animals, causing myiasis, and even in humans and may result in huge livestock losses. The horse bot fly *Gasterophilus intestinalis* lives inside horse stomachs as larvae.

Finally, Acalyptrata includes the leaf-miners Agromyzidae, the true fruit flies (Tephritidae, exemplified by *Ceratitis capitata* and *Anastrepha*), and the laboratory fruit flies of the geneticists, *Drosophila melanogaster* (Drosophilidae). Tephritidae are named true fruit flies because they usually attack fruits of living plants, whereas Drosophilidae actually are fungus feeders that acquired the name fruit fly because they feed on decaying fruit.

Most Brachycera larvae feed on decaying plants (Drosophilidae) or animals (Muscidae, Calliphoridae, Sarcophagidae), although they may be also predatory (Asilidae, robber flies), herbivorous or scavengers (Tabanidae), leaf and stem miners (Agromyzidae, leaf miner flies), parasites of insects (Tachinidae) or of mammals (Gasterophilidae), and inhabitants of fruits (Tephritidae). Brachycera adults may be predatory (Asilidae), pollen feeders (Syrphidae, the hover flies) and blood feeders (e.g., Tabanidae females, the horn fly *Haematobia irritans*, the stable fly *Stomoxys calcitrans*, and the tsetse fly *Glossina palpalis*), but most feed on nectar or liquids associated with decaying material (Daly et al. 1998).

1.4.4 *Lepidoptera*

Lepidoptera (butterflies and moths) amounts to about 13% of the insect species and originated in the Late Carboniferous (about 300 Ma) with mandibulate adults and larvae feeding externally on nonvascular land plants (Bryophites: hornworts, mosses, liverworts). In the Middle Triassic (about 241 Ma), a long coiled tube-like proboscis evolved, which allowed lepidopterans to acquire nectar from flowering plants (Kawahara et al. 2019). Hearing organs appeared independently in several lineages of moths, before the origin of bats, pre-adapting them to detect the bat sonar, which permitted them to avoid bat predation (Kawahara et al. 2019).

Extant species of Lepidoptera include basal lineages and the clade Ditrysia, containing the great majority of species. Ditrysia comprises several superfamilies like Tineoidea (e.g., *Tineola bisselliela*, the common clothes moth), Tortricoidea (e.g., the leaf rollers of the family Tortricidae), and Papilionoidea (butterflies) that is the sister group of all remaining moths superfamilies, exemplified by Pyraloidea (snout moths, which includes the Pyralidae *Ephestia kueniella* and the Crambidae *Diathraea saccharalis*); Noctuoidea (miller moths as *Spodoptera frugiperda*); Bombycoidea comprising the families Bombycidae (e.g., the familiar silkworm, *Bombyx mori*), Sphingidae (e.g., the cassava hornworm, *Erinnyis levis* and the tobacco hornworm, *Manduca sexta*); and Saturniidae (e.g., the giant silkworm moth, *Hyalophora cecropia*). Other details of the internal phylogenetic relationships of Lepidoptera are discussed by Kawahara et al. (2019). Adults usually feed on nectar, honeydew, or fermenting sap with their proboscis, whereas larvae as a rule are external feeders of higher plants, which nearly all are attacked by at least one species of Lepidoptera.

According to the patterns of insect evolution discussed above, the major selective pressures affecting insect guts identified were: (a) adaptations to deal with large amounts of dilute fluid food (see Chap. 13 under Hemiptera); (b) adaptations to digesting plant and fungal cells as a result of horizontal transfer of genes from microorganisms and recruitment of lysosomal proteins as digestive enzymes (see Chaps. 9 and 10); (c) adaptations to avoiding plant inhibitors by gene expansion and new functionalization (see Chap. 11); and (d) adaptations to avoiding prolonged exposure to natural enemies and to living in short-lived media by reduction of life span permitted by more efficient midguts (see Chap. 14).

References

Beckman N, Hurd LE (2003) Pollen feeding and fitness in praying mantids: the vegetarian side of a tritrophic predator. Environ Entomol 32:881–885

Buhtz A, Kolasa A, Arlt K et al (2004) Xylem sap protein composition is conserved among different plant species. Planta 219:610–618

Daly HV, Doyen JT, Purcell AH III (1998) Introduction to insect biology and diversity, 2nd edn. Oxford University Press, Oxford

Grimaldi D, Engel MS (2005) Evolution of the insects. Cambridge University Press, New York
Inward D, Beccaloni G, Eggleton P (2007) Death of an order: a comprehensive molecular phylogenetic study confirms that termites are eusocial cockroaches. Biol Lett 3:331–335
Johnson KP, Dietrich CH, Friedrich F et al (2018) Phylogenomics and the evolution of hemipteroid insects. Proc Natl Acad Sci U S A 115:12775–12780
Kawahara AY, Plotkin D, Espeland M et al (2019) Phylogenomics reveals the evolutionary timing and pattern of butterflies and moths. Proc Natl Acad Sci U S A 116:22657–22663
Kehr J (2006) Phloem sap proteins: their identities and potential roles in the interaction between plants and phloem-feeding insects. J Exp Bot 57:767–774
Lo N, Engel MS, Cameron S et al (2007) Save Isoptera: a comment on Inward et al. Biol Lett 3:562–563
Lopez-Millan AF, Morales F, Abadia A et al (2000) Effects of iron deficiency on the composition of the leaf apoplastic fluid and xylem sap in sugar beet. Implications for iron and carbon transport. Plant Physiol 124:873–884
McKenna DD, Shin S, Ahrens D et al (2019) The evolution and genomics basis of beetle diversity. Proc Natl Acad Sci USA. https://doi.org/10.1073/pnas19096551161-1
Misof B, Liu S, Meusemann K et al (2014) Phylogenomics resolves the timing and pattern of insect evolution. Science 346:763–767
Peters RS, Krogmann L, Mayer C et al (2017) Evolutionary history of the hymenoptera. Curr Biol 27:1013–1018
Weirauch C, Schuh RT, Cassis G et al (2019) Revisiting habitat and lifestyle transitions in Heteroptera (Insecta: Hemiptera): insights from a combined morphological and molecular phylogeny. Cladistics 35:67–105
Wigmann BM, Yeats DK (2017) Phylogeny of Diptera. In: Kirk-Spriggs AH, Sinclair BJ (eds) Manual of afrotropical Diptera vol 1: introductory chapters and keys to Diptera families, Suricata, vol 4. South African Biodiversity Institute, Pretoria, pp 253–265
Yeats DK, Wigmann BM, Courtney GW et al (2007) Phylogeny and systematics of Diptera: two decades of progress and prospects. Zootaxa 1668:565–590
Zhang C, Yu X, Ayre BG et al (2012) The origin and composition of cucurbit "phloem" exudate. Plant Physiol 158:1873–1882
Zhang H-L, Huang Y, Lin L-L et al (2013) The phylogeny of Orthoptera (Insecta) as deduced from mitogenomic gene sequences. Zool Stud 52:37
Zhang S-Q, Che L-H, Li Y et al (2018) Evolutionary history of Coleoptera revealed by extensive sampling of genes and species. Nature Com 9:205

Chapter 2
Overview of Insect Midgut Function

Abstract Food may be acquired by biting with chewing parts, sucking by pierce-sucking mouthparts, or by ingesting a pre-digested or pre-dispersed meal. Digestion is a stepwise process by which the molecules of food are hydrolyzed into components able to be absorbed. The first step of digestion is the initial digestion, when food polymers result in oligomers, followed by the second step, intermediary digestion, corresponding to the conversion of oligomers into dimers and, finally, the third step, final digestion, in which the dimers are cleaved into monomers that are absorbed. The gut morphology varies among insect taxa from the basal plan formed by a capacious crop followed by a midgut with anteriorly placed ceca, ending in a hindgut. The midgut has inside an anatomical chitin-protein film, the peritrophic membrane (PM), that separates two luminal compartments: endoperitrophic (inside PM) and ectoperitrophic (outside PM) spaces. In polyneopterans and lower holometabolans, the enzymes of initial and intermediary digestion move freely inside the midgut, whereas in higher holometabolans, only the enzymes of initial digestion traverse PM into the endoperitrophic space. The enzymes of intermediary digestion are retained in the ectoperitrophic space because they are larger than the PM pores (7–9 nm dia). There are midgut countercurrent fluxes caused by the secretion of fluid in a posterior region and its absorption in an anterior region. These countercurrent fluxes decrease the loss of enzymes by excretion as part of the enhancement of digestive efficiency caused by the compartmentalization of digestion. The sites of water and nutrient absorption are identified with the use of a non-absorbable dye and nutrients. The concentration of dye indicates water removal, its dilution indicates water secretion, whereas a change in the ratio of nutrient and dye indicates nutrient absorption.

W. R. Terra et al., *Molecular Physiology and Evolution of Insect Digestive Systems*, Entomology in Focus 7, https://doi.org/10.1007/978-3-031-39233-7_2

2.1 Food Handling and Ingestion

Digestion is the process by which the molecules in the food once ingested are hydrolyzed into smaller units that can be absorbed by the gut cells. The gut is responsible for all steps of food processing after ingestion: digestion, absorption, formation of feces, and their delivery.

Food must be prepared before ingestion, and this process depends on the kind of food, insect mouthparts, and insect habits. The most usual form of acquiring solid food is by biting with chewing mouthparts lubricated with saliva. In this case, the saliva usually is devoid of digestive enzymes. Occasionally, the saliva contains amylase and α-glucosidase (Walker 2003), and in rarer cases, as observed in cockroaches, it has also laminarinase and cellulase (Genta et al. 2003). It should be noticed, however, that salivary enzymes have only a subsidiary role in digestion (see Table 2.1 as an example), as most digestion is actually carried out by midgut enzymes. Another way of ingesting solid food is after extraoral digestion, which will be detailed below.

Liquid food usually is nectar, blood, or plant sap. Nectar is taken by lapping up (bees) or sucking (adult lepidopterans) mouthparts. Blood is acquired with piercing-sucking mouthparts in a fast and painless process to avoid host reaction thanks to the presence in the saliva of analgesics, vasodilators, and anticoagulants, but without digestive enzymes (Arca and Ribeiro 2018). Plant sap is taken by hemipterans with the aid of pierce-sucking mouthparts. Their mouthparts consist in a stylet that are used to penetrate the diet and have two channels: one for injecting salivary fluids and another for the inflow of the diet. Sap-sucking hemipterans have a kind of saliva that form a sheath surrounding the stylets and another containing enzymes to facilitate access to the plant conducting vessels (Walker 2003). This is usually known as the stylet-ensheath mechanism (see other details in Chap. 13). As this mechanism is one kind of extraoral digestion, this will be detailed below.

Extraoral (or preoral) digestion is the enzymatic treatment of food before ingestion. This strategy is used mostly by predaceous insects, which usually regurgitate midgut contents over the prey (reflux extraoral digestion) and after a while re-ingest the partially digested material from the prey (Cohen 1995; Canton and Bonning 2020). Enzymes involved in the reflux extraoral digestion include all enzymes found

Table 2.1 Representative digestive enzymes in salivary glands and different gut sites of *Abracris flavolineata* (Orthoptera) adults

		Foregut	Midgut		
Enzyme	Salivary gland	Crop	Caeca	Ventriculus	Hindgut
Amylase	5.3	40.5	39.2	12.9	7.4
Trypsin	0.03	30.4	40.3	21.4	7.9
Maltase	0.09	54.0	26.0	10.8	9.2
Aminopeptidase	0.11	23.0	49.0	20.9	7.1
Trehalase	11.1	20.5	17.0	32.0	20.7

Data according to Ferreira et al. (1990). Figures are % activity in relation to total gut in the midgut luminal contents, which are those responsible for initial and intermediary digestion. This was shown, for example, by stimulating predaceous larvae of *Pyrearinus termitilluminans* with balls of absorbent paper. After several attempts, the larvae attack the paper balls and regurgitate over them. The digestive enzymes recovered from the balls were qualitatively and quantitatively the same as those occurring in midgut contents. Enzymes carrying final digestion are restricted to midgut cells (Colepicolo-Neto et al. 1986).

Another type of preoral digestion (non-reflux extraoral digestion) is actually a dispersion of the tissues of the prey (either animal or vegetal) by digesting the intercellular cement with digestive fluids produced and secreted by specialized organs, usually the salivary glands (Cohen 1995; Canton and Bonning 2020). This kind of preoral digestion is found mainly among hemipterans. In the plant sap–feeding hemipterans, the enzymes usually employed to assist in plant penetration by breaking the extracellular matrices are mainly pectinases, glucanases, and cellulases (Sharma et al. 2014; Tan et al. 2016). Some metalloprotease are able to degrade the sieve tube proteins occur in aphid saliva, improving nutrition by complementing the usually poor phloem sap (Canton and Bonning 2020). Phytophagous Heteroptera are not sap feeders and they access their plant diet in a lacerate and flush mechanism, which consists of the repeated insertion and withdrawal of the stylet. The pierced plant tissue is then flushed with saliva that disperses it employing the same kind of enzymes as the sap feeders. Predaceous hemipterans disperse the tissues of their preys by a lacerate and flush mechanism with enzymes like salivary collagenase that degrades the fibers of collagen, one of the constituents of the extracellular matrix in animals. The once-ingested tissue fragments of preys (observed in midgut contents) are further digested in the midgut (Fialho et al. 2012).

The evolutionary advantage of the extraoral digestion in comparison with the acquisition of food by biting the prey and ingesting it in piece meals is that it avoids the ingestion of indigestible tissues as the arthropod exoskeletons and perhaps also provides a defense against pathogens and noxious chemicals in the diet. In the case of sap and phytophagous hemipterans, saliva may also contain proteins affecting plant defenses, like a calcium-binding protein that suppresses phloem sieve occlusion to prevent sap loss following injury (Will et al. 2007, 2009).

Finally, there is another form of extraoral digestion found among wood-feeding Siricidae (Hymenoptera). The larvae of these insects have modified mouth parts that press fungus-attacked xylem and squeeze out the nutritious material that is subsequently ingested and further digested (Thompson et al. 2014).

2.2 Gut Morphology

A generalized diagram of the insect gut is shown in Fig. 2.1. Foregut and hindgut cells are covered with a cuticle, which is non-permeable to most molecules in the foregut, but which is permeable to water, ions, and some organic compounds in the hindgut. Midgut cells lack a cuticle but are separated from the midgut contents by a

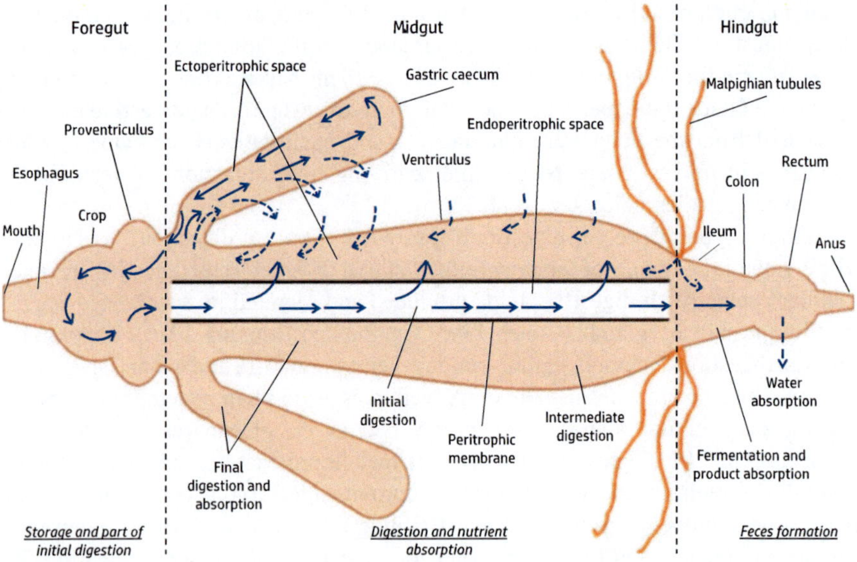

Fig. 2.1 Generalized diagram of the insect gut showing the sites of important digestive events and water fluxes (*dotted arrows*) and the circulation of digestive enzymes (*solid arrows*)

film-like anatomical structure named peritrophic membrane (PM). PM divides the midgut contents into endoperitrophic (inside PM) and ectoperitrophic (outside PM) spaces. The foregut begins with the mouth, followed by the cibarium (see below), pharynx, and esophagus, which is linked to the crop (a storage organ) and an organ (proventriculus), which is a triturating device in some insects, whereas in most insects it is only a valve controlling the entrance of food into the midgut. The midgut comprises a tube (ventriculus) from which may branch blind sacs (gastric or midgut ceca). Anteriorly placed midgut ceca are involved in digestion and water and nutrient absorption, whereas midgut ceca placed elsewhere are implicated in other functions (details below). In some insects, usually blood-feeders, the anterior midgut is dilated and stores food. In these cases, the anterior midgut is frequently named stomach. The Malpighian tubules (an excretory organ) branch from the gut at the region of the sphincter (pylorus) that separates the midgut from the hindgut, and in a few insects they are joined to form a ureter (Fig. 2.2b). The hindgut comprises the ileum, colon, and rectum and ends at the anus, but may be modified in a fermentative chamber harboring microorganisms that may assist cellulose degradation (see Chaps. 10 and 13). The gut epithelium is surrounded by longitudinal and circular muscles that propel the food bolus along the gut by peristalsis that are wave-like contractions of the circular muscles. The gut is oxygenated by the tracheal system and is more innervated at the fore- and hindgut than in the midgut. Visceral muscles connect the gut to the body wall and act as dilators of the gut. In the foregut, those muscles form a pump mainly developed in fluid feeders (cibarial pump). In chewing

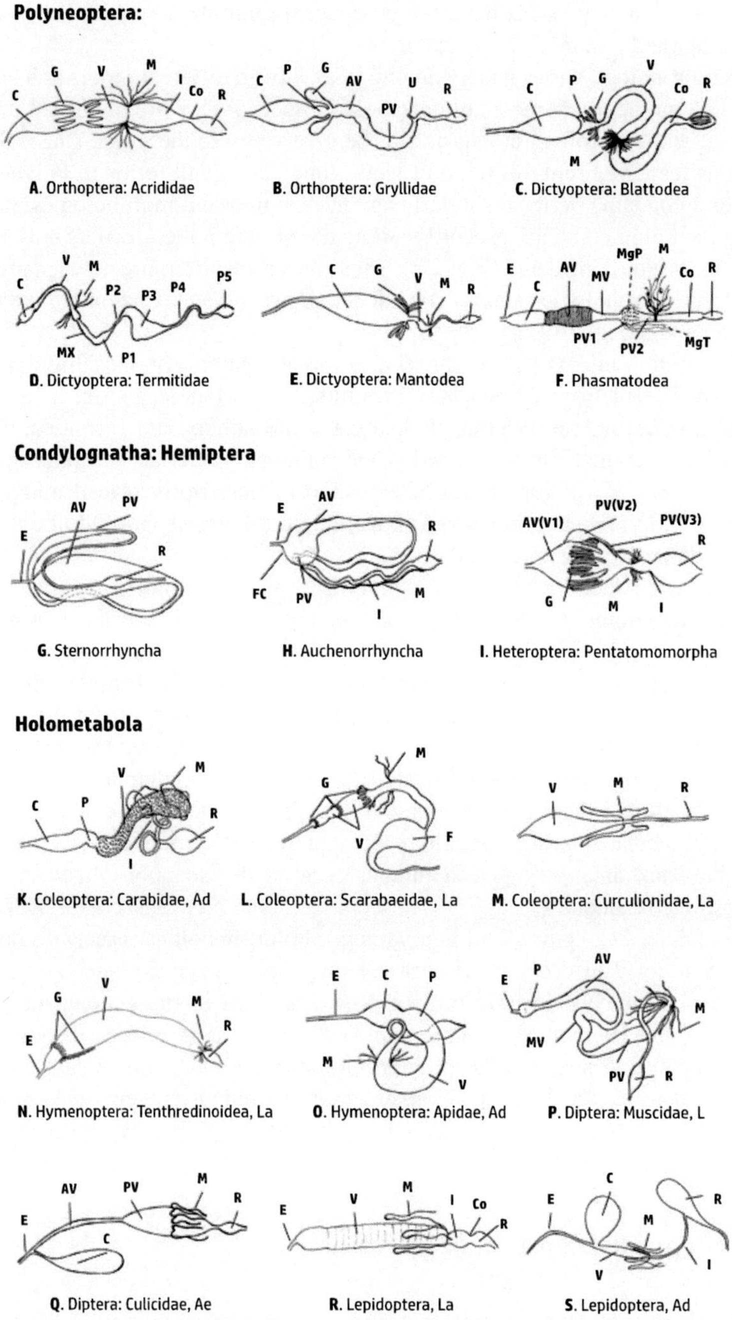

Fig. 2.2 Major insect gut types. *Ad* adult, *AV* anterior ventriculus (midgut), *C* crop, *Co* colon, *E* esophagus, *F* fermentation chamber, *FC* filter chamber, *G* midgut (gastric) ceca, *I* ileum, *La* larva, *M* Malpighian tubules, *MgP* midgut protuberances, *MgT* midgut tubules, *Mx* Mixed segment, *P* proventriculus, *Pa* paunch, *P1-5* hindgut regions in termites, *PV* posterior ventriculus, *R* rectum, *U* ureter, *V* ventriculus. Not drawn to scale. (Based partly in Terra WR (1988). Physiology and biochemistry of insect digestion: An evolutionary perspective. *Braz J Med Biol Res* 21: 675–734)

insects, the dilator muscles form the pharyngeal pump that enables the insects to drink water and pump air during the molts.

Gut morphology varies widely among insects from different orders and between the larvae and adults in most holometabolan insects. As we are interested in insect digestion and absorption, emphasis will be given only to the insect stages that are notorious feeders. From this point of view, young and adult forms of Polyneoptera and Condylognatha perform similarly and have similar gut morphologies, whereas among the holometabolan, except for some blood- and pollen-feeders only the larval forms will be considered (Fig. 2.2). Those interested in a more detailed presentation of gut morphologies should consult Snodgrass (1935), Chapman (1985), and Simpson and Douglas (2013).

Polyneoptera insects have a capacious crop and anteriorly placed midgut ceca, which are absent from Dictyoptera Termitidae and Mantodea (Fig. 2.2a–f). The enlargement of hindgut structures in Dictyoptera Blattodea and Termitidae that are involved in dealing with wood and other cellulosic materials is noticeable (see Chaps. 10 and 13). Phasmatodea have a kind of posteriorly placed midgut ceca resembling Malpighian tubules known as midgut tubules supposed to be active in midgut alkalization (Monteiro et al. 2014) (Fig. 2.2a–f).

Hemipterans lack PM, crops, and anteriorly placed midgut ceca (Fig. 2.2g–i). Some Pentatomomorpha have posteriorly placed midgut ceca in contact with the anterior parts of the ventriculus (Fig. 2.2i) and are assumed to have a function similar to the filter chamber of Auchenorrhyncha and Sternorrhyncha insects (Goodchild 1963). The filter chamber is a structure formed by the apposition of the posterior midgut and the proximal ends of the Malpighian tubules to the anterior midgut, thus facilitating the passage of water directly from the anterior midgut to the Malpighian tubules, resulting in the concentration of dilute food (Le Cahérec et al. 1997).

Holometabola guts are variable. Megaloptera and primitive Coleoptera have crops but lack anteriorly placed midgut ceca, with exceptions like Coleoptera Dermestidae (Caldeira et al. 2007). Except for some adult insects (Hymenoptera Apidae, Diptera Culicidae, and Lepidoptera), holometabolous insects do not have crops. Anteriorly placed midgut ceca are only found in mosquitoes and midges. Coleoptera Scarabaeidae have midgut ceca organized in rings along the midgut (Fig. 2.2l) and Hymenoptera Tenthredinoidea have a ring of ceca disposed in U format in the ventral side of the anterior midgut (Fig. 2.2n). In both cases, the ceca are thought to be involved in the alkalization of midgut contents (see details in Chap. 6).

2.3 Stepwise Digestion of Nutrients

Most food molecules are polymers and must be reduced to monomers to be absorbed. Digestion of polymers (proteins and starch) occurs in three phases: initial, intermediate, and final (Fig. 2.3). Initial digestion is carried out by enzymes that cleave the internal bonds of the polymers (e.g., endopeptidases in the case of proteins and

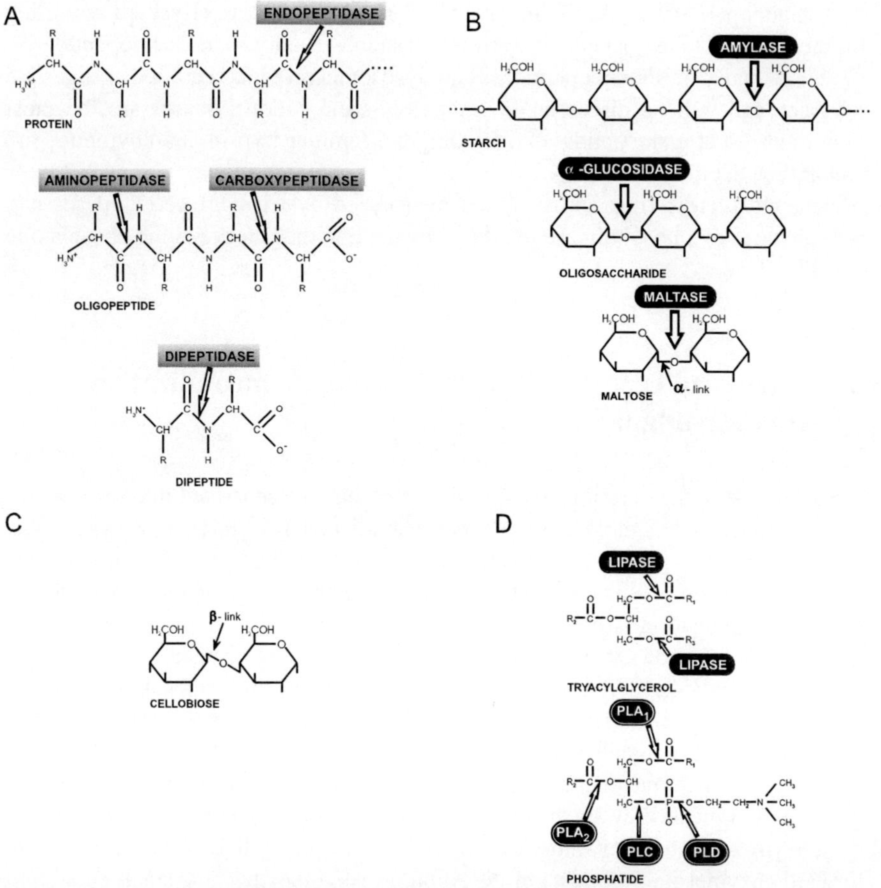

Fig. 2.3 Stepwise digestion of nutrients. Arrows point to the chemical bonds hydrolyzed by the identified enzymes. (**a**) Protein digestion; R, different amino acid moieties; (**b**) starch digestion; (**c**) β-linked glucoside; (**d**) lipid digestion; *PL* phospholipase, *R* fatty acyl moieties. (Reprinted with permission from Terra WR, Ferreira C, 2012. Biochemistry and molecular biology of digestion., in: Gilbert, L.I. (Ed.), Insect Molecular Biology and Biochemistry. Academic Press/Elsevier, London, pp. 365–418)

amylase for starch). During intermediate digestion, oligomers (oligopeptides and oligosaccharides) are split into dimers by, respectively, carboxypeptidases and amylases. In sequence, during the final digestion, the dimers (dipeptides and maltase) are hydrolyzed by dipeptidases and maltases. Aminopeptidases are active on small oligopeptides and also on tri- and dipeptides. Thus, aminopeptidases are active in the final digestion of proteins in combination or in replacement of dipeptidases. Likewise, α-glucosidases act on small oligosaccharides and disaccharides and, hence, they are responsible for the final digestion of starch. When the α-glucosidases hydrolyze only maltose they are named maltase.

The endopeptidases usually involved in insect digestion are classified according to their active site groups in serine-, cysteine-, and aspartic-endopeptidases. Cysteine- and aspartic-endopeptidases are lysosomal proteins recruited as digestive enzymes and will be discussed in Chap. 9. Serine endopeptidases are the most important insect endopeptidases, including the familiar trypsin and chymotrypsin, and will be detailed in Chap. 4.

Triacylglycerols (fats and oils) are hydrolyzed by triacylglycerol lipases and phosphatides by phospholipase identified by the link they cleave in the phosphatide (see Fig. 2.3).

2.4 Phases of Digestion and Their Compartmentalization in the Midgut

We discussed in the anterior item that digestion may be separated into primary (or initial), intermediary, and final digestion accomplished by different enzymes. In this chapter, we will begin to relate the phases of digestion with the gut compartments to provide a complete picture of the spatial organization of digestion. We will see that this organization is closely associated with the phylogenetic position of the insect. Table 2.1 shows that except for the final digestion of proteins (indicated by a high aminopeptidase activity) that occurs in the midgut, most digestion is carried out in the crop of *Abracris flavolineata* (Polyneoptera: Orthoptera). The participation of the salivary glands is negligible. As the enzymes are stable in midgut contents along their movement across the gut, the low enzyme activities recovered in the hindgut indicate they are excreted at a low rate. This low excretory rate of digestive enzymes is caused by midgut countercurrent fluxes of fluid. These fluxes move forward enzymes and products of digestion, as soon they became small enough to pass through PM into the ectoperitrophic space, thus recovering the enzymes before excretion and enhancing digestion efficiency. This endo-ectoperitrophic circulation of digestive enzymes is named enzyme recycling (details in Chap. 6). Secretory and absorptive regions are qualitatively inferred with the use of non-absorbable dyes, like amaranth. Secretory regions accumulate dyes injected into the hemolymph on their hemal side, whereas absorptive regions accumulate dyes on their luminal side. This will be discussed in detail in Chap. 6. A similar arrangement of the digestive process was observed in several other Polyneoptera insects (see Chap. 14). It should be remarked that it is not necessary to be functional in enzyme recycling that the countercurrent flux of fluid occurs from the posterior to the anterior midgut. It suffices a countercurrent flux taking place from a posterior and to an anterior site, both in the posterior midgut, as exemplified in detail for the fly larvae (see Chap. 6).

Among Holometabola, the lower Coleoptera, exemplified by *Pheropsophus aequinoctialis* (Coleoptera: Adephaga: Carabidae), the spatial organization of digestion is similar to Polyneoptera insects (see details in Chap. 14). In higher Coleoptera, like *Tenebrio molitor* (Coleoptera: Polyphaga: Tenebrionidae), as they have no crops, most digestion occurs in midgut contents, with the final digestion taking place in midgut cells (Table 2.2). As before, the excretion of digestive enzymes is very low.

Compartmentalization of digestion in higher Holometabola (Panorpoidea) is more complicated. This was shown for the first time with the lower Diptera *Rhynchosciatra americana* (Sciaridae). When the activities of the digestive enzymes determined in midgut cells and contents were summed up, the figures for trehalase and other enzymes were much less than the activities determined in whole midguts. This means that some enzyme-containing compartment is being lost. Soon it became clear that what is being lost was the ectoperitrophic contents that in these larvae include the contents of their large midgut ceca (as depicted in Fig. 2.1). When the ectoperitrophic fluid was collected by puncturing the midgut ceca with a capillary and the enzyme activities determined in it were added to those of midgut contents (material enclosed by PM) and midgut cells, the sum was similar to activities of the whole midgut (Table 2.2; Terra et al. 1979). This means that the ectoperitrophic space contains enzymes that although secreted by midgut cells do not traverse PM. This is highlighted by a comparison of the enzyme-specific activities (enzyme units per mg protein). As is shown in Table 2.2, the specific activities of acetylglucosaminidase, carboxypeptidase, aminopeptidase, and trehalase are significantly higher in the ecto- than in the endoperitrophic space. Thus, in this insect, initial digestion takes place in the endoperitrophic space (PM contents), intermediate

Table 2.2 Representative digestive enzymes in midgut compartments of insects pertaining to three orders

	Coleoptera midgut		Diptera midgut			Lepidoptera midgut		
Enzyme	Contents	Cells	Ecto contents	Endo contents	Cells	Ecto contents	Endo contents	Cells
Amylase	94.4	3.8	31.3 (150)	39.5 (144)	29.2 (50)	1.5 (570)	96.7 (4900)	1.8 (20)
Trypsin	83.2	9.5	61.8 (50.5)	34.4 (37)	12.4 (4)	4.2 (70)	94.5 (200)	1.3 (0.7)
Acetylglucosaminidase	–	–	13.5 (79)	6.3 (17)	80.2 (31)	36.5 (34)	1.5 (0.21)	62 (2.1)
Maltase	95.0	3.6	Trace	Trace	88 (30)	1.7 (70)	3.4 (20)	94.9 (130)
Carboxypeptidase A	–	–	17.7(121)	10.4 (54)	71.9 (150)	(50)	(17)	(23)
Aminopeptidase	10.9	73.0	29.7 (1348)	13.5 (465)	56.8 (700)	12.7 (700)	7.8 (60)	80 (150)
Trehalase	72.0	15.3	74.8 (345)	24.8 (7.1)	18 (30)	–	–	–

Data according to Terra et al. (1985) (Coleoptera), Terra et al. (1979) (Diptera), and Ferreira et al. (1994) (Lepidoptera). Figures are % activity of total midgut and in parentheses are specific activities (mUnits/ mg protein)

Ecto contents ectoperitrophic contents, *endo contents* endoperitrophic contents; – not determined digestion in the ectoperitrophic space, and final digestion at midgut cells. A similar conclusion may be arrived at by analyzing data (Table 2.2) for Lepidoptera, another panorpoid insect.

Most insects do not have midgut ceca and hence it is not possible to obtain uncontaminated ectoperitrophic fluid by puncturing them. The ectoperitrophic fluid in the midgut is contained in the ectoperitrophic space, which in this region is almost virtual (see space between the cell microvilli and PM in Fig. 2.4). However, it is possible to prepare a sample enriched with the ectoperitrophic fluid by washing the luminal surface of midgut tissues. This technique provides good results supported by the finding that enzymes involved in intermediate digestion have specific activities higher in the washings than in PM contents (Table 2.2, Ferreira et al. 1994). With this technique, ectoperitrophic proteins were identified by proteomic techniques by Liu et al. (2018). But as their study did not include PM contents, it was not possible to know which of the proteins were retained in the ectoperitrophic space, and which were en route to PM contents. More recently, a proteomic study with samples of ectoperitrophic fluid and PM contents identified many proteins involved in intermediary digestion, like carboxypeptidases and aminopeptidase that do not pass through PM (Fuzita et al. 2021), confirming the enzymological data previously discussed. The spatial organization of digestion described above was also found in other Holometabola (see Chap. 15).

An interesting question arises now. Why do some enzymes pass and others do not pass through the PM? The most obvious answer is their size. As expected, the

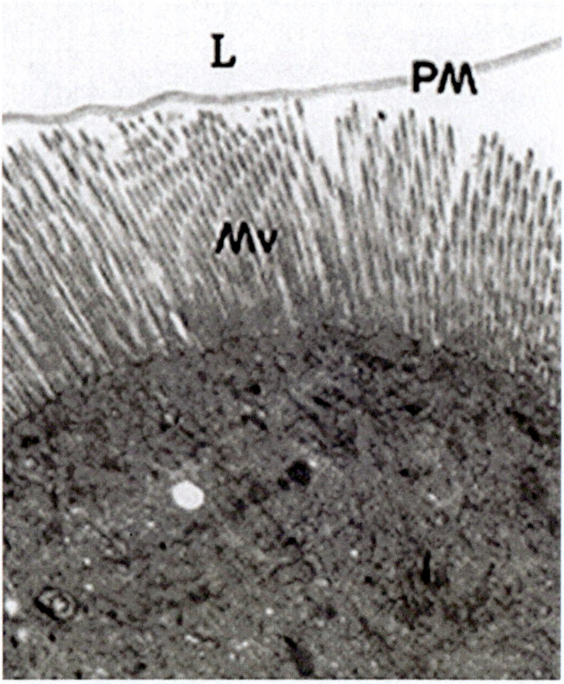

Fig. 2.4 Electron micrograph showing the very small space between PM and microvilli (ectoperitrophic space) of a posterior midgut cell of *M. domestica*. L, lumen; Mv, microvilli; PM, peritrophic membrane. Magnification 7500X. (Reprinted with permission from Terra WR, Ferreira C (2009) Digestive system. In: Resh, V.H., Cardé, R.T. (Eds.), Encyclopedia of Insects.2nd ed., Academic Press, San Diego, CA, pp.273–281; © Elsevier)

molecular weights of the enzymes retained in the ectoperitrophic space are larger than those passing through the PM (Table 2.3; Santos and Terra 1986; Terra and Ferreira 1983). It is possible to estimate the radius and hence the diameter of an enzyme given its molecular weight. This is done by interpolation in a plot of log molecular weight against Stokes' radius of known proteins. Stokes' radius is the radius of hydrate polymers, like a protein, in solution (Atkins and de Paula 2010). Tabulation of molecular weight and Stokes' radius of several proteins may be found in Brewer et al. (1974) and in La Verde et al. (2017).

Data in Table 2.3 suggest that PM pores have 7–8 nm of diameter and that carboxypeptidase A does not traverse PM because it occurs as hexamers that dissociate during electrophoresis. PM pores sizes will be discussed in detail in Chap. 7. For now, it suffices to conclude that the sophistication of midgut compartmentalization apparently is a consequence of using larger or oligomeric enzymes for intermediate and final digestion or restricting those enzymes on the midgut cell surface (see Chap. 5). This kind of compartmentalization seems to increase the efficiency of digestion, thus supporting faster growth rates in panorpoid insects. This becomes apparent when one considers that the life span of a fly or a butterfly is about 6 weeks in comparison to that of beetles of about 12 months (Sehnal 1985). This increase in efficiency is also observed in the relative growth rates, RGR (increase in biomass per initial biomass per day), and relative food consumption rates, RCR (dry weight of food ingested per dry weight of biomass). Thus, Orthoptera on grass has: RGR, 0.07; RCR, 0.35, whereas the parameters averaged for all nutrient classes are (RGR, RCR): Coleoptera, 0.07, 0.6; Lepidoptera, 0.3, 1.8 (Slansky and Scriber 1985).

Hemipterans are expected to have lost the compartmentalization of digestion, as they evolved from insects adapted to suck plant sap, usually rich in sugars and poor in polymeric nutrients. Thus, they only have to hydrolyze sugars (usually sucrose). However, the hemipterans evolved to feed on saps rich in proteins and to use other

Table 2.3 Molecular weights and estimated molecular diameters of selected digestive enzymes that pass or do not pass through the midgut peritrophic membrane of *Erinnyis ello* (Lepidoptera: Sphingidae) and *Rhynchosciara americana* (Diptera: Sciaridae)

		Lepidoptera			Diptera		
Enzymes	Pass through PM	MW (kDa)C	MW (kDa)E	Dia (nm)	MW (kDa) C	MW (kDa)E	Dia (nm)
Amylase	Y	56	40	6.2 (mean)	65	–	6.8
Trypsin	Y	61	48	6.5 (mean)	–	40	6.2
Acetylglucosaminidase	N	134	149	8.6 (mean)	141	–	8.6
Carboxypeptidase A	N	90	17	7.6, 3.9 (respectively)	–	–	–
Aminopeptidase	N	–	–	–	111	–	8.1
Trehalase	N	106	100	7.9 (mean)	–	122	8.2

Lepidoptera data according to Santos and Terra (1986) and Diptera data according to Terra and Ferreira (1983). *C* ultracentrifugation data, *E* electrophoresis data, *Dia* hydrated diameter of the molecule calculated as explained in the text, *MW*, molecular weight, – not determined nutrient sources. This led to evolutionary pressures to compartmentalize digestion, which, due to their evolutionary course, resulted in a peculiar compartmentalization. Sternorrhyncha insects, exemplified by the aphid *Acyrthosiphon pisum*, have cell digestive enzymes associated with the cells, thus avoiding their loss with honeydew (Cristofoletti et al. 2003). There are not enough data on Auchenorrhyncha, but it is probable that they are similar to Sternorrhyncha. Digestion compartmentalization in Heteroptera, exemplified by the Cimicomorpha *Rhodnius prolixus* (Terra et al. 1988) and the Pentatomomorpha *Dysdercus peruvianus* (Silva and Terra 1994), have initial digestion in luminal contents and intermediary and final digestion in cells.

2.5 Identification of the Sites of Transport of Water, Ions, and Nutrients Along the Gut

In this section, the physiological techniques to identify the sites of absorption of water, ions, and nutrients will be overviewed. The molecular processes underlying these phenomena will be postponed to Chap. 6, whereas particular cases will be described in Chap. 13.

The first attempt to identify the sites of absorption of substances along the insect gut used histological techniques. With this approach, the identification in a gut region of the storage of a substance, like glycogen and oil drops, was considered to reveal the sites of active absorption. However, the results obtained with this technique are not reliable, as revealed by quantitative data on absorption events. These quantitative data were obtained by feeding (or injecting into) insects with a solution containing a non-absorbable dye, like amaranth, and radioactively labeled compounds, such as glucose, amino acids, and tripalmitin. A concentration of dye in a gut region indicates water removal from that region. The change in the ratio of the concentration of radioactively labeled substances and that of dye, as the solution containing them moves along the gut, permits to calculate the amount of absorption occurring in each gut region (Treherne 1958a, b, 1959). Instead of using radiolabeled substances, it is possible to determine them with sensible analytical procedures (see, e.g., Silva and Terra 1994). Movements of ions in the insect gut are usually determined by the quantification of ions in the hemolymph and gut contents and with electrophysiological methods as reviewed by Treherne (1967), Dow (1986), and Klein et al. (1996).

The data on substances moved along the gut are the bases of the present knowledge on the functions of the different insect gut regions. According to that data, the crop is impermeable to water, ions, and nutrients; the midgut ceca and anterior ventriculus (midgut) are the major sites of nutrient and water absorption; the anterior hindgut (ileum) absorbs some water and ions and some organic compounds, whereas the posterior hindgut (rectum) is the region where most water and ions are absorbed before the dry feces are expelled (Fig. 2.1) (Simpson and Douglas 2013).

The organization of the absorptive processes along the insect gut described above was derived mainly from studies performed with cockroaches and locusts. The organization may vary somewhat among the different insect taxa. These differences will be dealt with in Chap. 13.

References

Arca B, Ribeiro JMC (2018) Saliva of hematophagous insects: a multifaceted toolkit. Curr Opin Insect Sci 29:102–109

Atkins P, de Paula J (2010) Physical chemistry, 9th edn. W. H. Freeman Co., New York

Brewer JN, Pesce AJ, Ashworth RB (1974) Experimental techniques in biochemistry. Prentice-Hall, Englewood Cliffs

Caldeira W, Dias AB, Terra WR et al (2007) Digestive enzyme compartmentalization and recycling and sites of absorption and secretion along the midgut of *Dermestes maculatus* (Coleoptera) larvae. Archs Insect Biochem Physiol 64:1–18

Canton PE, Bonning BC (2020) Extraoral digestion: outsourcing the role of the hemipteran midgut. Curr Opin Insect Sci 41:86–91

Chapman RF (1985) Structure of the digestive system. In: Kerkut GA, Gilbert LI (eds) Comprehensive insect physiology, biochemistry and pharmacology, vol 4. Pergamon Press, Oxford, pp 165–211

Cohen AC (1995) Extra-oral digestion in predaceous terrestrial Arthropoda. Annu Ver Entomol 40:85–103

Colepicolo-Neto P, Bechara EJH, Ferreira C et al (1986) Evolutionary considerations of the spatial organization of digestion in the luminescente predaceous larvae of Pyrearinus termitilluminans (Coleoptera: Elateridae). Insect Biochem 16:811–817

Cristofoletti PT, Ribeiro AF, Deraison C et al (2003) Midgut adaptation and digestive enzyme distribution in a phloem feeding insect, the pea aphid *Acyrthosiphon pisum*. J Insect Physiol 49:11–24

Dow JAT (1986) Insect midgut function. Adv Insect Physiol 19:187–328

Ferreira C, Oliveira MC, Terra WR (1990) Compartmentalization of the digestive process in *Abracris flavolineata* (Orthoptera: Acrididae). Insect Biochem 20:267–274

Ferreira C, Capella NA, Sitnik R et al (1994) Digestive enzymes in midgut cells, endo- and ecto-peritrpohic contentes and peritrophic membranes of *Spodoptera frugiperda* (Lepidoptera) larvae. Arch Insect Biochem Physiol 26:299–313

Fialho MCQ, Moreira NR, Zanuncio JC et al (2012) Prey digestion in the midgut of the predatory bug *Podisus nigrispinus* (Hemiptera: Pentatomidae). J Insect Physiol 58:850–856

Fuzita FJ, Palmisano G, Pimenta DC et al (2021) A proteomic approach to identify digestive enzymes, their exocytic and microapocrine secretory routes and their compartmentalization in the midgut of *Spodoptera fugiperda*. Comp Biochem Physiol B 257:110670

Genta FA, Terra WR, Ferreira C (2003) Action pattern, specificity, lytic activities, and physiological role of five digestive β-glucanases isolated from *Periplaneta americana*. Insect Biochem Mol Biol 33:1085–1097

Goodchild AJP (1963) Studies on the functional anatomy of the intestines of Heteroptera. J Zool 141:851–910

Klein U, Koch A, Moffett DF (1996) Ion transport in Lepidoptera. In: Lehane MJ, Billingsley PF (eds) Biology of the insect gut. Chapman & Hall, London, pp 236–264

La Verde V, Dominici P, Astegno A (2017) Determination of hydrodinamic radius of proteins by size exclusion chromatography. Bio Protoc 7(8):e2230

Le Cahérec F, Guillam MT, Beuron F et al (1997) Aquaporin-related proteins in the filter chambre of homopteran insects. Cell Tissue Res 290:143–151

Liu I, Qu M, Yang J et al (2018) Physiological differentiation along the midgut of *Bombyx mori*-inspirations from proteomics and gene expression. Patterns of the secreted proteins in the ecto-peritrophic space. Insect Mol Biol 27:247–259

Monteiro EC, Tamaki FK, Terra WR et al (2014) The digestive system of the "stick bug"Cladomorphus phyllinus (Phasmida: Phasmatidae): a morphological, physiological and biochemical analysis. Arthropod Struct Dev 43:123–134

Santos CD, Terra WR (1986) Distribution and characterization of oligomeric digestive enzymes from *Erinnyis ello* larvae and inferences concerning secretory mechanisms and the permeability of the peritrophic membrane. Insect Biochem 16:691–700

Sehnal F (1985) Growth and life cycles. In: Kerkut A, Gilbert LI (eds) Comprehensive insect physiology, biochemistry and pharmacology, vol 2. Pergamon, Oxford, pp 1–86

Sharma A, Khan NA, Subrahmanyam et al (2014) Salivary proteins of plant-feeding hemipteroids. Implication in phytophagy. Bull Entomol Res 104: 117–136

Silva CP, Terra WR (1994) Digestive and absorptive sites along the midgut of the cotton seed sucker bug *Dysdercus peruvianus* (Hemiptera: Pyrrhocoridae). Insect Biochem Mol Biol 24:493–505

Simpson SJ, Douglas AE (eds) (2013) RF Chapman the insects. Structure and function, 5th edn. Cambridge University Press, Cambridge

Slansky F, Scriber JM (1985) Food consumption and utilization. In: Kerkut A, Gilbert LI (eds) Comprehensive insect physiology, biochemistry and pharmacology, vol 4. Pergamon Press, Oxford, pp 87–163

Snodgrass RE (1935) Principles of insect morphology. McGraw-Hill, New York

Tan X, Xu X, Gao Y et al (2016) Levels of salivary enzymes of *Apolygus lucorum* (Hemiptera: Miridae) from 1st instar nymph to adult, and their potential relation to bug feeding. PLoS One 11:e0168848

Terra WR, Ferreira C (1983) Further evidence that enzymes involved in the final stages of digestion by *Rhynchosciara* do not enter the endoperitrophic space. Insect Biochem 13:143–150

Terra WR, Ferreira C, De Bianchi AG (1979) Distribution of digestive enzymes among the endo- and ectoperitrophic spaces and midgut cells of Rhynchosciara and its physiological significance. J Insect Physiol 25:487–494

Terra WR, Ferreira C, Bastos F (1985) Phylogenetic considerations of insect digestion. Disaccharidases and the spatial organization of digestion in the *Tenebrio molitor* larvae. Insect Biochem 15:443–449

Terra WR, Ferreira C, Garcia ES (1988) Origin, distribution, properties and functions of the major *Rhodnius prolixus* midgut hydrolases. Insect Biochem 18:423–434

Thompson BM, Bodart J, McEwen C et al (2014) Adaptations for symbiont-mediated external digestion in *Sirex noctilio* (hymenoptera: Siricidae). Annals Entomol Soc Amer 107:453–459

Treherne JE (1958a) The absorption of glucose from the alimentary canal of the locust *Schistocerca gregaria* (Forsk.). J Exp Biol 35:297–306

Treherne JE (1958b) The digestion and absorption of tripalmitin in the cockroach, *Periplaneta americana* L. J Exp Biol 35:611–625

Treherne JE (1959) Amino acid absorption in the locust (Schistocerca gregaria Forsk.). J Exp Biol 36:533–545

Treherne JE (1967) Gut absorption. Rev Ent 12:43–58

Walker GP (2003) Salivary glands. In: Cardé RT (ed) Resh VH. Encyclopedia of Insects, San Diego, pp 1011–1017

Will T, Tjallingii WF, Thönnessen A et al (2007) Molecular sabotage of plant defence by aphid saliva. Proc Natl Acad Sci U S A 104:10536–10541

Will T, Kornemann SR, Furch ACV et al (2009) Aphid watery saliva counteracts sieve-tube occlusion: a universal phenomenon? J Exp Biol 212:3305–3312

Chapter 3
Types and Chemistry of Diets

Abstract The knowledge of diets and their chemical components demanding digestion is a prerequisite to understanding the evolution of insect midguts. The clade Insecta is a hyper-diverse group, and this diversity extends to an impressive number of different feeding habits. Therefore, we can only summarize some of the most conspicuous and representative types of diets used by insects in their natural environments. The diversity of chemical compounds found in these diets is also staggering. This makes insects excellent models for investigating the comparative physiology of digestion among animals. The following diets will be discussed: detritus, plant exudates, other plant material, animal flesh, blood, fungi, lichens, bacteria, and nectar. In this chapter, we will discuss some of the major types and chemical compositions of diets used by insects. Our focus will be on chemical constituents that require some special type of digestion or that offer some challenge that leads to adaptations in the insect digestive system.

3.1 Initial Considerations

Insects are involved as major players in most of the trophic interactions in terrestrial and lake communities (Rainford and Mayhew 2015; Roman-Palacios et al. 2019). According to the different types of diets, insect feeding habits can be classified as detritivory, fungivory, herbivory, parasitism, parasitoidism, predation, and nonfeeding (aphagous insects). In this book, we consider herbivory as the habit of feeding on autotrophs, including marginal feeding habits such as xylophagy (wood feeding), pollenivory (feeding on pollen), and feeding on plant exudates. Some insects combine different forms of food and are therefore called omnivores, while specialists or oligophagous species, by definition, make use of a restricted number of food resources available in their ecological community.

Another important aspect that must be considered is that the dominant insects with the greatest number of species (Holometabola) are those whose immature forms usually have different feeding habits than the adult forms. Therefore, we must

W. R. Terra et al., *Molecular Physiology and Evolution of Insect Digestive Systems*, Entomology in Focus 7, https://doi.org/10.1007/978-3-031-39233-7_3

expect that immature and adult forms are adapted to deal with different diets and do not compete for the same food resource. On the other hand, in insect species that have an incomplete metamorphosis life cycle, the immature forms (nymphs) usually feed on the same type of diet used by adults. To get an idea of how the different types of diets are distributed among the insect families, in larval forms, herbivory predominates with about 51% of the species, 30% being detritivorous, 30% predatory species, 19% fungivory, 13% of parasitism, and less than 1% of parasitism. The numbers are different for the adult forms, where eating habits are dominated by liquid-feeding species, with approximately equal proportions of detritivory, herbivory, and predatorism and less representation of fungivory, non-feeding species, and hematophagy (blood-feeding) (Rainford and Mayhew 2015; Simon et al. 2015; Roman-Palacios et al. 2019).

Herbivorous insects and their host plants represent more than 50% of all known species on Earth (Futuyma and Agrawal 2009; Simon et al. 2015). Insects like other animals require several macro- and micronutrients to grow, develop, and reproduce. Herbivorous insects feeding on live or dead plant tissues like wood are especially challenged to acquire vital nutrients. It is important to note that herbivorous insects face a large stoichiometric imbalance between the adequate amounts of chemical elements needed for full fitness such as nitrogen (N), phosphorus (P), potassium (K), calcium (Ca), magnesium (Mg), manganese (Mn), iron (Fe), sodium (Na), copper (Cu), sulfur (S), and zinc (Zn), and those found in plant tissues, which are adaptatively biased on just carbon (C), hydrogen (H), and oxygen (O). On the other hand, insects that feed on the bodies of other animals or that feed on detritus that has been somehow processed by microorganisms have access to more balanced diets from a nutritional point of view. In the case of nutritionally poor diets, such as certain types of detritus or plant saps, insects must assume one of two possibilities or a combination of both: to live slowly and process a large amount of plant material or have their diet supplemented in some way, usually with the participation of bacteria or fungi.

Table 3.1 summarizes the types of insect diets and their key chemical components.

3.2 Detritus

3.2.1 Types of Detritus

Detritus are the remains of living beings as exuvias, feces, urine, feathers, hairs, leaves, trunks, and whole dead bodies. Detritus is a source of nutrients that supports trophic food chains in several ecosystems and the animals that consume them contribute to the recycling process of organic matter. Among detritivores, innumerable species of insects play a prominent role in the decomposition of leaf litter, bodies of different types of living beings, or in the use of feces as a food source. Some types of detritus, such as leaf litter or fallen logs, are nearly continuously available, while

Table 3.1 Diet items and key chemical components that are digested by insects

Types of diets	Chemical component	Types of chemical bonds to be hydrolyzed
Plant material, such as litter, leaves, seeds, roots, pollen grains, etc	Cellulose	Glc-β(1,4)-Glc; Galacturonic acids linked by α(1,4)-glycosidic bonds; [-α-D-GalA-α(1,2)-L-Rha-α(1,4)-]
Plant material, such as litter, leaves, seeds, roots, pollen grains, etc	Hemicellulose (arabinoxylan, glucomannan, glucuronoxylan, xylan, etc.)	Glc-β(1,4)-Xyl; β(1,4)-linked D-mannose and D-glucose residues with a branching through β(1,6)-glycosidic bonds
Plant material, such as litter, leaves, seeds, roots, pollen grains, etc	Pectin (homogalacturonan, rhamnogalacturonan-I, rhamnogalacturonan-II)	Linear polymer of GalA linked through α(1,4)-glycosidic bonds; disaccharide repeat [-α-D-GalA-α(1,2)-L-Rha-α(1,4)-]; a backbone of α(1,4)-linked GalA residues decorated at varying intervals with 12 different types of monosaccharides in over 20 different glycosidic bonds
Fungi cell wall; cuticle of invertebrate exoskeletons	Chitin	Linear polymer of GlcNAc linked through β(1,4)-glycosidic bonds
Fungi cell wall	Chitosan	Deacetylated chitin
Plant intracellular storage polymer found mainly in leaves, seeds, and roots	Starch	Branched homopolymer made of Glc residues linked through α(1,4) or α(1,6)-glycosidic bonds
Intracellular storage polymer from animals	Glycogen	Branched homopolymer made of Glc residues linked through α(1,4) or α(1,6)-glycosidic bonds
Thylakoid and lamellar membranes from chloroplasts	Galactolipids	Gal residue linked through a β(1,4)-glycosidic bond to diacylglycerol esterified with two fat acids
Nectar, fruits.	Sucrose	D-Glc-α(1,2)β-D-Fru
Found in G(+) bacterial cell walls	Peptidoglycans	Their sugar component consists of alternating GlcNAc-β(1,4)-NAM attached to an oligopeptide chain made of three to five amino acid residues
Lichens	Lichenin	Homopolymer made of repeating glucose units linked by β(1,3) and β(1,4) glycosidic bonds
Lichens	Thamnolan	Heteroglycan composed of a core of rhamnose residues with two different galactofuranosyl chains, β (1,3) Galf and β (1,5) Galf
Structural protein making up claws, feathers, hairs, horns, nails, scales, and skins of vertebrates	Keratin	Keratins have a large amount of the amino acid cysteine, required for the disulfide bridges that confer strength and rigidity to the protein

Ara arabinose, *Fru* fructose, *Gal* galactose, *Galf* galactofuranosyl, *Glc* glucose, *GlcNAc* N-acetylglucosamine, *NAM* N-acetylmuramic acid carcasses and animal feces are much more ephemeral and difficult to predict where they will appear and for how long they will be available. The quantity, quality, and availability of key nutrients needed for the growth of detritivores is another important difference between detritus types. Litter-eating insects deal with a carbon-rich food source that can be both low in nitrogenous nutrients and difficult to digest.

Plant-derived detritus such as leaf litter and decaying logs are food resources that contain appreciable amounts of energy sequestrated in the structure of polysaccharides, although these resources are not easily accessible due to their anatomical and chemical nature. The very nature of the main chemical components that give structural stability to plant bodies sequester a great part of precious elements like C, P, S, and N, which are elements present in the monomers that make up plant polymers such as proteins, cellulose, hemicelluloses, polyphenols, pectin, lignin, etc. Even after the death or partial digestion of plant tissues, most of the carbon and nitrogen compounds are in the form of supramolecular structures that are intrinsically recalcitrant to digestion by non-specialized insects. Animals that feed on already dead plant or animal tissues are called saprophagous. There are three main groups of saprophagous insects: first, those that feed on dead plant matter; second, those insects that feed on feces; and third, those insects that feed on dead animals. Many insects are important decomposers especially species from Blattodae (especially cockroaches and termites), Coleoptera (especially scarabs and dermestids), Diptera (especially flies), and Hymenoptera (especially ants and wasps).

3.2.2 Litter Chemical Composition

The main anatomical cellular component responsible for the structural stability of plants is the plant cell wall, which can be divided into two types: the primary cell wall and the secondary plant cell wall. Initially, the assembly of the primary cell wall takes place, whose main chemical components are cellulose, hemicellulose, and pectin. With the development of the plant, there is a posterior deposition of lignin for the formation of the secondary cell wall. Cellulose, hemicellulose, and lignin combine with each other to form lignocellulose, a highly complex polymeric structure that is recalcitrant to digestion, requiring complete degradation of a consortium of glycoside hydrolases, glycosyl transferases, polysaccharide lyases, carbohydrate esterases, and other activities from enzymes such as lignin-modifying enzymes, lignin-degrading auxiliary enzymes, and lytic polysaccharide monooxygenases (Lodha 1974; Popper and Fry 2007; Mohnen 2008; Caffall and Mohnen 2009) (Fig. 3.1) (see also Chap. 10).

Cellulose is by far the largest source of potential dietary carbon present in plant tissues. Chemically speaking, cellulose is a linear polymer of D-glucose in which glucose residues are linked together by β-(1,4)-glycosidic bonds. While most cellulose persists for long periods after the plant death or from the falling of the leaves, a considerable amount of the other types of carbon-rich compounds rapidly decreases in a matter of days. The large amounts of glucose residues trapped in the

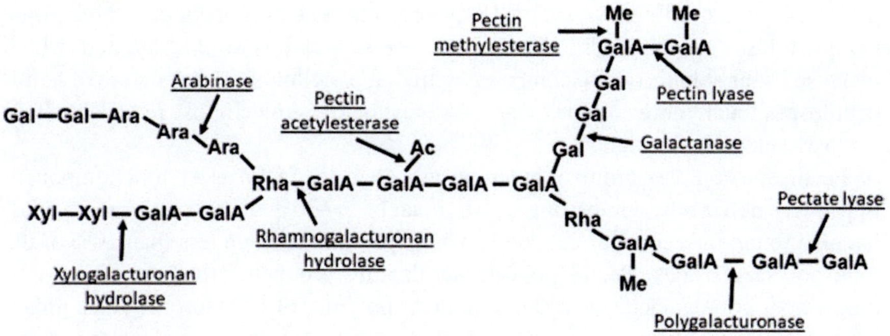

Fig. 3.1 Schematic representation of a pectic polysaccharide structure showing the action points of the major enzymes responsible for the hydrolysis of this complex carbohydrate found in plant cell walls. *Ara* arabinose, *Gal* galactose, *GalA* galacturonic acid, *Me* methyl ester, *OAc* acetyl ester, *Rha* rhamnose, *Xyl* xylose

β-(1,4) glycosidic bonds have become a selection pressure for the emergence of insects capable of digesting cellulose in different degrees (Shelomi et al. 2020). Nearly 80 insect species from 20 families representing eight orders are capable of digesting cellulose. The digestive efficiency of this glucose homopolymer varies between 40% and 90%. The digestion of cellulose involves synergistic activities of three types of enzymes: endoglucanases, exoglucanases, and β-glucosidases.

Hemicellulose is a highly heterogeneous polysaccharide from the point of view of its monomer identities, the substitutions observed in these monomers, and the types of glycosidic bonds that unite them. The hemicellulose structures also vary between different types of plant cell walls. Based on the chemical composition of their main linear chain, hemicelluloses were grouped into mannans, xyloglucan, and xylans, whether the predominant monomers are mannose, xylose/glucose, or xylose, respectively. Xyloglucans are major components in primary cell walls, whereas hemicelluloses containing polymerized xylan or glucomannan are the main hemicellulose types in secondary cell walls. Xylan is the second most abundant polysaccharide after cellulose and contributes to the structural stability of the cell wall in wood. Xyloglucans consist of a linear backbone of glucose joined by β-(1,4)-glycosidic bonds with up to 75% of the glucose residues being linked at the C6 position to residues of the pentose xylose. This basic structure is common for all the xyloglucans, but additional residues can be linked depending on the plant source, conferring both structural diversity and new physicochemical properties to the cell wall. The mannan polysaccharides are further subclassified into glucomannans, galactomannans, galactoglucomannans, and mannans. The linear backbone of glucomannans consists of both mannose and glucose linked through β-(1,4)-glycosidic bonds in a non-repeating pattern. The glucomannans in angiosperms are formed by β-(1,4)-linked D-mannose and D-glucose residues with a branching level of 8% through β-(1,6)-glycosidic bonds. The linear backbone of galactomannans and galactoglucomannans commonly found in gymnosperms are formed of β-(1,4)-linked mannose or mannose and glucose repeating units with a single unit of galactose side chain linked via α-(1,6)-glycosidic bonds at various intervals, while the mannans contain a linear chain of β-(1,4)-bonded repeating mannose units. Some hemicelluloses can be modified with galactose side chains or O-acetylated, which increases their solubility. So, compared with that of cellulose, the structure of hemicellulose is much more complex, and the digestion of hemicellulose requires a more extensive enzymatic arsenal.

Pectin is a complex group of heteropolysaccharides found as a major component in primary cell walls, comprising approximately 25–30% of their polysaccharides, but also found in secondary cell walls, where pectin makes up less than 10% of the total polysaccharides. Pectin polysaccharides are grouped into three major subtypes: first, homogalacturonan (HG), a linear polymer of galacturonic acids linked by α-(1,4)-glycosidic bonds, which is methyl esterified at the C6 position to varying degrees depending on the plant source. Homogalacturonan is the most abundant pectic polysaccharide comprising up to 65% of pectin. Second, rhamnogalacturonan-I (RG-I), which contains a backbone of the disaccharide repeat [−α-D-GalA-α(1,2)-L-Rha-α(1,4)-], is also an abundant pectic polysaccharide. RG-I has a high degree of type and number of monosaccharides, linear oligosaccharides, and branched oligosaccharides attached to its main chain, whereas some galacturonic acid residues in the RG-I backbones are acetyl esterified. The level of variation in RG-I structures and mode of deposition in different plant tissues suggest diverse functional specialization and impose a challenge to its complete degradation. The third subtype of pectic polysaccharides contains rhamnogalacturonan-II (RG-II). This group is the most complex polysaccharide in plant cell walls and makes up 10% of pectin. Its structure has a backbone of α-1,4-linked galacturonic acid residues decorated at varying intervals with side branches consisting of 12 different types of monosaccharides in over 20 different glycosidic bonds (Albersheim et al. 2011). Experimental evidence suggests that pectic polysaccharides are covalently crosslinked with each other or with other types of polysaccharides including xyloglucans and xylans in the plant cell walls (Ishii and Matsunaga 1996; Ishii et al. 2001; Mohnen 2008; Albersheim et al. 2011). Figure 3.1 and Table 3.1 illustrate the general structure of polysaccharides found in plant cell walls and the points of attack by enzymes that break down these polymers as digestive enzymes in insects.

Another major polymeric structural macromolecule found in plant cell walls is lignin, the second most abundant terrestrial organic material after cellulose. Lignin is a major constituent in secondary cell walls showing a complex hydrophobic polymeric structure consisting of p-hydroxyphenylpropane derivatives known as lignols crosslinked by ether or carbon–carbon bonds in diverse ways. The lignols are of three main subtypes: first, paracoumaryl alcohol or 4-hydroxyphenylpropane (H); second, guaiacyl alcohol or 4-hydroxy-3-methoxyphenylpropane (G); and third, syringyl alcohol or 3,5-dimethoxy-4-hydroxyphenylpropane (S). The lignin polymer is structured in an amorphous way giving rise to an encrusting material in which the cellulose microfibrils and hemicellulose are embedded. So, lignin is a mixture of polymeric phenolic compounds that add complexity and resistance to the amalgama of polysaccharides that make up the plant cell walls.

It is worth mentioning that lignin is a member of numerous classes of biomolecules commonly found in plant tissues called plant phenolics. Structurally, they are characterized by an aromatic ring with one or more hydroxyl groups. The group of phenolic compounds encompasses a heterogeneous set of secondary metabolites widely distributed in plant tissues and it consists of three subtypes: first, the polyphenols (lignin and tannin are examples); second, the oligophenols (flavonoids and coumarins are members of this group); and third, the monophenols (benzoic acid, cinnamic acid, gallic acid, salicylic acid, and vanillin are examples of this group). Plant polyphenols are multifunctional chemical entities involved in a myriad of functions in plant tissues. As a significant part of the plant phenolics remains present after the fall of branches and leaves or after the death of plants, phenolics may affect the digestive system in litter-consuming insects. For example, simple phenolic acids can affect the redox status of the midgut lumen during digestion or polyphenols can bind to proteins, decreasing the digestibility of ingested detritus.

Tannins are a class of plant polyphenols of particular interest due to their multiple activities, including their ability to precipitate alkaloids, amino acids, and proteins, and also due to their antioxidant activity. There are three major subtypes of tannins based on the identity of their monomers: first, the hydrolyzable tannins, whose monomers are gallic acids; second, phlorotannins, formed by phloroglucinol monomers and found in brown algae; and third, condensed tannins, which are polymers of heavily hydroxylated flavans. Plant tannins are mainly deposited in the vacuoles of plant cells, but they can also be found on the surface of leaves, seeds, pericarp, and roots. Softwoods may have a lower tannin content than hardwoods and the tannin contents decrease considerably during the fruit ripening process. There is evidence in the literature that condensed tannins bind non-specifically to dietary plant proteins lowering their digestibility and to proteins directly involved in digestion, such as digestive enzymes and nutrient transporters in non-adapted insects. Therefore, the presence of tannins in the diet requires adaptations that minimize these antinutritional effects. These include the presence of surfactants and high luminal pH values in the midgut.

3.2.3 Litter and Wood Quality

The associations of detritus and microorganisms assure direct contact between detritivores and numerous bacteria, fungi, and protozoans, which initiate the chemical degradation of the recalcitrant components and aid in alleviating nutritional imbalances in insects. Frequently it is the colonized detritus that would serve as an enriched food source for detritivores. Dead plant tissues like litter and wood are rich in carbohydrates and can serve as an energy source if digested, but they are poor in other macronutrients, impacting on the fitness of detritivorous insects. Bacterial N fixation and enzymatic secretion partially reduce the C/N and lignin/N ratios in the litter, whereas fungal activity can rearrange nutrients from the environment since the early stages of wood decay. Fungal hyphae are connected to nutritional sources of organic matter and minerals in rocks present nearby. The acquired nutrients are then translocated to the decaying leaves and wood through the fungal mycelium.

This microbial enrichment of dead plant tissues creates a better nutritional resource for saproxylophagous insects that allows them to grow, develop, and reproduce.

The consumption of plant parts before deposition or death also has an important impact on the chemical composition and quality of the litterfall or wood generated. Insect herbivory can alter nitrogen and phosphorus concentrations in living foliage and induce the production of secondary defensive metabolites such as alkaloids and tannins. These changes can affect litter quality parameters such as protein precipitation activity or Carbon/Nitrogen, Carbon/Phosphorus, and lignin/Nitrogen ratios (Chapman et al. 2003; Łukowski et al. 2021).

3.2.4 *Feces*

As a source of nutrients, fecal pellets are not very different in terms of chemical composition from decaying plant tissues. Compared to litter, feces have more favorable conditions for rapid colonization by microbes, such as a higher proportion of fragmented plant tissues, a higher surface/volume ratio, which facilitates retaining more moisture, having higher pH values, and exhaling volatile substances that attract detritivores. Colonization by bacteria and fungi not only improves parameters such as C/N and lignin/N ratios, but the bacteria and fungi themselves are an enriched food source compared to the original plant material. Ultimately, feces result in a more balanced diet for saproxylophagous insects compared to the diet of insects that deal directly with plant tissues without the prior processing that occurs in transit through the animals' intestinal tract. In addition to offering macro- and micronutrients per se, fecal pellets are also a source of microbial enzymes, microbial metabolites, and mutualistic microbiota that can help further nutrient extraction (Cammack et al. 2021).

3.2.5 *Carcasses*

The total biomass of vertebrate animals that is deposited as cadavers in terrestrial ecosystems is estimated to be approximately 1%, much less than 99% attributed to the deposition of plant-derived detritus (Carter et al. 2007). However, when animal carcasses are deposited, they represent a concentrated point of nutrients that attract a large community of scavengers, among which necrophilic insects have great ecological and forensic importance (Carter et al. 2007). It is worth mentioning that unlike what happens with feces, in carcass decomposition, bacterial action is more important than the action of fungi. These bacterial communities are formed by the preexisting resident microbiota of the animal and by the environmental microbiota existing where the carcass was deposited and is primarily responsible for attracting or repelling insects through the emission of volatile compounds. The development of scavenger insects also has a strong effect on bacterial growth, shaping microbial community formation and ecological succession in decaying carrion.

There are three different phases during animal body decomposition: early, middle, and late stages of decomposition (Stiegler et al. 2020). The succession patterns of bacteria and necrophagous insects on carcasses have been shown to be useful in estimations for "postmortem interval" (PMI) (Carter et al. 2007; Stiegler et al. 2020).

3.3 Plant Materials

3.3.1 Leaves

Leaves are characterized by their intense metabolic activity, being directly responsible for carrying out photosynthesis, which ultimately allows plants to be autotrophic organisms. For this reason, leaf tissues are among the most mechanically and chemically protected plant tissues. First, the cuticle, which is an acellular structure rich in wax and other hydrophobic compounds that coats the epidermis on both surfaces of the leaf. Second, the epidermis, which is a cellular layer located just below the cuticle on the upper and lower surfaces of the leaf lamina. Third, the mesophyll, which is organized as palisade parenchyma (vertically elongated columnar cells juxtaposed and oriented perpendicular to the leaf surface) and spongy parenchyma (formed by cells of irregular shape and not so compacted). Four, the veins which form the leaf vascular system and are in the spongy layer of the mesophyll. A vein is made up of a vascular bundle. At the core of each bundle, there are clusters of two distinct types of conductive cells: xylem cells that bring water and minerals from the roots to the leaf, and phloem cells that generally move sap, with assimilated compounds (products of photosynthesis, such as sucrose) from the leaf to the rest of the plant. The xylem cells are usually located closer to the external surface of the plant than phloem cells. Both xylem and phloem elements are embedded in a dense parenchyma tissue, called a sheath, which usually has more lignin deposits.

Host leaf nutritional quality is determinant for the fitness of leaf-feeding insects, and mechanical structures in plant leaves can limit foliage digestion. Usually, larval stages select their host leaves based on factors such as leaf nitrogen, water content, digestibility, leaf hardness and thickness, cuticle thickness, and chemical defenses, whereas during the adult stage, nitrogen content and tannin contents of the leaves seem to be more important (Rainford and Mayhew 2015; Tokuda 2019). Leaf anatomy can also influence its digestibility by determining the size and shape of particles after the insect mastication movements. Generally, the cuticle is not digested by insects, but instead it remains intact and binds to fragments of vascular tissues and other nondegraded cell wall components. The antinutritional role of lignin has been identified as a key factor limiting the digestibility of leaves, and lignin deposition acts dramatically to limit the degradation of certain leaf tissues. As leaves mature, sites of lignification increase. The deposition of lignin is also particularly dependent on stressful biotic or abiotic factors. Leaves with highly lignified tissues are the most resistant to digestion, but other foliar tissues that are not so rigid are sometimes also refractory to digestion due to antinutritional components such as toxic secondary metabolites. So, a combination of the insect mouth apparatus and leaf anatomic structures interact to shape the fragments of leaves, thereby promoting the passage of leaf residues through the intestinal tract and influencing access to nutrients. Normally, the fragments after insect digestion of leaves contain large proportions of the epidermis, sclerenchyma, and xylem cells, which are tissues generally resistant to digestion by non-adapted insects. The accumulation of certain minerals in leaf cuticles can also act as a feeding barrier to herbivorous insects. Silica and calcium contents absorbed by the roots can be deposited in the leaves of several plant species and impose difficulties in the digestion of leaf-feeding insects (Lanning et al. 1980; Prychid et al. 2004; Boldt and Altland 2021).

Leaves have organelles called plastids, exemplified by the chloroplast, which is responsible for carrying out photosynthesis. Other plastids are: chromoplasts, which store carotenoids; amyloplasts, which store starch granules; proteinoplasts, which store proteins; and elaioplasts which are specialized in storing lipids. Three main classes of lipids appear in plant cell membranes: glycerolipids, sphingolipids, and sterols. Of these, by far the most abundant and important are glycerolipids, which can be subdivided into: galactolipids (GL), phospholipids (PL), sulfolipids (SL), and triacylglycerols (TAG) (Reszczyńska and Hanaka 2020). In leaf photosynthetic membranes, however, the main constituents are galactolipids. These do not have phosphorus in their constitution and are divided into two classes: the monogalactosyldiacylglycerol (MGDG) and the digalactosyldiacylglycerol (DGDG). MGDG and DGDG have one or two galactose residues as polar headgroups and both are naturally enriched in 18 or 16 carbon polyunsaturated fatty acids (PUFAs) at the sn-1 and sn-2 positions. MGDG and DGDG can account for 50% and 25% of total thylakoid lipids, respectively. Galactolipids are much more abundant in the plant kingdom than TAG and they are also found in seeds and latex. In leaf-feeding insects, galactolipids may represent the main source of dietary fatty acids both as building blocks of membranes and as a source of energy.

3.3.2 *Plant Exudates*

Some vascular plants are characterized by two anatomical canal systems that are different from the prominent vascular systems formed by the xylem and phloem elements in the vascular bundles. These complementary tubular systems are called laticifers and secretory ducts. Both the secretory ducts and the laticifers are associated, at some point, with vascular bundles, from where the exchange of chemical substances and water can occur. The most obvious difference between laticifers and the secretory ducts is that latex is stored inside the laticifer cells, while in the secretory ducts, the synthesized material that can generate resin or gum is located extracellularly in the duct lumen (Dussourd 2021). A common feature of all liquefied material transported by the vascular systems of plants is that they can leak, flow, gel, dry, and solidify when there is a perforation in vascular bundles and ducts. The material that runs off from mechanical damage to plant tissue is known as exudates.

Plant exudates can be produced by different organs, such as roots, seeds, leaves, stems, and bark from trees or shrubs in response to mechanical injuries caused by several stressors, such as insect chewing or piercing, infections by bacteria or fungi, and abiotic stresses like drought (Kora 2021). Depending on their physical properties, chemical composition, and mode of production and storage by plant tissues, exudates can be classified as resins, gums, latex, or saps.

Resins are solid or semisolid amorphous exudates, which can be produced in almost any organ or tissue in the plant, depending on the species. Resins are insoluble in water and are formed from a complex mixture of low and high molecular weight organic compounds as pectic polysaccharides. Many resins are characterized by their contents of terpenes and mixtures of essential oils. Resins are stored under pressure in resin ducts and are released when these ducts are broken by herbivores. When dried, resins can seal the wounding site and in some cases, the aggressor can be intoxicated, be entrapped, or have its mouth parts immobilized by the sticky resin.

Gums exude from the site of damage to plant tissue initially as a sticky liquid that flows like tears, but which upon exposure to air dry into translucent, amorphous bodies. In the same way as resins, gums can trap attacking insects and help in the healing process of plant tissue. They consist of a complex mixture of pectic polysaccharides, glycoproteins, and some salts of Ca, K, and Mg. The most common of these polysaccharides are composed of glucuronic or galacturonic acid residues and derivatives with peptides. Gums are not soluble in organic solvents, but they are either water soluble or swell and change to a gel when soaked in water.

Latex, also known as milk sap, is a cytoplasmic exudate of specialized living cells, called laticifers that synthesize and accumulate it (Konno et al. 2004; Konno 2011). Not all latex has a milky appearance, although a sticky consistency is common. Latex contains inorganic ions and a variety of low and high molecular weight organic compounds, such as alkaloids, cardenolides, terpenoids, carbohydrates, amino acids, volatile compounds as well as various proteins and enzymes, such α-amylase and peptidase inhibitors, chitinases, glycosyl hydrolases, and peptidases, which can act as defensive compounds against insects (Konno 2011). Like resins, latex is also stored under pressure and is released in response to a mechanical damage to the laticifers. Besides chemical defense, latex can also coagulate when exposed to air and entrap feeding insects or immobilize their mouth parts (Dussourd and Eisner 1987; Dussourd and Hoyle 2000; Zalucki et al. 2001).

Plant saps are fluids transported in xylem cells or in phloem sieve tube elements, as described before. Despite their defensive roles (Zalucki et al. 2001; Agrawal and Konno 2009; Konno 2011; Dussourd 2021), except for xylem saps, the other types of exudates represent important sources of nutrients as polysaccharides, proteins, sugars, organic acids, free amino acids, and peptides. Many insect species have developed strategies to take advantage of these food sources, despite some large nutritional imbalances such as C/N and C/P ratios. Even the unlikely niche of using xylem sap as food is occupied by highly specialized insects like Auchenorrhyncha (see Chap. 14). Some exudates are so nutritious that they can sustain herbivory by sucking stink bugs followed by facilitated feeding of opportunist wasps and ants (Scaccini and Pozzebon 2021).

3.3.3 Pollen

Pollen grains consumed as food by insects are typically coated with a sticky, lipid-rich material called pollenkitt that covers the indigestible thick and sculptured outer wall exine. The inner wall intine is composed of cellulose and pectin (normally the pectin content is higher compared to other cell walls), and the pollen's protoplasm contains most of the nutrients available in this type of food. Numerous species from different orders of insects feed on pollen grains, including Collembola, Diptera, Coleoptera, Orthoptera, Thysanoptera, and Hymenoptera. Pollen grains are a nutrient-rich source for pollinators, providing carbohydrates (mainly starch); lipids; proteins; vitamins; sterols; minerals such as K, P, Fe, S, Ca, Cu, Mg, Zn, Na, and Mn; and other micronutrients. The protein/lipid ratios and contents of other nutrients in pollen grains differ considerably between plant species and environmental conditions (Ziska et al. 2016; Vaudo et al. 2020). Some plant species like conifers and grasses produce protein-poor pollens, whereas the protein contents in pollen can range from 2.5% to 61%, with an average of 25–45% in nectary-containing plants (Roulston and Cane 2000). Certain types of pollen have multiple layers of exine, which makes it more difficult to be broken down and release the nutrients contained in the protoplasm. The other defense front relies on the accumulation of secondary metabolites such as sesquiterpene lactones and antinutritional proteins, such as peptidase inhibitors (Roulston and Cane 2000; Vanderplanck et al. 2020).

3.3.4 Nectar

Nectar is a sugar-rich liquid produced by plants in glands called nectaries, either within the flowers to attract pollinators (floral nectar) or by extrafloral nectaries (extrafloral nectar), which provide a nutrient source to feed mutualists, which in turn provide protection from herbivores (González-Teuber and Heil 2009; Bogo et al. 2021). The chemical composition of nectar when secreted is dominated by the concentrations of three main sugars, the disaccharide sucrose and the hexoses glucose and fructose. These sugars can be present in different proportions, and in many nectars, glucose and fructose predominate and only trace amounts of sucrose are found. In addition to carbohydrates, different types of nectar may contain lower concentrations of proteins, canonic amino acids, non-protein amino acids, lipids, and a diversity of secondary metabolites, of which alkaloids and phenols are the most common (Adler 2000; Nepi 2014). Sucrose-rich nectars are generally preferred by Lepidoptera and long-tongued bees, while nectars rich in free hexoses are generally preferred by short-tongued bees and flies (Baker and Baker 1990; Bogo et al. 2021). Contamination due to floral visitors appears to be the main cause of major modifications of the nectar chemical composition (González-Teuber and Heil 2009; Bogo et al. 2021).

3.3.5 Seeds

The investment of plants in the resources allocated to seeds is large to ensure the propagation of the species genes. For this reason, concentrated amounts of nutrients are usually deposited in the seeds that guarantee the necessary resources for germination and the initial stages of seedling establishment. The combination of concentrated nutrients and the urgent need to avoid damage that compromises seed viability makes seeds among the most protected plant materials against herbivores and pathogens (Harborne 1993; War et al. 2012; Whitehead et al. 2013, 2021; Dalling et al. 2020).

Seed lines of defense are physical and chemical in nature and can be anywhere in the seed anatomy. Obviously, there is a wide variety of seed types, mainly in terms of size and concentration of potentially toxic compounds, ranging from secondary metabolites to poisonous proteins. Basically, seed coats are the seed's first defense front and may vary in chemical composition, hardness, thickness, and surfaces that make it difficult for pathogens and herbivores to access the precious embryo (Janzen 1977; De Sá et al. 2014, 2018). Protected by the seed coat, it is in the cotyledons or endosperm that the greatest nutrient reserves are deposited to be mobilized by the embryo during germination. Both cotyledons and endosperms are connected directly to the embryo via vascular tissue (xylem and phloem), allowing the passage of nutrients to the embryo during germination. In addition to their function as a store of carbohydrates, lipids, proteins, and micronutrients, cotyledons and endosperms also contribute to the seed's physical and chemical defenses. Some types of cotyledons and endosperms may have such a high hardness or such low humidity that they are difficult for ingestion or invasion by herbivores, including insects (Kiltie 1982; Dalling et al. 2020).

From the point of view of chemical composition, in addition to the lignocellulosic material present in the seed coat layers, there is also deposition of polyphenols, toxic proteins, and other secondary metabolites that can deter insect penetration through the seed coat (De Sá et al. 2014, 2018). The embryo and its cotyledons or endosperm contain a more diverse set of nutrients compared with the seed coat. Chemical composition analyses of several edible and non-edible seeds indicated that seeds are typically low in moisture and high in proteins and can be rich in lipids or starch (Chunhieng et al. 2004; Grosso et al. 2000). Generally, when the seed is very rich in lipids, it accumulates little energy in the form of starch and, vice versa, when a seed stores a lot of starch, it has moderate levels of lipids. There are also seeds with balanced levels of starch and lipids. The protein content can vary from 7.50% to 21%. Total soluble sugars range from 0.55% to 4%, whereas the starch content can reach 40% (w/w). Seeds are also rich sources of vitamins and minerals.

Seed proteins contain almost all the essential amino acids required by several herbivores, but normally seeds are deficient in lysine, methionine, threonine, and tryptophan. The diversity of proteins found in seeds is very large, ranging from storage proteins that can be multifunctional, such as vicilins in legume seeds (Fabaceae), to proteins with remarkable biological activities, such as abrin, arcelins, lectins, enzymes such as chitinases, glucanases, urease, inhibitors of peptidases, amylase inhibitors, and polygalacturonase inhibitors (Dalling et al. 2020). Some seeds also have peptides with insecticidal activity, such as defensins, which are peptides containing 44 to 54 amino acid residues and a three-dimensional structure reinforced by disulfide bonds, or cyclotides, which are cyclic peptides with about 30 residues and high insecticidal activity against Lepidoptera and Coleoptera and microorganisms (Pelegrini et al. 2007; Craik 2010). As mentioned above, seeds can be very rich in different types of carbohydrates, ranging from cellulose microfibers, starch granules, to different types of pectins. Some seeds contain in their endosperm high percentages of mucilage, in which the main component is galactomannans. Seed mucilage can form up to 30% of the seed composition with a corresponding decrease in the amount of starch stored in these seeds.

Plants that invest heavily in secondary metabolites in their leaves may also allocate the same compounds in seeds. Potentially toxic compounds of seeds can be as concentrated as those in leaves and roots or even more concentrated in certain cases. The diversity of secondary metabolites already described in seeds is very large. To illustrate part of this diversity, we mention some classes of compounds that can negatively or positively affect seed-feeding insects, such as vitamins on the positive side and non-protein amino acids such as L-canavanine, a structural analogue of arginine and very toxic to non-adapted insects; phytate, an anti-nutritional compound that affects the absorption of iron; alkaloids; toxic glycosides; tannin; furanocoumarins; and saponins. In many cases, the metabolic cost is high, as many of these compounds are synthesized from amino acids, reflecting the importance of investing in the successful propagation of the species through the seeds.

3.4 Bacteria

The ingestion of bacteria as a source of nutrients by insects has already been mentioned indirectly earlier in this chapter, for example, in the consumption of bacteria that grow in feces and mainly on decaying fruits and corpses. These microorganisms grow quickly in these environments and promote the bioconversion of original tissues of animal or plant origin into bacterial mass. Adapted insects are tolerant or immune to toxins secreted by bacteria and can break through the bacterial cell wall to gain access to the nutrients inside, which are better digested and absorbed than the original dead or infected material. The N-containing compounds of bacterial cell walls are mainly phospholipid membranes and peptidoglycans. Peptidoglycan architecture is the basis of Gram classification, by which bacteria are divided into either Gram-positive or Gram-negative. Gram-positive bacteria possess a thick multilayered peptidoglycan cell wall that is exposed to the cell exterior, whereas cell walls of Gram-negative bacteria have a thin, predominantly monolayered peptidoglycan covered by an outer lipid bilayer. Once the cell wall is broken, macro- and micronutrients from the cytoplasmic fraction may be rapidly taken up and efficiently used to build new insect biomass. Other details of bacteria as food of insects are found in Chap. 12.

3.5 Fungi

Fungal cells are formed by the plasma membrane, cytoplasm with organelles typical of heterotrophs, and a cell wall quite different in chemical composition from those of plant or bacterial cell walls. This cell wall is composed of multilayers and is basically made up of heteropolysaccharides and glycoproteins. Among the polysaccharides, glycans [α-(1,3)-glycans, β-(1,3)-glycans and β-(1,6)-glycans] and chitin (β-(1,4)-N-acetylglucosamine) are among the most common in fungal cells. The glycoproteins found are rich in mannose residues, forming the so-called mannoproteins. Other polymeric constituents that can be found in fungal walls in smaller amounts are chitosans (deacetylated chitin) and galactomannans. Lipids and melanin can also be found in fungal cell walls. Despite this generalization of wall constituents, the chemical components can vary among species of fungi (Free 2013; Biedermann and Vega 2020). Other details on the role of fungi in insect nutrition may be found in Chaps. 10 and 12.

3.6 Blood

Vertebrate blood is an example of a liquid diet that is very rich in protein by dry weight, but that is deficient in other nutrients such as carbohydrates, lipids, and vitamins. Whole blood demands adaptations from insects that have adapted to exploit this food resource. Nutrients are present in the two fractions of blood, which are: first, plasma, the liquid part that makes up about 55% of the volume. Second, the cellular fraction, composed mainly of red blood cells or erythrocytes with leukocytes and platelets in smaller amounts, making up 45% of the blood volume. Plasma is a source of proteins, such as fibrinogen, globulins, and serum albumin, which is a major blood protein, while erythrocytes are full of hemoglobin, the most abundant protein in the diet of hematophagous insects. Therefore, for the complete process of digestion, blood cells must be broken down for nutrients to be released. So, erythrocyte membrane lysis (hemolysis) is important for releasing large amounts of hemoglobin available for digestion. When hemoglobin is digested, a large amount of the free heme group is released. Free heme is potentially toxic due to its ability to generate reactive oxygen species, which requires adaptations to the high intake of this pro-oxidant molecule (Graça-Souza et al. 2006; Gandara et al. 2021).

3.7 Lichen

Lichens are composite organisms that arise from the symbiotic relationship between microalgae or cyanobacteria (known as photobionts) living among filaments (hyphae) of certain fungi species (the mycobionts). The part of a lichen that is not involved in reproduction is called the *thallus*. Most lichens develop a stratified thallus divided into the upper cortex, photobiont layer, medulla, and lower cortex. Many lichen species are eaten by insects such as some larvae of moths and termites due to their nutritional properties. Lichens are rich in carbohydrates and a good source of proteins and minerals, being relatively poor in lipids. Lichens have a unique chemical constitution with respect to secondary metabolism compounds. Most often, fungi are responsible for the synthesis and extracellular deposition of secondary metabolites in the filamentous part of the lichen, but some compounds are synthesized by the algae or synergistically by all partners (Zhao et al. 2021).

All lichen species investigated so far produce considerable amounts of polysaccharides (up to 50% of dry weight), and some authors have reported the immunostimulatory activity of these polymers. The lichen polysaccharides can be linear or branched homopolysaccharides (α- and β-glucans) or pectic heteropolysaccharides, mainly galactomannans and heteroxylans. Examples of these polysaccharides include chitin, cellulose, lichenin, isolichenin, thamnolan, and pustulan. The nutritional value of lichen is mainly due to their low-fat content, high carbohydrate content, abundant mineral elements, and good protein and vitamin sources.

References

Adler LS (2000) The ecological significance of toxic nectar. Oikos 91:409–420. https://doi.org/10.1034/j.1600-0706.2000.910301.x

Agrawal AA, Konno K (2009) Latex: a model for understanding mechanisms, ecology, and evolution of plant defense against herbivory. Annu Rev Ecol Evol Syst 40:311–331. https://doi.org/10.1146/annurev.ecolsys.110308.120307

Albersheim P, Darvill A, Roberts K et al (2011) Plant cell walls. Garland Science, New York

Baker HG, Baker I (1990) The predictive value of nectar chemistry to the recognition of pollinator types. Isr J Bot 39:157–166

Biedermann PHW, Vega FE (2020) Ecology and evolution of insect–fungus mutualisms. Annu Rev Entomol 65:431–455. https://doi.org/10.1146/annurev-ento-011019-024910

Bogo G, Fisogni A, Rabassa-Juvanteny J et al (2021) Nectar chemistry is not only a plant's affair: floral visitors affect nectar sugar and amino acid composition. Oikos 130:1180–1192. https://doi.org/10.1111/oik.08176

Boldt JK, Altland JE (2021) Petunia (*Petunia xhybrida*) cultivars vary in silicon accumulation and distribution. HortScience 56:305–312. https://doi.org/10.21273/HORTSCI15486-20

Caffall KH, Mohnen D (2009) The structure, function, and biosynthesis of plant cell wall pectic polysaccharides. Carbohydr Res 344:1879–1900. https://doi.org/10.1016/j.carres.2009.05.021

Cammack JA, Miranda CD, Jordan HR et al (2021) Upcycling of manure with insects: current and future prospects. J Insects Food Feed 7:605–619. https://doi.org/10.3920/JIFF2020.0093

Carter DO, Yellowlees D, Tibbett M (2007) Cadaver decomposition in terrestrial ecosystems. Naturwissenschaften 94:12–24. https://doi.org/10.1007/s00114-006-0159-1

Chapman SK, Hart SC, Cobb NS, Whitham TG, Koch GW (2003) Insect herbivory increases litter quality and decomposition: an extension of the acceleration hypothesis. Ecology 84:2867–2876. https://doi.org/10.1890/02-0046

Chunhieng T, Petritis K, Elfakir C et al (2004) Study of selenium distribution in the protein fractions of the Brazil nut (*Bertholletia excelsa*). J Agric Food Chem 52:4318–4322. https://doi.org/10.1021/jf049643e

Craik DJ (2010) Discovery and applications of the plant cyclotides. Toxicon 56:1092–1102. https://doi.org/10.1016/j.toxicon.2010.02.021

Dalling JW, Davis AS, Arnold AE et al (2020) Extending plant defense theory to seeds. Annu Rev Ecol Evol Syst 51:123–141. https://doi.org/10.1146/annurev-ecolsys-012120-115156

De Sá LFR, Wermelinger TT, Ribeiro ES et al (2014) Effects of *Phaseolus vulgaris* (Fabaceae) seed coat on the embryonic and larval development of the cowpea weevil *Callosobruchus maculatus* (Coleoptera: Bruchidae). J Insect Physiol 60:50–57. https://doi.org/10.1016/j.jinsphys.2013.10.004

De Sá LFR, Ventury KE, Machado OLT et al (2018) Toxic effect of *Canavalia ensiformis* seed coat on larval development of *Callosobruchus maculatus*. J Pest Sci 91:313–326. https://doi.org/10.1007/s10340-017-0895-3

Dussourd DE (2021) Does secretory canal architecture determine the sabotage behaviors of insect folivores? Arthropod Plant Interact 15:71–81. https://doi.org/10.1007/s11829-020-09798-x

Dussourd DE, Eisner T (1987) Vein-cutting behavior: insect counterploy to the latex defense of plants. Science 237:898–901

Dussourd DE, Hoyle AM (2000) Poisoned plusiines: toxicity of milkweed latex and cardenolides to some generalist caterpillars. Chemoecology 10:11–16. https://doi.org/10.1007/PL00001810

Free SJ (2013) Fungal cell wall organization and biosynthesis. Adv Genet 81:34–71. https://doi.org/10.1016/B978-0-12-407677-8.00002-6

Futuyma DJ, Agrawal AA (2009) Macroevolution and the biological diversity of plants and herbivores. Proc Natl Acad Sci U S A 106:18054–18061. https://doi.org/10.1073/pnas.090410610

Gandara AC, Dias FA, de Lemos PC et al (2021) Urate and NOX5 control blood digestion in the hematophagous insect *Rhodnius prolixus*. Front Physiol 12:633093. https://doi.org/10.3389/fphys.2021.633093

González-Teuber M, Heil M (2009) Nectar chemistry is tailored for both attraction of mutualists and protection from exploiters. Plant Signal Behav 4:809–813. https://doi.org/10.4161/psb.4.9.9393

Graça-Souza AV, Maya-Monteiro C, Paiva-Silva GO et al (2006) Adaptations against heme toxicity in blood-feeding arthropods. Insect Biochem Molec Biol 36:322–335. https://doi.org/10.1016/j.ibmb.2006.01.009

Grosso NR, Nepote V, Guzman CA (2000) Chemical composition of some wild peanut species (*Arachis* L.) seeds. J Agric Food Chem 48:806–809. https://doi.org/10.1021/jf9901744

Harborne JB (1993) Introduction to ecological biochemistry; Elsevier Academic Press: London, ISBN 9780080918587

Ishii T, Matsunaga T (1996) Isolation and characterization of a boron-rhamnogalacturonan-II complex from cell walls of sugar beet pulp. Carbohydr Res 284:1–9. https://doi.org/10.1016/0008-6215(96)00010-9

Ishii T, Matsunaga T, Hayashi N (2001) Formation of rhamnogalacturonan II-borate dimer in pectin determines cell wall thickness of pumpkin tissue. Plant Physiol 126:1698–1705. https://doi.org/10.1104/pp.126.4.1698

Janzen DH (1977) How southern cowpea weevil larvae (Bruchidae: *Callosobruchus maculatus*) die on nonhost seeds. Ecology 58:921–927

Kiltie RA (1982) Bite force as a basis for niche differentiation between rain forest peccaries (*Tayassu tajacu* and *T. pecari*). Biotropica 14:188–195

Konno K (2011) Plant latex and other exudates as plant defense systems: roles of various defense chemicals and proteins contained therein. Phytochemistry 72:1510–1530. https://doi.org/10.1016/j.phytochem.2011.02.016

Konno K, Hirayama C, Nakamura M et al (2004) Papain protects papaya trees from herbivorous insects: role of cysteine proteases in latex. Plant J 37:370–378. https://doi.org/10.1046/j.1365-313X.2003.01968.x

Kora AJ (2021) Exudate tree gums: properties and applications. In: Inamuddin MIA, Boddula R, Altalhi T (eds) Polysaccharides: properties and applications. Scrivener Publishing LLC

Lanning FC, Hopkins TL, Loera JC (1980) Silica and ash content and depositional patterns in tissues of mature *Zea mays* plants. Ann Bot 45:549–554

Lodha BC (1974) Decomposition of digested litter. In: Dickinson CH, Pugh GJP (eds) Biology of plant litter decomposition. Academic, London, pp 213–241

Łukowski A, Giertych MJ, Żmuda M et al (2021) Decomposition of herbivore-damaged leaves of understory species growing in oak and pine stands. Forests 12(3):304. https://doi.org/10.3390/f12030304

Mohnen D (2008) Pectin structure and biosynthesis. Curr Opin Plant Biol 11:266–277. https://doi.org/10.1016/j.pbi.2008.03.006

Nepi M (2014) Beyond nectar sweetness: the hidden ecological role of non-protein amino acids in nectar. J Ecol 102:108–115. https://doi.org/10.1111/1365-2745.12170

Pelegrini PB, Quirino BF, Franco OL (2007) Plant cyclotides: an unusual class of defense compounds. Peptides 28:1475–1481. https://doi.org/10.1016/j.peptides.2007.04.025

Popper ZA, Fry SC (2007) Xyloglucan–pectin linkages are formed intra-protoplasmically, contribute to wall-assembly, and remain stable in the cell wall. Planta 227:781–794. https://doi.org/10.1007/s00425-007-0656-2

Prychid CJ, Rudall PJ, Gregory M (2004) Systematics and biology of silica bodies in monocotyledons. Botanic Rev 69:377–440. https://doi.org/10.1663/0006-8101(2004)069[0377:SABOSB]2.0.CO;2

Rainford JL, Mayhew PJ (2015) Diet evolution and clade richness in Hexapoda: a phylogenetic study of higher taxa. Am Nat 186:777–791. https://doi.org/10.5061/dryad.6f75v

Reszczyńska E, Hanaka A (2020) Lipids composition in plant membranes. Cell Biochem Biophys 78:401–414. https://doi.org/10.1007/s12013-020-00947-w

Roman-Palacios C, Scholl JP, Wiens JJ (2019) Evolution of diet across the animal tree of life. Evol Lett 3:339–347. https://doi.org/10.1002/evl3.127

Roulston TH, Cane JH (2000) Pollen nutritional content and digestibility for animals. Pl Syst Evol 222:187–209. https://doi.org/10.1007/BF00984102

Scaccini D, Pozzebon A (2021) Invasive brown marmorated stink bug (Hemiptera: Pentatomidae) facilitates feeding of European wasps and ants (hymenoptera: Vespidae: Formicidae) on plant exudates. Eur J Entomol 118:24–30. https://doi.org/10.14411/eje.2021.003

Shelomi M, Wipfler B, Zhou X et al (2020) Multifunctional cellulase enzymes are ancestral in Polyneoptera. Insect Mol Biol 29:124–135. https://doi.org/10.1111/imb.12614

Simon J-C, d'Alençon E, Guy E et al (2015) Genomics of adaptation to host plants in herbivorous insects. Brief Funct Genom 14:413–423. https://doi.org/10.1093/bfgp/elv015

Stiegler J, Von Hoermann C, Müller J et al (2020) Carcass provisioning for scavenger conservation in a temperate forest ecosystem. Ecosphere 11:e03063. https://doi.org/10.1002/ecs2.3063

Tokuda G (2019) Plant cell wall degradation in insects: recent progress on endogenous enzymes revealed by multi-omics technologies. Adv Insect Physiol 57:97–136. https://doi.org/10.1016/bs.aiip.2019.08.001

Vanderplanck M, Gilles H, Nonclercq D et al (2020) Asteraceae paradox: chemical and mechanical protection of *Taraxacum* pollen. Insects 11:304. https://doi.org/10.3390/insects11050304

Vaudo AD, Tooker JF, Patch HM et al (2020) Pollen protein: lipid macronutrient ratios may guide broad patterns of bee species floral preferences. Insects 11:132. https://doi.org/10.3390/insects11020132

War AR, Paulraj MG, Ahmad T et al (2012) Mechanisms of plant defense against insect herbivores. Plant Signal Behav 7:1306–1320. https://doi.org/10.4161/psb.21663

Whitehead SR, Jeffrey CS, Leonard MD et al (2013) Patterns of secondary metabolite allocation to fruits and seeds in *Piper reticulatum*. J Chem Ecol 39:1373–1384. https://doi.org/10.1007/s10886-013-0362-4

Whitehead SR, Schneider GF, Dybzinski R et al (2021) Fruits, frugivores, and the evolution of phytochemical diversity. Oikos 00:1–16. https://doi.org/10.1111/oik.08332

Zalucki MP, Malcolm SB, Paine TD et al (2001) It's the first bites that count: survival of first-instar monarchs on milkweeds. Aust Ecol 26:547–555. https://doi.org/10.1046/j.1442-9993.2001.01132.x

Zhao Y, Wang M, Xu B (2021) A comprehensive review on secondary metabolites and health-promoting effects of edible lichen. J Funct Foods 80:104283. https://doi.org/10.1016/j.jff.2020.104283

Ziska LH, Pettis JS, Edwards J et al (2016) Rising atmospheric CO_2 is reducing the protein concentration of a floral pollen source essential for North American bees. Proc R Soc B Biol Sci 283:20160414. https://doi.org/10.1098/rspb.2016.0414

Chapter 4
Ordinary Digestive Enzymes

Abstract Procedures to obtain reliable enzyme kinetic parameters are discussed. The classification of enzymes regarding their function (substrate specificity and mechanism, Enzyme Commission Nomenclature) and protein families (MEROPS, CAZy, and Brenda platforms) is presented. The mechanism of the enzymes acting on the major nutrient substrates is described. The substrates discussed are carbohydrates (starch, oligosaccharides, and disaccharides); peptides (proteins, oligopeptides, and dipeptides); and acylglycerol esters (triacylglycerols and phospholipids). The enzymes discussed in detail are: serine endopeptidases (trypsins, chymotrypsins, elastases); collagenases; carboxypeptidases; aminopeptidases; amylases; α-glucosidases; β-glucosidases; β-fructosidases; β-N-acetylhexosaminidases; α-mannosidases; α-galactosidases; myrosinases; trehalases; triacylglycerol lipases; and phospholipases. Lysosomal proteins recruited as digestive enzymes (cysteine- and aspartic-endopeptidases) and enzymes involved in the degradation of cell walls are the subject of specific chapters.

4.1 Introduction

Enzymes are classified into hydrolases, oxidoreductases, transferases, lyases, isomerases, ligases, and translocases depending on the type of reaction they catalyze and are numbered according to the Nomenclature Committee (Enzyme Commission, EC) of the International Union of the Biochemistry and Molecular Biology Societies (IUBMB). Digestive enzymes are hydrolases, except laccases which are oxidoreductases of the group of multicopper oxidases.

Hydrolases are separated according to the type of chemical bond they cleave. Peptidases (peptide hydrolases, EC 3.4) split peptide bonds between amino acids, glycosyl hydrolases cleave glyosidic links, and lipases broke ester bonds, all of which are diagrammatically represented in Fig. 2.3. General information for all enzymes may be found in the website Brenda (www.brenda-enzymes.org).

W. R. Terra et al., *Molecular Physiology and Evolution of Insect Digestive Systems*, Entomology in Focus 7, https://doi.org/10.1007/978-3-031-39233-7_4

Peptidases are divided into endopeptidases (EC 3.4.21-29) and exopeptidases (EC 3.2.4.11-19). Endopeptidases attack internal bonds in peptides and are classified into subclasses according to the catalytic mechanism. Serine endopeptidases (EC 3.4.21) have a serine and a histidine in the active site; cysteine endopeptidase (EC 3.4.22), a cysteine; aspartic endopeptidases, an aspartic acid; and, finally, metalloendopeptidases need a metal ion in the catalytic mechanism. Exopeptidases comprise enzymes that remove single (occasionally two) amino acids from the C-terminus (carboxypeptidases, EC 3.4.16-18) or from the N-terminus (aminopeptidases, EC 3.4.4.11) of the peptide chain and the enzymes specific for dipeptides (dipeptide hydrolases, EC 3.4.13) (see Fig. 2.3). Peptidases are also classified in families according to their structures, exemplified by S1, family of serine endopeptidases and M14, for metallocarboxypeptidases. Details of this kind of classification may be found on the website "merops.sanger.ac.uk" (Rawlings et al. 2018). In this chapter, we will discuss only the exopeptidases and the serine endopeptidases like trypsin, chymotrypsin, and elastase. The other endopeptidases will be dealt in Chap. 9.

Glycosyl hydrolases (EC 3.2) are grouped depending on their substrate specificities. Endopolisaccharidases are named from their substrates, as exemplified by amylase, cellulase, pectinase, and chitinase. Only amylase is the subject of this chapter. The other endopolysaccharidases will be described in Chap. 10. Oligosaccharidases and disaccharidases are classified according to the monosaccharide that gives the reducing group to the glycosidic link and on bond configuration (α or β) (see Fig. 2.3). Glycosyl hydrolases are also classified based on their structures into families (Glycosyl hydrolases families, GHF). Details on this classification are available in CAZy (Carbohydrate active enzymes database, www.cazy.org) (Dunla et al., 2022). Glycoside hydrolases use two catalytical mechanisms, one retaining and one inverting the configuration of the anomeric carbon of the substrate. In retaining glycosidases there is a glycosyl-enzyme intermediate, and after the hydrolysis of this intermediate, the resulting product has the same configuration of the anomeric carbon of the substrate. In inverting glycosidases, hydrolysis occurs in one step, and the configuration of the anomeric carbon of the substrate is changed. The active sites of glycosyl hydrolases are divided into subsites, numbered −1, −2, −3, and so on from the point of cleavage toward the nonreducing end of the substrate and +1, +2, +3, and so on from this point toward the reducing end (Davies et al. 1997). The negative subsite(s) is also called the glycone site and the positive(s) one(s) the aglycone site.

Enzymes that hydrolyze lipids are varied because lipids are a large and heterogenous group insoluble in water, but easily solubilized in apolar solvents. These enzymes are: (a) carboxylic ester hydrolases (EC 3.1.1), exemplified by lipases, esterases, and phospholipases A and B; (b) phosphoric diester hydrolases (EC 3.1.4) that include phospholipases C and D; and (c) phosphoric monoester hydrolases (EC 3.1.3) and phosphatases (Fig. 2.3).

Before advancing to analyzing the specific digestive enzymes, we will discuss the procedures to assure reliable enzyme assays. Enzyme assays are relatively simple, but are somewhat tricky and may easily produce results impossible to interpret or compare to others in the literature.

4.2 Reliable Enzyme Assays

Enzymes are proteins (with few exceptions as some RNA) that are able to transform one or more substances (named enzyme substrates) into other(s), which are named enzyme products.

The amount of enzyme is measured by its activity. One unit of enzyme activity corresponds to the amount of enzymes that are able to change one μmol of substrate per minute in defined conditions of pH and temperature. For this, we maintain the enzyme with its substrate in an appropriate buffer for a period of time and then calculate the amount of substrate transformed by minute. The assay conditions must be chosen to assure that the enzyme activity does not change during the incubation period. The observed constant velocity of enzyme action is called initial velocity. As the reaction advances, several factors may influence the enzyme activity, which begins to decline. The time during which the enzymatic reaction conserves the initial velocity depends on the reaction itself and on the conditions it is maintained. A practical way to know if the reaction is occurring in initial velocity is to verify if the substrate consumption (or product formation) is constant during the elapsed time. Thus, substrate will decrease (and product will increase) linearly with time.

The linearity of product increase may be verified using assays with continuous recording of product formation or using test tubes with identical compositions that will be incubated for different periods of time before product detection.

Incubating enzymes for different periods of time has other advantages besides showing the integrity of the enzyme activity. Some substrates may be contaminated with a certain amount of product, and depending on the assay done, the enzyme preparation can also bring this contamination to the assay. This occurs, for example, when the enzyme is producing glucose and we are using cell or tissue homogenates that contain glucose. The enzyme activity is calculated from the slope of the straight line formed by the product versus time, discounting the amount of product found at time zero.

As mentioned before, the initial velocity is the constant velocity found for some time, from the start of the reaction on, and it is the higher velocity that can be achieved in the conditions of the assay. It is worth differentiating initial velocity from maximal velocity, that is the enzymatic velocity achieved when the enzyme is saturated with its substrate (i.e., the enzyme is almost always bound to the substrate). From the definition of enzyme activity, it is clear that how many units of enzyme are present in a sample can be calculated, even if the enzyme is not the sole protein present in the assay.

4.3 Peptidases

4.3.1 Serine Endopeptidases

Serine endopeptidases are homologous to chymotrypsin in having the catalytic triad (serine, histidine, and aspartic acid) in the active site. Because of that, the numbering of the amino acid residues mentioned below is referred to as bovine chymotrypsin (Barrett et al. 2012).

Trypsin (EC 3.4.21.4) is an endopeptidase that preferentially breaks the peptide bond of the protein chain on the carboxyl side of L-amino acids such as Arg or Lys. Trypsins that have been purified from many insects have molecular weights in the range 20–35 kDa and usually have alkaline pH optima (Terra and Ferreira 2012). Trypsins are found in all insect midguts, except hemipterans, although some of them have trypsin in the salivary glands (Zeng et al. 2002).

Trypsin is synthesized as proenzymes, depending on activation (removal of the activation peptide), and has conserved N-terminal residues IUGG, three pairs of cysteine residues, and, in a pocket, the residue Asp189 that determines its substrate specificity (Terra and Ferreira 2012). Despite the overall similarity with mammalian trypsins, insect trypsins differ from them in not being activated or stabilized by calcium ions, and in some preferences for substrates. For example, whereas most mammalian and insect trypsins have Arg as primary specificity, lepidopteran trypsins prefer Lys. Furthermore, by analyzing the hydrolysis data of 60 peptides, differing in residues around the primary site of hydrolysis, it was found that there is a trend of insect trypsins of more evolved insects to have active sites more hydrophobic, which has as a consequence an increase in the resistance to plant protein inhibitors with their hydrophilic binding sites (Lopes et al. 2006; see also Chap. 11). Another conclusion from those sets of data is that trypsins from different insects bind their substrates or the transition states (high energy intermediate of reaction) with different strengths. Therefore, trypsin subsites differently favor substrate binding or catalysis (Marana et al. 2002). Trypsins are coded by polygenic families, and how this affects their interactions with plant inhibitors is the subject of Chap. 11.

Chymotrypsins are endopeptidases that preferentially break protein chains at the carboxyl side of aromatic amino acids (Barrett et al. 2012). They have been purified from many insect sources and have their molecular weight and sequence features similar to trypsin, except for their specificity pocket, which has Ser/Gly/Tyr 189, instead of Asp 189 (Terra and Ferreira 2012). The chymotrypsin of the fire ant had its 3D structure solved and the structure found was similar to the one of vertebrates. However, it differs from that of the vertebrates in the activation mechanism and in the structure of the subsites around the cleaving site (Botos et al. 2000). In agreement with this, many differences were found among insect and mammalian chymotrypsins, based on kinetics data obtained with several peptides differing in the amino acid residues around the preferential breaking bond (Sato et al. 2008). Insect chymotrypsin structural characteristics that may affect their interaction with inhibitors are discussed in Chap. 11.

Elastase (EC 3.4.21.36) preferentially cleaves peptide bonds at the carboxyl side of amino acids with small hydrophobic lateral groups (e.g., Ala) (Barrett et al. 2012). This enzyme has been characterized in the cricket *Teleogryllus commodus* (Christeller et al. 1990), in the gypsy moth (Valaitis 1995), and in the fire ant *Solenopsis invicta*, from which the enzyme was sequenced and shown to be more similar to chymotrypsin than to known elastases (Whitworth et al. 1998). However, more work is needed regarding demanding for this enzyme in insects.

4.3.2 Collagenases

Collagen and hyaluronic acid are the major components of the extracellular matrix that maintain together the cells in animal tissues (Alberts et al. 2008). The hydrolysis of collagen or hyaluronic acid destroys the intercellular cement setting free the cells. Collagenases are enzymes that hydrolyze bonds in the triple helix of collagen. The best-known insect collagenase is the one of the endoparasitic cattle fly *Hypoderma lineatum*, which enters through the skin of the host, digesting its extracellular matrix. The sequence of this enzyme is homologous to the serine endopeptidases of the chymotrypsin family (Lecroisey et al. 1979, 1987). Another kind of insect collagenase is the collagenase of family 9 of the metallopeptidases found in the saliva of the predator hemipteran *Podisus nigrispinus*, which is used to disperse the host tissue cells before ingestion (Fialho et al. 2012).

4.3.3 Carboxypeptidases

Carboxypeptidases remove one amino acid residue from the C-terminal of the peptide chain. Taking into account their mechanisms of action they are divided into two major groups: metallocarboxypeptidases (EC 3.4.17) and serine carboxypeptidases (EC 3.4.16). The insect digestive carboxypeptidases are mainly metallocarboxypeptidases from the M14 family (Rawlings et al. 2018; merops.sanger.ac.uk) that need a divalent metal for activity. Their molecular masses are in the range 20–50 kDa and their optimal pH is alkaline (Terra et al. 2019). Carboxypeptidases are secreted as pro-enzymes that can be activated by trypsin action.

Digestive M14 carboxypeptidases are classified into carboxypeptidase A, which releases amino acids with aromatic or branched hydrocarbon chains, and carboxypeptidase B, which removes Lys or Arg from the C-terminus of their substrates. Another kind of midgut M14 carboxypeptidase was described in the midgut of *Helicoverpa armigera* and named carboxypeptidase MC (Bown and Gatehouse 2004). The enzyme removes Glu from the C-terminus of the peptide chain. The specificity of M14 carboxypeptidases is mainly determined by one amino acid residue that is present at the bottom of the pocket, where the terminal residue of the substrate binds. This amino acid is neutral (polar or hydrophobic) in carboxypeptidase A, Asp or Glu in carboxypeptidase B, and Arg or Lys in carboxypeptidase MC. The specificities of carboxypeptidases A and B allow these enzymes to continue the digestion of peptides resulting after the action of chymotrypsin and trypsin, respectively.

Carboxypeptidase A from *Helicoverpa armigera* (Estébanez-Perpiñá et al. 2001) had its 3D structure revealed. It is similar to the mammalian M14 carboxypeptidases. However, it has a Ser residue in the S1'subsite (see nomenclature of peptide subsites in Chap. 9, Fig. 9.1), instead of Ile or Asp, found in human carboxypeptidases A and B, respectively, and also has an enlarged substrate preference.

The crystal structure of *Aedes aegypti* carboxypeptidase B was determined (Gavor et al. 2021), and the interaction of this sensitive enzyme with potato carboxypeptidase inhibitor was compared to that in the insensitive *Helicoverpa zea* carboxypeptidase B (Bayés et al. 2005). The data revealed that

loop regions in the active site region of *Aedes aegypti* carboxypeptidase B are replaced by alpha-helices in *Helicoverpa zea* carboxypeptidase B, thus hampering the binding of the inhibitor into the active site (Gavor et al. 2021). *Aedes aegypti* carboxypeptidase B has a role in Dengue virus development (Gavor et al. 2021). The enzyme binds the virus, inhibiting its maturation and reducing its liberation into insect hemolymph.

The gene coding for the most expressed carboxypeptidase A of *S. frugiperda* was cloned and a recombinant enzyme was produced and used to raise antibodies in a rabbit (Ferreira et al. 2015). By immunocytolocalization with the antibodies, the enzyme was found inside small vesicles in the anterior midgut cells and inside large vesicles in the cells from the posterior midgut. These results indicate that enzyme secretion occurs by a microapocrine mechanism (see secretory mechanisms in Chap. 5) in the anterior midgut and by classical exocytosis in the posterior region. The results also showed that most carboxypeptidase A is secreted in the posterior midgut (Ferreira et al. 2015).

A prolyl carboxypeptidase (EC 3.4.16.2) was described in the anterior midgut lumen of *T. molitor* larvae (Goptar et al. 2013). It is a serine carboxypeptidase from family S28 that removes one amino acid residue bound to the carboxyl group of a Pro residue. It is active in acidic pH and is found in lysosomes. *T. molitor* has in its midgut lumen several enzymes that are present in lysosomes from other organisms (see Chap. 9). This insect feeds on cereals; and the role of this prolyl carboxypeptidase may be the digestion of gliadin, a protein present in cereals that is rich in Pro residues.

Some insects, such as hemipterans and bruchid beetles, do not have M14 carboxypeptidases in their midguts (Ferreira et al. 2015). They have instead carboxypeptidases from family S10, usually found in lysosomes. Hemipterans are known to have lost serine proteinases on adapting to sap-feeding and on returning to high protein diets recruited lysosomal endopeptidases (Chap. 9). Similarly, two S10 carboxypeptidases expressed only in the midgut are found in the gut lumen of *D. peruvianus* (Ferreira et al. 2015). They are predicted to be one CPC, hydrolyzing predominantly C-terminal hydrophobic amino acid residues and CPD, hydrolyzing mainly C-terminal basic amino acid residues. With these specificities, they are able to continue the digestion initiated by cathepsin D and L, respectively, which are major endopeptidases in these insects (see Chap. 9).

4.3.4 Aminopeptidases

Aminopeptidases are metalloenzymes that sequentially hydrolyze amino acid residues from the N-terminus of peptides. Aminopeptidases preferentially remove Ala and Leu (Aminopeptidase N, EC 3.4.11.2) or Asp (Aminopeptidase A, EC 3.4.11.7) from peptides (Norén et al. 1986).

Aminopeptidases in less evolved insects (e.g., Orthoptera, Hemiptera, and Coleoptera Adephaga) have large amounts of soluble aminopeptidases in midgut contents. In more evolved insects (Coleoptera Polyphaga, Diptera, and Lepidoptera), aminopeptidases are chiefly bound to the microvillar membranes of midgut cells (Terra and Ferreira 1994). Insect aminopeptidases are aminopeptidases N, except in *Rhynchosciara americana*, which have in addition an aminopeptidase A (Klinkowstrom et al. 1994). Insect aminopeptidases N from different sources were sequenced and all of them have a signal peptide, a zinc-binding motif, and a C-terminal sequence with a phosphatidyl inositol anchor (Terra and Ferreira 2012). The active site of some insect aminopeptidases was studied with the use of multiple inhibition analysis, protection against EDTA inactivation, the effect of other enzyme inactivators, etc. The data showed that the active site of the *R. americana* microvillar aminopeptidases has a hydrophobic domain which, once occupied, causes conformational changes associated with the catalytic step, and a polar subsite (Ferreira and Terra 1986). The active site of *T. molitor* aminopeptidase N includes a metal ion, a carboxylate, and a deprotonated imidazole that are necessary for catalysis, which is influenced by an arginine in the neighborhood of the active site. Furthermore, a phenol and carboxylate group are associated with substrate binding (Cristofoletti and Terra 2000).

Di-, tri-, and pentapeptides are hydrolyzed by insect aminopeptidases N, although most of them prefer tripeptides (Baker and Woo 1981; Cristofoletti and Terra 1999; Cristofoletti et al. 2006; Bozic et al. 2008). Aminopeptidases N may account for up to 55% of the proteins in the midgut microvillar membranes of *T. molitor* (Cristofoletti and Terra 1999). Probably because of that, those enzymes are targets of the *Bacillus thuringiensis* δ-endotoxins. These endotoxins, after binding to aminopeptidases or another receptor protein, form channels through which cell contents extrude, causing insect deaths (Knight et al. 1995). Except for dipterans and lepidopterans, all insects seem to have a single aminopeptidase N. Dipterans have several aminopeptidases N with different substrates specificities (see above). The large diversity of aminopeptidases N in lepidopterans is associated with the resistance against *B. thuringiensis* endotoxins (Hughes 2014), despite the lack of studies on the substrate specificities of lepidopteran aminopeptidases N.

We saw that oligopeptides of different sizes, including dipeptides, are substrates for insect aminopeptidases N. The released amino acids may be absorbed by amino acid transporters, whereas oligopeptides may be absorbed with the aid of the H^+-oligopeptide symporter in the midgut of insects (see Chap. 6), thus it is not unexpected that midgut dipeptidases have low activity or are non-existent in insects (Terra and Ferreira 2012).

4.4 Glycosyl Hydrolases

4.4.1 Amylases and α-Glucosidases

α-Amylases (EC 3.2.1.1), hereafter named simply amylases, catalyze the break of internal bonds in linear α-1,4-glucan chains, like starch and glycogen, resulting in maltose, maltotriose, and branched maltodextrins. These products will be further digested under the action of maltases (α-glucosidases, see below). The major part of the insect amylases has molecular weights in the range 48–60 kDa, K_m values around 0.1% for soluble starch, and pH optima corresponding to the pH of the midgut contents from where the amylases were isolated (Terra and Ferreira 1994, 2012; D'Amico et al. 2000).

Insect amylases need calcium ions for activity or stability, are activated by chloride ions with the displacement of pH optimum (lepidopteran amylases are exceptions) (Terra and Ferreira 1994, 2012), and frequently are not very active on intact starch granules (Celinska et al. 2015), making mastication necessary (Silva et al. 2001).

The action pattern of amylases is a description of the pattern of products they form. The average number of links hydrolyzed after the first bond is broken is named degree of multiple attack. Liquefying amylases are those that have no multiple attack, that is, they perform a single round of catalysis each time they bind to the substrate. Saccharifying (or processive) amylases carried out multiple catalytic events every time it attacks the substrate. The salivary gland amylase from *Periplaneta americana* is liquefying, whereas its midgut amylase is saccharifying (Tamaki et al. 2014). Also saccharifying are the midgut amylases from *Rhynchosciara americana, Sitophilus granarius, S. zeamais, S. oryzae,* and *Bombyx mori* (Terra and Ferreira 2012).

The only insect amylase whose 3D structure was solved is that one of *T. molitor* larvae (Strobl et al. 1998). The enzyme substrate–binding site is located between domains A and B; the calcium ion site is placed in domain B, and domain C occurs in the C-terminal end of the enzyme. The chloride-binding site is in domain A. A flexible loop protruding close to the catalytic cleft, thought to be an ancestral characteristic, is absent from many insect amylase sequences (Da Lage 2018).

Muscomorpha insects have, in addition to amylases, a paralog with amylolytic activity, amyrel, which can be chloride-dependent or chloride independent. In chloride-dependent amyrel, the addition of chloride does not modify its pH optimum nor the enzyme affinity for starch. The physiological role of amyrel is still unknown (Da Lage et al. 1998; Claisse et al. 2016).

Insect amylases pertain to the glycosyl hydrolase subfamily GH13_15 and those of vertebrates belong to GH13_24 (Stam et al. 2006). Amylase genes vary from one (e.g., honeybees) to more than 12 (some mosquitoes) and are absent from the blood-sucker *Pediculus humanus* and the sap-feeding aphid *Acyrthosiphon pisum*, but unexpectedly are present in the blood-feeding *Rhodnius prolixus* (Da Lage 2018). A possibility in the last case is that the amylase was recruited to another function associated with hematophagy, as occurred with the α-glucosidase that in *R. prolixus* catalyzes the formation of hemozoin from the heme of hemoglobin (Da Lage 2018).

α-Glucosidases (EC 3.2.1.20) remove sequentially α-1,4 linked glucoside residues from the terminal nonreducing end of disaccharides or oligosaccharides. Because α-glucosidases hydrolyze maltose (glucose α-1,4 glucose), they are frequently named maltases. Insect maltases hydrolyze α-1,4 and α-1,6 linkages from the nonreducing ends of oligosaccharides produced by amylases, including the residual branched maltodextrins (Takewaki et al. 1993). It should be stressed that the terminal digestion of starch in mammals is very different from insects. Oligosaccharides produced by mammalian amylases are digested by maltase-glucoamylase (EC 3.2.1.13) and sucrase-isomaltase (EC 3.2.1.10) (Boron and Boulpaed 2017), which coding genes are absent from insect genomes. Mammalian maltases pertain to family GH 31, whereas insect maltases belong to glycosyl hydrolase family GH13, as amylase (Gabriško and Janeček 2011). All insect maltase sequences have the invariant residues pertaining to the active site of the amylase family of enzymes (Darboux et al. 2001) and three other residues Gly 69, Pro 77, and Gly 323 *(C. pipiens* maltase numbering), supposed to have a structural function in some α-glucosidases (Janecek 1997).

Insect maltases also hydrolyze sucrose, which is α-1,2- β fructoside, from the glucoside moiety, whereas fungi hydrolyze fructose from the fructoside moiety. Fungi β-fructosidase is usually named sucrase. Thus, the insect enzyme hydrolyzing sucrose should not be called sucrase, but maltase. It is interesting to remark that there is an increasing number of reports on insect midgut β-fructosidases (see next item). Insect maltases usually prefer as substrate oligosaccharides up to at least maltopentaose (Terra and Ferreira 1994). However, the digestive maltase from *D. peruvianus* (Silva and Terra 1995) and *Quesada gigas* (Fonseca et al. 2010) prefer oligosaccharides up to maltotriose and the enzyme from bee, up to maltotriose (Nishimoto et al. 2001).

Insect digestive maltases may be soluble in the midgut contents or immobilized at the midgut cell surface by entrapping in the cell glycocalyx or by binding in the cell microvillar (Terra and Ferreira 1994), perimicrovillar (Silva and Terra 1995), or in the modified perimicrovillar (Cristofoletti et al. 2003) membranes. The binding of insect maltases in cell membranes may use unorthodox anchors. The MdMal-A2 of *M. domestica* anchor seems to be a prolongation of the signal peptide (Pimentel et al. 2018) and, in the case of the maltase from *A. pisum,* apparently it is a predominantly hydrophobic C-terminal region, but which is not predicted to be a peptide anchor (Price et al. 2007).

4.4.2 β-Fructosidases

In the majority of insect midguts, sucrose is hydrolyzed by an α-glucosidase, but in more recent years reports of the presence of a β-fructosidases in the midgut of Lepidoptera and Coleoptera increased (Santos and Terra 1986; Daimon et al. 2008; Carneiro et al. 2004; Pedezzi et al. 2014). β-Fructosidases (EC 3.2.1.26) are hydrolases that recognize and remove one β-fructosyl residue from the nonreducing end of a di-, oligo-, or polysaccharide. They have sucrose, raffinose, nistose, kestose, and inulin as substrates.

Insect β-fructosidases are members of the family 32 of Glycoside hydrolases. They are retaining enzymes with an Asp and Glu residue as catalytical groups and have a five-fold propeller at the N-terminal end, where the active site is located, and a β-sandwich at the C-terminal end (Verhaest et al. 2005).

As the insect β-fructosidase sequences branch with the sequences of bacterial enzymes, it is believed that the enzyme was acquired from these organisms by horizontal gene transfer (Daimon et al. 2008). Accordingly, the structure of the β-fructosidase from *B. mori* was resolved and it is more similar to the bacterial enzymes than to the eukaryotic ones (Miyazaki et al. 2020).

An immunohistochemistry study detected the presence of β-fructosidase in the goblet cell cavity of *B. mori* larval midgut (Daimon et al. 2008). This result may be taken with care because, as discussed for trehalase below, histochemistry caused technical artifacts.

The role of β-fructosidase was initially presumed to enable *B. mori* larvae to eat on mulberry leaves (Daimon et al. 2008). These leaves have sugar mimic alkaloids that inhibit more α-glucosidases (responsible for sucrose hydrolysis in the majority of insects) than β-fructosidases (Miyazaki et al. 2020). Nevertheless, other Lepidoptera species are not resistant to the compounds present in the mulberry leaves (Konno et al. 2006), although β-fructosidases have been found in all Lepidoptera insects, doubting the role described for the enzyme. Unquestionably, β-fructosidase is important in *B. mori* metabolism. RNAi was used to decrease the enzyme activity in *B. mori* larvae, resulting in a smaller body size and reduced amount of trehalose and glycogen in the fat body (Gan et al. 2018).

Fructans are present as a reserve of energy in more than 36,000 plant species (Quin et al. 2023) and are found as stock substance in 15% of Angiosperms (Hendry 1993). In many Coleoptera and in Lepidoptera insects, where β-fructosidases were detected, the enzyme may have some role in digesting other fructans, besides sucrose, present in the food. Anyhow, further studies are necessary to elucidate the role of β-fructosidases in insects.

4.4.3 *β-N-Acetylhexosaminidases, α-Mannosidases, and α-Galactosidases*

β-N-Acetylhexosaminidases or β-N-acetylglucosaminidase (EC 3.2.1.52) catalyze the hydrolysis of terminal nonreducing β-N-acetylhexosamine residues from β-N-acetylhexosaminides, like chitooligosaccharides and from glycoproteins with terminal N-acetylglucosamine. In insects, those enzymes are from family GH20 and were found in the molting fluid, hemolymph, and gut, where they finish chitin degradation started by chitinases (Muthukrishnan et al. 2012). A search in many insect genomes showed that there are four groups of them. Group I includes the enzymatically well-characterized β-N- acetylglucosaminidases; Group II does not differ much from Group I; Group III corresponds to enzymes involved in intracellular N-glycan processing; and, finally, Group IV comprises the β-N-acetylglucosaminidases with a substrate specificity wider than the usual β-N-acetylglucosaminidases, and to remark this difference they are named as β-N-Acetylhexosaminidases (Mark et al. 2003; Leonard et al. 2006; Hogenkamp et al. 2008).

The suppression of the expression of β-N-acetylglucosaminidase coding genes of *T. castaneum* showed that they are all associated with the degradation of cuticle in the moltings. However, as some of these genes are also highly expressed in gut tissues, their coded enzymes were thought to play a role in peritrophic membrane turnover (Hogenkenkamp et al., 2008). This role is clear in *Aedes aegypti*, whose peritrophic membrane is synthesized following blood-feeding, accompanied by an increase in β-N-acetylglucosaminidase synthesis (Filho et al. 2002). Another possibility, at least for seed beetles which feed on fungi-contaminated flour, is that the β-N-acetylglucosaminidase finishes the digestion of fungi cell wall started by a chitinase devoid of chitin-binding domain, presumably for not attacking the peritrophic membrane (Genta et al. 2006).

α-Mannosidases (EC 3.2.1.24) are enzymes that hydrolyze the nonreducing residue of mannose from glycoconjugates. Animal α-mannosidases are separated into Class I α-mannosidases, which are of family GH47, and Class II α-mannosidases, pertaining to family GH38. Class I α-mannosidases hydrolyze α-1,2 mannose links in the endoplasmic reticulum or Golgi Complex. Class II α-mannosidases cleave α-1,2, α-1,3, and α-1,6 mannose bonds. Golgi Class II α-mannosidases cleave α-1,6 and α-1,3 linked mannose residues, whereas the lysosomal ones hydrolyze α-1,2, α-1,3, and α-1,6 linked mannose residues (Gonzalez and Jordan 2000). Data on insect recombinant α-mannosidases of family GH47 (Ren et al. 1995; Kawar and Jarvis 2001) and from both Golgi and lysosomal α-mannosidases of family GH38 are available (Nemcovicova et al. 2012).

Insect digestive α-mannosidases, now known to be similar to the GH38 lysosomal α-mannosidases, have been described in several insects, mainly in those with diets with a high raffinose content (*Dysdercus peruvianus*, Silva and Terra 1994) or in insects having fungi-contaminated food (*Tenebrio molitor*, Terra et al. 1985; Moreira et al. 2015). As described earlier, *T. molitor* midgut has a laminarinase that degrades fungal cell walls (Genta et al. 2009). α-Mannosidases are thought to remove mannose units from the α-mannosides occurring in the outermost layer of fungi cell walls (Martinez-Esparza et al. 2006).

α-Galactosidases (EC 3.2.1.22) catalyze the hydrolysis of α-D-galactosidic bonds in the nonreducing terminus of galactooligosaccharides, like melibiose, raffinose, stachiose, and galactolipids, exemplified by digalactosyldiacylglycerol (Dey and Pridham 1972). Galactooligosaccharides are common in lipid-rich seeds and galactolipids that are widespread in the chloroplast membranes of leaves (Harwood 1980). Insect digestive α-galactosidases have been partially purified. The enzymes can be more active on raffinose than on melibiose, like that of *D. peruvianus* (Silva and Terra 1997) or the contrary as in *Abracris flavollineata* (Ferreira et al. 1999). Several different α-galactosidases were partially purified from *T. molitor* and *Spodoptera frugiperda* (Grossmann and Terra 2001).

4.4.4 β-Glycosidases

β-Glycosidases (EC 3.2.1.21) are enzymes that hydrolyze the terminal, nonreducing residue of glucose or another monosaccharide from the substrate. They are able to hydrolyze di- and oligosaccharides, alkylglucosides, and toxic glucosides produced by plants and, in some cases, also hydrolyze polysaccharides (Ferreira et al. 2001). β-glycosidases are important enzymes for the insects, because they are responsible for the intermediary and final digestion of cellulose and hemicelluloses, which are abundant in plant cell walls, releasing glucose and oligomers. Those enzymes are also important in insect-plant interactions, since plants synthesize many different toxic β-glycosides.

The number of β-glycosidases present in the insect midgut may vary, with some being able to hydrolyze a wide range of substrates, and others a few types of β-glycosides (Ferreira et al. 1998, 2001, 2003; Azevedo et al. 2003). Digestive β-glucosidases prefer β-glucosides or β-galactosides. Some have high activity against substrates with hydrophobic moieties in the aglycone, whereas others are more active against substrates with hydrophilic aglycone moieties (Terra and Ferreira 1994). The physiological role of β-glycosidases acting on hydrophobic substrates may be the hydrolysis of galactolipids that are present in high amount in plant tissues. The prevailing galactolipids in plants are 2,3-diacyl β-galactoside D-glycerol (monogalactosyl diglyceride) and 2,3-diacyl 1-alpha galactosyl 1,6 β-galactosyl D-glycerol (digalactosyl diglyceride) (Harwood,1980). These enzymes may act after the removal of the galactose residue by α-galactosidase. In agreement with this hypothesis, β-Gly47 from *S. frugiperda* hydrolyzes ceramide, although with low activity (Marana et al. 2000).

Insect β-glycosidases best characterized have molecular masses from 30 to 150 kDa, 4.5 to 6.5 optimum pH, pI values from 3.7 to 6.8, and Km values for cellobiose or p-nitrophenyl β-glucoside from 0.2 to 2 mM. Free energy relationships

(Withers and Rupitz 1990) can show how similar active sites are, indicating the types and strength of interactions among substrate and amino acid residues in the active site. This approach is used to compare the enzyme specificities, and Azevedo et al. (2003) showed that β-glycosidases found in the midgut of a Coleoptera (*T. molitor)* and a Lepidoptera (*D. saccharalis)* are more similar to each other than to enzymes from the same insect. These results indicate that some β-glycosidases have important properties that are conserved in different orders.

The enzymes hydrolyzing di- and oligosaccharides, which active sites were studied, have four subsites for glucose binding. One is the glycone-binding site (+1) and the other three are in the aglycon-binding site (−1, −2, and −3) (Ferreira et al. 2001, 2003; Marana et al. 2001; Azevedo et al. 2003). Enzymes with high activity against disaccharides have a high affinity to glucose in the −1 and + 1 subsites. The ones hydrolyzing mainly oligosaccharides have similar affinity to glucose in the −1, +1, and +2 subsites (Ferreira et al. 2001, 2003).

Insect β-glycosidases are also named cellobiase, because they hydrolyze cellobiose, the disaccharide resulting from cellulose digestion. Nevertheless, many midgut enzymes are more active against laminaribiose (glucose β-1,3 glucose), indicating that they participate in the final digestion of β-1,3 glucans of plants and fungi (Terra and Ferreira 1994, 2012). Insect midgut β-glycosidases are mainly from Glycoside hydrolases family 1 (Drula et al. 2022) that pertain to Clan A, which has proteins with a (β/α)8 barrel structure. They are retaining enzymes with two catalytical Glu residues. A comparison of β-glycosidases from Family 1 revealed that catalysis-related amino acid residues and the ones present in the glycone-binding site are more conserved than the amino acid residues present at the aglycone-binding site (Tamaki et al. 2016). This may be a consequence of a high diversity of aglycones found in nature, which are substrates of the GH1 β-glycosidases. As a consequence of this structure, mutations in one amino acid residue located at the aglycone-binding site have a lower effect in Km and kcat than a mutation in a catalysis-related residue or present at the glycone-binding site (Tamaki et al. 2016).

One mechanism by which plants defend themselves from herbivorous insects is producing toxic- β-glucosides. They generally have glucose bound to a varied aglycone. Toxic β-glucosides may be present in concentrations such as 0.5% to 1% of the vegetal organ weight, are widespread in plants, and only the cyanogenic ones are produced by 11% of plants (Schoonhoven et al. 2005) corresponding to more than 2500 species (Zagrobelny et al. 2004). Midgut insect β-glucosidases have different specificities and can hydrolyze a great diversity of these β-glucosides that have in most cases deleterious effects. Insects avoid the problem caused by the release of the aglycone by detoxifying them (Spencer 1988) or by decreasing only the activity of the specific β-glycosidase that hydrolyzes the toxic β-glycoside, without affecting the other β-glycosidases (Ferreira et al. 1997; Pankoke et al. 2012). This is an advantage for those insects that have many β-glucosidases with different specificities in the gut.

The crystal structures of the β-glycosidases from the termite *Neotermes koshunensis* (Jeng et al. 2011) and the Lepidoptera *S. frugiperda* (Tamaki et al. 2016) were also solved and are similar to other β-glycosidases from GH1 family.

4.4.5 *Myrosinase*

Myrosinase (EC 3.2.147) is a member of the glycosyl hydrolase family 1 that hydrolyzes glucosinolates, which are β-D-thioglucosides. The aphid *Brevicoryne brassicae* has a midgut myrosinase and its encoding cDNA was cloned and sequenced (Jones et al. 2002) and its 3D structure resolved (Husebye et al. 2005). The enzyme has two catalytic Glu residues (plant myrosinases have only one) and its sequence is more similar to animal β-O-glucosidases than to plant myrosinases. Perhaps insect myrosinases should be considered β-glucosidases with specificities to both O- and thio-glucoside bonds.

4.4.6 *Trehalases*

Trehalase (EC 3.2.1.28) hydrolyzes the disaccharide trehalose into two glucose molecules. The enzyme is important in fungi, nematodes, and insects but, in spite of this, it is not extensively studied. Trehalose is the main circulating sugar in insects and due to this, every tissue must have a trehalase to provide an energy source for cellular metabolism.

Insect trehalases are members of Family 37 of glycosyl hydrolases that includes only trehalases. This family has an (alpha/alpha)6 barrel and are inverting enzymes. Insect midguts have a soluble and a membrane-bound trehalase. The presence of a membrane-bound enzyme in the basal membranes of midgut cells was suggested by cell fractionation (Capella et al. 1997). The membrane-bound enzyme can hydrolyze trehalose from hemolymph, delivering glucose to the cells.

Membrane-bound trehalases have a predicted transmembrane region in its C-terminal end. In cladograms, membrane-bound trehalases branch together but are separated from the soluble enzymes in the majority of insect orders (Gomez et al. 2013). This separation indicates that the trehalase gene duplicated and diverged before Condylognatha+Phthyraptera and Holometabola were set apart. The membrane-bound trehalases, which are supposed to perform the basic role of delivering glucose to cells from hemolymph trehalose, presumably give rise to the soluble ones after losing the C-terminal transmembrane portion. The soluble and membrane-bound Diptera trehalases branched together in cladograms. Apparently in evolution, the Diptera lost the soluble trehalase and regained it after duplication and divergence of the membrane-bound gene (Gomez et al. 2013).

Mitsumasu et al. (2005), using immunohistochemistry, reported the localization of soluble trehalase in the goblet cell cavity in the midgut of *Bombyx mori* larvae. Silva et al. (2009) raised antibodies against *Spodoptera frugiperda* soluble trehalase in rabbits and used them to immunocytolocalize the enzyme. Several control media, including saline-buffer solution, followed by immunofluorescence visualization showed fluorescence in goblet cells. When the anti-trehalase serum was added, fluorescence was detected in goblet cells plus columnar cell microvilli regions. When immunogold labeling was analyzed under an electron microscope, trehalase labeling was found inside small vesicles in anterior columnar cells, inside large vesicles in posterior columnar cells, and associated with columnar cells microvilli. No labeling was detected in controls or in goblet cells. These results indicate that trehalase, like many other midgut enzymes, is secreted only by columnar cells and its detection in Lepidoptera goblet cells is due to technical artifacts.

S. frugiperda trehalase structure was modeled by Silva et al. (2010) using the only trehalase with 3D structure available at the time, the trehalase from *Escherichia coli* (Gibson et al. 2007). D322 and E520 (*S. frugiperda* numbering) were confirmed as, respectively, the basic and acid catalyst using site-directed mutagenesis. The trehalase mutants D322A and E520A loss activity and showed the disappearance of, respectively, the ionizable group with higher or lower pKa, which in plots of activity versus pH imply in the absence of the right or left branch of the bell-like curve of activity versus pH. The authors also showed that three Arg residues, present in the active center, are as important as the catalytic residues for *S. frugiperda* activity, since the production of the mutants R169A, R222A, and R287A led to enzymes with kcat values as low as the ones having the catalytical carboxyl residues mutated. The removal of Arg residues from the enzyme conducted to changes in the pK values of the catalytical groups.

Trehalases have two presumed subsites for glucose binding, identified as subsite −1 and +1 by the structural study of *E. Coli* trehalase (Gibson et al. 2007). Based on protection against amino acid residues, chemical modification (Silva et al. 2004, 2010) showed that in *S. frugiperda* trehalase, the subsite −1 contains the basic catalyst (nucleophile) and binds the competitive inhibitor methyl α-glucoside and mandelonitrile. The subsite +1 contains the acid catalyst (proton donor) and binds a Tris molecule. Based on the fact that Tris inhibition occurs at pH 9 (where Tris have a positive charge), but not at pH 6 (where Tris is a neutral molecule), subsite +1 might have a positive charge.

Kinetical studies performed with *S .frugiperda* trehalase by Silva et al. (2004) showed that the enzyme undergoes conformational changes after ligand binding. The His residue reactant diethyl pyrocarbonate does not change trehalase activity, but in the presence of the competitive inhibitor methyl α-glucoside, the activity decreases and the pKa of the proton donor changes. This result indicates a conformational modification after inhibitor binding that exposed the His residue to the reactant. Gibson et al. (2007) solved the 3D structure of *E. coli* trehalase complexed with inhibitors. They showed that the compounds were buried within the structure, indicating that important conformational changes would be necessary to bind the substrate and release the products.

Soluble trehalase from *S. frugiperda* was modeled with *E. coli* periplasmic trehalase, and the results indicated that the N- and C-terminal ends of the insect enzyme could have loops responsible for trehalase motility (Silva et al. 2010, 2015). Silva et al. (2015), trying to identify regions responsible for this mobility in *S. frugiperda* trehalase, produced two truncated enzymes, one with 101 N-terminal and one with 101 N-terminal plus 31 C-terminal residues missing. Comparison of wild type with these truncated enzymes showed that Km and Ki for competitive inhibitors were the same, whereas kcat values decreased. These results indicate that the position of the catalytical residues, but not of the substrate binding residues, changed after mutation. The study also showed that modification of an His residue only occurs if the truncated enzymes are incubated with a competitive inhibitor, as demonstrated to occur with the wild type, indicating that conformational changes still take place after the removal of N- and C- terminal portions. Spyro Orange dye increases fluorescence when it binds to hydrophobic regions of the protein, and was used to detect conformational changes in the wild type and mutated enzymes after additions of substrate or inhibitors (Silva et al. 2015). The results showed that the wild type was less affected by ligands than the truncated enzymes, indicating that the conformational changes occur mainly near the active site.

Adhav et al. (2019) studied the structure of *Enterobacter cloacae* and compared structures of trehalases obtained with and without ligands. They concluded that there are three regions, called hood domain, side loop, and lid hoop (comprising, respectively, 65, 13, and 14 amino acid residues), that might undergo conformational changes during the catalytical cycle. Both *S. frugiperda* truncated mutant trehalases (Silva et al. 2015) lack the region corresponding to the *E. cloacae* hood domain, but contain the regions corresponding to the side and loop domains, which could be responsible for *S. frugiperda* trehalase structural changes.

One *Drosophyla melanogaster* partial trehalase sequence that does not contain the section corresponding to the hood domain and the lid loop was expressed (Shukla et al. 2016). Since the enzyme is still active, the lid loop rich in glycine residues is not essential for enzyme motility. Further studies are necessary to clarify the role of amino acid residues involved in trehalase conformational changes.

4.4.7 Lipases and Phospholipases

Important components of many insect diets are triacylglycerols (TAGs), such as oils and fats, and phospholipids, like phosphatides and galactolipids, occurring in cell membranes (Vance and Vance 2008). TAG lipases remove fatty acids from tri- and diacylglycerols. Phospholipase A1 hydrolyzes the fatty acid from position 1 of the phosphatide, phospholipase A2, from position 2 and, finally, phospholipase B, from both positions (see Fig. 2.3). The phosphatide depleted of one acyl ester becomes a lysophosphatide, which has detergent properties (Canavoso et al. 2001). It should be noticed that phospholipase A2 is found among insects, but belongs to a family different from the other lipases (Stanley 2006) and will not be treated here. Galactolipids are mostly mono- and digalatosyl diacylglycerols abundant in the chloroplast. After the removal of galactose residues by the sequential action of α-galactosidase and β-galactosidase, the resulting diacylglycerol is substrate for TAG lipases.

TAG lipases belong to different families (Derewenda 1994), from which only the neutral (or pancreatic) and acid (or gastric) lipases are of interest in the case of insects (Horne et al. 2009). Neutral TAG lipases of mammals have a domain, named PLAT domain, that binds colipase and enhances TAG activity by facilitating TAG association with the enzyme (Miled et al. 2000). The β-loop and lid are sequences near the active site that are associated with substrate specificity. Enzymes with long sequences are active only on TAGs. Those enzymes with short sequences hydrolyze only phosphatides and, finally, the enzymes with long β-loops and lid sequences hydrolyze phosphatides and TAGs (Aoki et al. 2007; Horne et al. 2009). Acid lipases differ from neutral lipases in having a core domain and a Cap domain containing the lid structure covering the active site (Miled et al. 2000).

Insect genomes do not have genes coding for colipases and, in accordance with that, insect neutral lipases lack the PLAT domain and have short β-loops and short lid sequences, showing they are phospholipases (Horne et al. 2009). Christeller et al. (2010), based on the expression of different lipases from lepidopterans feeding diets with a varied lipid composition, advanced the proposal that neutral lipases hydrolyze phosphatides, whereas acid lipases, TAGs. Favoring this view is the finding of Barroso et al. (2021) that neutral lipases are expressed in the anterior midgut and acid lipases in the middle and posterior midgut. According to them, the neutral lipases (actually phospholipases A1) hydrolyze phosphatides in the anterior midgut, releasing lysophosphatides that emulsify fats, which make them easier to be digested by acid lipases in the middle and posterior midgut.

Work done with crude or partially purified TAG lipases preparations showed that insect lipases preferentially remove unsaturated fatty acids from the external positions of the TAG and are activated by calcium (Bollade et al. 1970; Tsuchida and Wells 1988; Majerowicz and Gondim 2013).

References

Adhav A, Harne S, Bhide A et al (2019) Mechanistic insights into enzymatic catalysis by trehalase from the insect gut endosymbiont *Enterobacter cloacae*. FEBS J 286:1700–1716

Alberts B, Johnson A, Lewis J, Raff M, Roberts K, Walter P (2008) Molecular biology of the cell, 5th edn. Garland Science, New York, p 1268p

Aoki J, Inoue A, Makide K et al (2007) Structure and function of extracellular phospholipase A1 belonging to the pancreatic lipase gene family. Biochimie 89:197–204

Azevedo TR, Terra WR, Ferreira C (2003) Purification and characterization of three β-glycosidases from midgut of the sugar cane borer, *Diatraea saccharalis*. Insect Biochem Mol Biol 33:81–92

Baker JE, Woo SM (1981) Properties and specificities of a digestive aminopeptidase from larvae of *Attagenus megatoma* (Coleoptera: Dermestidae). Comp Biochem Physiol B 69:189–193

Barrett AJ, Rawlings ND, Wosner JF (2012) Handbook of proteolytic enzymes, 3rd edn. Elsevier Academic Press, London

Barroso IG, Cardoso C, Ferreira C et al (2021) Transcriptomic and proteomic analysis of the underlying mechanisms of digestion of triacylglycerols and phosphatides and absorption and fate of fatty acids along the midgut of *Musca domestica*. Comp Biochem Physiol D 100826

Bayés A, Comellas-Bigler M, de la Vega MR et al (2005) Structural basis of the resistance of an insect carboxypeptidase to plant protease inhibitors. Proc Natl Acad Sci U S A 102:16602–16607

Bollade D, Paris R, Moulins M (1970) Origine et mode d'action de la lipase intestinale chez les blattes. J Insect Physiol 16:45–53

Boron WF, Boulpaed EL (2017) Medical physiology. Elsevier, Philadelphia

Botos I, Meyer E, Nguyen M et al (2000) The structure of an insect chymotrypsin. J Mol Biol 298:895–901

Bown DP, Gatehouse JA (2004) Characterization of a digestive carboxypeptidase from the insect pest corn earworm (*Helicoverpa armigera*) with novel specificity towards C- terminal glutamate residues. Eur J Biochem 271:2000–2011

Bozic N, Ivanovic J, Nenadovic V et al (2008) Purification and properties of a major leucyl aminopeptidase of *Morimus funereus* (Coleoptera: Cerambycidae) larvae. Comp Biochem Physiol B 149:454–462

Canavoso LM, Jouni ZE, Karnas KJ et al (2001) Fat metabolism in insects. Ann Rev Nut 21:23–46

Capella AN, Terra WR, Ribeiro AF et al (1997) Cytoskeleton removal and characterization of the microvillar membranes isolated from two midgut regions of *Spodoptera frugiperda* (Lepidoptera). Insect Biochem Mol Biol 27:793–801

Carneiro CN, Isejima EM, Samuels RI et al (2004) Sucrose hydrolases from the midgut of the sugarcane stalk borer *Diatraea saccharalis*. J Insect Physiol 50:1093–1101

Celinska E, Bialas W, Borkowska M et al (2015) Cloning, expression, and purification of insect *(Sitophilus oryzae)* alpha-amylase, able to digest granular starch, in *Yarrowia lipolytica* host. Appl Microbiol Biotech 99:2727–2739

Christeller JT, Laing WA, Shaw BD et al (1990) Characterization and partial purification of the digestive proteases of the black field cricket, *Teleogryllus commodus* (Walker): elastase is a major component. Insect Biochem 20:157–164

Christeller JT, Poulton J, Markwick NM et al (2010) The effect of diet on the expression of lipase genes in the midgut of the lightbrown apple moth (*Epiphyas postvittana* Walker; Tortricidae). Insect Molec Biol 19:9–25

Claisse G, Feller GMB, Da Lage J-L (2016) A single amino-acid substitution toggles chloride dependence of the alpha-amylase paralog amyrel in *Drosophila melanogaster* and *Drosophila virilis* species. Insect Biochem Mol Biol 75:70–77

Cristofoletti PT, Terra WR (1999) Specificity, anchoring, and subsites in the active center of a microvillar aminopeptidase purified from *Tenebrio molitor* (Coleoptera) midgut cells. Insect Biochem Mol Biol 29:807–819

Cristofoletti PT, Terra WR (2000) The role of amino acid residues in the site of a midgut microvillar aminopeptidase from the beetle *Tenebrio molitor*. Biochem Biophys Acta 1479:185–195

Cristofoletti PT, Ribeiro AF, Deraison C et al (2003) Midgut adaptation and digestive enzyme distribution in a phloem feeding insect, the pea aphid *Acyrthosiphon pisum*. J Insect Physiol 49:11–24

Cristofoletti PT, Mendonça de Sousa FA, Rahbe Y et al (2006) Characterization of a membrane-bound aminopeptidase purified from *Acyrthosiphon pisum* midgut cells: a major binding site for toxic mannose lectins. FEBS J 273:5574–4488

D'Amico S, Gerday C, Feller G (2000) Structural similarities and evolutionary relationships in chloride-dependent α-amylases. Gene 253:95–105

Da Lage J-L (2018) The amylases of insects. Int J insect Sci 10:1–14

Da Lage J-L, Renard E, Chartois F et al (1998) *Amyrel*, a paralogous gene of the amylase gene family in *Drosophila melanogaster* and the *Sophophora subgenus*. Proc Natl Acad Sci U S A 95:6848–6853

Daimon T, Taguchi T, Meng Y et al (2008) Beta-fructofuranosidase genes of the silkworm, *Bombyx mori:* insights into enzymatic adaptation of *B.mori* to toxic alkaloids in mulberry latex. J Biol Chem 283:15271–15279

Darboux I, Nielsen-LeRoux C, Charles JF et al (2001) The receptor of *Bacillus sphaericus* binary toxin in *Culex pipiens* (Diptera: Culicidae) midgut: molecular cloning and expression. Insect Biochem Mol Biol 31:981–990

Davies GJ, Wilson KS, Henrissat B (1997) Nomenclature for sugar-binding subsites in glycosyl hydrolases. Biochem J 321:557–559

Derewenda ZS (1994) Structure and function of lipases. Adv Protein Chem 45:1–52

Dey PM, Pridham JB (1972) Biochemistry of α-galactosidases. Adv Enzymol 36:91–130

Drula E, Garron M-L, Dogan S et al (2022) The carbohydrate-active enzyme database: functions and literature. Nucleic Acids Res 50:D571–D577

Estébanez-Perpiñá E, Bayés A, Vendrell J et al (2001) Crystal structure of a novel midgut procarboxypeptidase from the cotton pest *Helicoverpa armigera*. J Mol Biol 313:629–638

Ferreira C, Terra WR (1986) Substrate specificity and binding loci for inhibitors in an aminopeptidase purified from the plasma membrane of midgut cells of an insect *Rhynchosciara americana* larva. Arch Biochem Biophys 244:478–485

Ferreira C, Parra JRP, Terra WR (1997) The effect of dietary plant glycosides on larval midgut β-glycosidases from *Spodoptera frugiperda* and *Diatreaea saccharalis*. Insect Biochem Mol Biol 27:55–59

Ferreira C, Torres BB, Terra WR (1998) Substrate specificities of midgut β-glycosidases from insects of different orders. Comp Biochem Physiol B 119:219–225

Ferreira C, Marana SR, Silva C (1999) Properties of digestive glycosidases and peptidases and the permeability of the peritrophic membranes of *Abracris flavolineata* (Orthoptera: Acrididae). Comp Biochem Physiol B 123:241–250

Ferreira AHP, Marana SR, Terra WR et al (2001) Purification, molecular cloning, and properties of a β-glycosidase isolated from midgut lumen of *Tenebrio molitor* (Coleoptera) larvae. Insect Biochem Mol Biol 31:1065–1076

Ferreira AHP, Terra WR, Ferreira C (2003) Characterization of a β-glycosidase highly active on disaccharides and of a β-galactosidase from *Tenebrio molitor* midgut lumen. Insect Biochem Mol Biol 33:253–265

Ferreira C, Rebola KGO, Cardoso C et al (2015) Insect midgut carboxypeptidases with emphasis on S10 hemipteran and M14 lepidopteran carboxypeptidases. Insect Mol Biol 24:222–239

Fialho MCQ, Moreira NR, Zanuncio JC (2012) Prey digestion in the midgut of the predatory bug *Podisus nigrispinus* (Hemiptera: Pentatomidae). J Insect Physiol 58:850–856

Filho BP, Lemos FJ, Secundino NF et al (2002) Presence of chitinase and beta-N-acetylglucosaminidase in Aedes aegypti. A chitinolytic system involving peritrophic matrix formation and degradation. Insect Biochem Mol Biol 32:1723–1729

Fonseca FV, Silva JR, Samuels RI et al (2010) Purification and partial characterization of a midgut membrane-bound α-glucosidase from *Quesada gigas* (Hemiptera: Cicadidae). Comp Biochem Physiol B 155:20–25

Gabriško M, Janeček S (2011) Characterization of maltase clusters in the genus Drosophila. J Mol Evol 72:104–118

Gan Q, Zhang X, Zhang D et al (2018)BmSUC1 is essential for glycometabolism modulation in the silkworm, *Bombyx mori*. 1861:543-553

Gavor E, Choong YK, Jobichen C et al (2021) Structure of *Aedes aegypti* carboxypeptidase B1-inhibitor complex uncover the disparity between mosquito and non-mosquito insect carboxypeptidase inhibition mechanism. Protein Sci 30:2445–2456

Genta FA, Blanes L, Cristofoletti PT et al (2006) Purification, characterization and molecular cloning of the major chitinase from *Tenebrio molitor* larval midgut. Insect Biochem Mol Biol 36:789–800

Genta FA, Bragatto I, Terra WR et al (2009) Purification, characterization and sequencing of the major β-1,3-glucanase from the midgut of *Tenebrio molitor* larvae. Insect Biochem Mol Biol 39:861–874

Gibson RP, Gloster TM, Roberts S et al (2007) Molecular basis for trehalase inhibition revealed by the structure of trehalase in complex with potent inhibitors. Angew Chem Int Ed Engl 46:4115–4119

Gomez A, Cardoso C, Genta FA et al (2013) Active site characterization and molecular cloning of *Tenebrio molitor* trehalase and comments on their insect homologs. Insect Biochem Mol Biol 43:768–780

Gonzalez DS, Jordan IK (2000) The a-mannosidases: phylogeny and adaptive diversification. Mol Biol Evol 17:292–300

Goptar IA, Shagin DA, Shagina IA (2013) A digestive prolyl carboxypeptidase in *Tenebrio molitor* larvae. Insect Biochem Mol Biol 43:501–509

Grossmann GA, Terra WR (2001) α-Galactosidases from the larval midgut of *Tenebrio molitor* (Coleoptera) and *Spodoptera frugiperda* (Lepidoptera). Comp Biochem Physiol B 128:109–122
Harwood JL (1980) Plant acyl lipids: structure, distribution and analysis. In: Stumpf PK (ed) The biochemistry of plants: a comprehensive treatise, Lipids: structure and function, vol 4. Academic, New York, pp 1–56
Hendry JAF (1993) Evolutionary origins and natural functions of fructans – a climatological, biogeographic and mechanistic appraisal. New Phytol 123:3–14
Hogenkamp DG, Arakane Y, Kramer KJ et al (2008) Characterization and expression of the β-N-acetylhexosaminidase gene family of *Tribolium castaneum*. Insect Biochem Mol Biol 38:478–489
Horne I, Hariotos VS, Oakeshott JG (2009) Comparative and functional genomics of lipases in holometabolous insects. Insect Biochem Molec Biol 39:547–567
Hughes AL (2014) Evolutionary diversification of aminopeptidase N in Lepidoptera by conserved clade-specific amino acid residues. Mol Phylogenet Evol 76:127–133
Husebye H, Artz S, Burmeister WP et al (2005) Crystal structure at 1.1-Å resolution of an insect myrosinase from *Brevicorine brassicae* shows its close relationship to β-glucosidases. Insect Biochem Mol Biol 35:1311–1320
Janecek S (1997) Alpha-amylase family: molecular biology and evolution. Progr Biophys Mol Biol 67:67–97
Jeng W-Y, Wang N-C, Lin M-H et al (2011) Structural and functional analysis of three β-glucosidases from bacterium *Clostridium cellulovorans*, fungus *Trichoderma reesei* and termite *Neotermes koshunensis*. J Struct Biol 173:46–56
Jones AME, Winge P, Bones AM et al (2002) Characterization and evolution of a myrosinase from the cabbage aphid *Brevicorine brassicae*. Insect Biochem Mol Biol 32:275–284
Kawar Z, Jarvis DL (2001) Biosynthesis and subcellular localization of a lepidop- teran insect alpha 1,2-mannosidase. Insect Biochem Mol Biol 31:289–297
Klinkowstrom AM, Terra WR, Ferreira C (1994) Aminopeptidase A from *Rhynchosciara americana* (Diptera) larval midguts: properties and midgut distribution. Arch Insect Biochem Physiol 27:301–315
Knight PJK, Knowles BH, Eller DJ (1995) Molecu-lar cloning of an insect aminopeptidase N that serves as a receptor of *Bacillus thuringiensis* Cry 1a (c) toxin. J Biol Chem 270:17765–17770
Konno K, Ono H, Nakamura M et al (2006) Mulberry látex rich in antidiabetic sugar-mimic alkaloids forces dieting on caterpillars. Proc Nat Acad Sci 103:1337–1341
Lecroisey A, Boulard C, Keil B (1979) Chemical and enzymatic characterization of the collagenase from the insect *Hypoderma lineaturn*. Eur J Biochem 101:385–393
Lecroisey A, Gilles AM, de Wolf A et al (1987) Complete amino acid sequence of the collagenase from the insect *Hypoderma lineaturn*. J Biol Chem 262:7548–7551
Leonard R, Rendic D, Rabouille C et al (2006) The *Drosophila* fused lobes gene encodes an *N*-acetylglucosaminidase involved in *N*-glycan processing. J Biol Chem 281:4867–4875
Lopes AR, Sato PM, Terra WR (2006) Insect chymotrypsins:chloromethyl ketone inactivation and substrate specificity relative to possible coevolutional adaptation of insects and plants. Arch Insect Biochem Physiol 70:151–216
Majerowicz D, Gondim KC (2013) Insect lipid metabolism: insights into gene expression regulation. In: Mandal SS (ed) Recent trends in gene expression. Nova Science Publishers, Inc., Hauppauge, New York, pp 147–190
Marana SR, Terra WR, Ferreira C (2000) Purification and properties of a β-glycosidase purified from midgut cells of *Spodoptera frugiperda* (Lepidoptera) larvae. Insect Biochem Mol Biol 30:1139–1146
Marana SR, Jacobs-Lorena M, Terra WR et al (2001) Amino acid residues involved in substrate binding and catalysis in an insect digestive β-glycosidase. Biochem Biophys Acta 1545:41–52
Marana SR, Lopes AR, Juliano L et al (2002) Subsites of trypsin active site favor catalysis or substrate binding. Biochem Biophys Res Commun 290:494–497
Mark BL, Mahuran DJ, Cherney MM et al (2003) Crystal structure of human betahexosaminidase B: understanding the molecular basis of Sanhoff and Tay-Sachs disease. J Mol Biol 327:1093–1109

Martinez-Esparza M, Sarazin A, Jouy N et al (2006) Comparative analysis of cell wall surface glycan expression in *Candida albicans* and *Saccharomyces cerevisae* yeasts by flow cytometry. J Immunol Methods 314:90–102

Miled N, Canaan S, Dupuis L et al (2000) Digestive lipases: from three-dimensional structure to physiology. Biochimie 82:973–986

Mitsumasu K, Azuma M, Niimi T et al (2005) Membrane-penetrating trehalase from silkworm *Bombyx mori*. Molecular cloning and localization in larval midgut. Insect Mol Biol 14:501–508

Miyazaki T, Oba N, Park EY (2020) Structural insight into the substrate specificity of *Bombyx mori* β-fructofuranosidase belonging to the glycoside family 32. Insect Biochem Mol Biol 127:103494

Moreira NR, Cardoso C, Ribeiro AF et al (2015) Insect midgut a-mannosidases from family 38 and 47 with emphasis on those of *Tenebrio molitor*. Insect Biochem Mol Biol 67:94–104

Muthukrishnan S, Merzendorfer H, Arakane Y et al (2012) Chitin metabolism in insects. In: Gilbert LI (ed) Insect molecular biology and biochemistry. Academic Press, London, pp 193–235

Nemcovicova I, Nemcovic M, Sestak S et al (2012) Expression, purification and preliminary crystallographic analysis of *Drosophila melanogaster* lysosomal a-mannosidase. Acta Crystallogr 68:965–970

Nishimoto M, Kubota M, Tsuji M et al (2001) Purification and substrate specificity of honeybee, *Apis mellifera* L., alpha-glucosidase III. Biosci Biotechnol Biochem 65:1610–1616

Norén O, Sjostrom H, Danielsen EM et al (1986) The enzymes of the enterocyte plasma membrane. In: Desnuelle P, Sjostrom H, Norén O (eds) Molecular and cellular basis of digestion. Elsevier, Amsterdam p, pp 355–365

Pankoke H, Bowers MD, Dobler S (2012) The interplay between toxin-releasing β-glucosidase and plant iridoid glycosides impairs larval development in a generalist caterpillar, *Grammia incorrupta* (Arctiidae). Insect Biochem Mol Biol 42:426–434

Pedezzi R, Fonseca FPP, Santos Junior CD et al (2014) A novel β-fructofuranosidase in Coleoptera: characterization of a β-fructofuranosidase from the sugarcane weevil *Sphenophorus levis*. Insect Biochem Mol Biol 55:31–38

Pimentel AC, Barroso IG, Ferreira JMJ et al (2018) M olecular machinery of starch digestion and glucose absorpsion along the midgut of *Musca domestica*. J Insect Physiol 109:11–20

Price DRG, Karley AJ, Ashford DA et al (2007) Molecular characterisation of a candidate gut sucrase in the pea aphid, *Acyrthosiphon pisum*. Insect Biochem Mol Biol 37:307–317

Quin Y-Q, Wang L-Y, Yang X-Y et al (2023) Inulin: properties and health benefits. Food Func 14:2948–2968

Rawlings ND, Barrett AJ, Thomas PD et al (2018) The MEROPS database of proteolytic enzymes, their substrates and inhibitors in 2017 and a comparison with peptidases in the PANTHER database. Nucleic Acids Res 46:D624–D632

Ren J, Bretthauer RK, Castellino FJ (1995) Purification and properties of a golgi- derived (a1,2)-mannosidase-I from Baculovirus-infected lepidopteran insect cells (IPLB-SF21AE) with preferential activity toward mannose$_6$-N-acetylglucosamine$_2$. Biochemist 34:2489–2495

Santos CD, Terra WR (1986) Midgut alpha-glucosidase and beta-fructosidase from *Erinnyis ello* larvae and imagóes: physical and kinetic-properties. Insect Biochem 16:819–824

Sato PM, Lopes AR, Juliano L et al (2008) Subsite substrate specificity of midgut insect chymotrypsins. Insect Biochem Mol Biol 38:628–633

Schoonhoven LM, van Loon JJA, Dicke M (2005) Insect-plant biology. 2nd ed Oxford university press p440

Shukla E, Thorat L, Bhavnani V et al (2016) Molecular cloning and *in silico* studies of physiologically significant trehalase from *Drosophila melanogaster*. Int J Biol Macromol 92:282–292

Silva CP, Terra WR (1994) Digestive and absorptive sites along the midgut of the cotton seed sucker bug *Dysdercus peruvianus* (Hemiptera: Pyrrhocoridae). Insect Biochem Mol Biol 24:493–505

Silva CP, Terra WR (1995) An α-glucosidase from perimicrovillar membranes of *Dysdercus peruvianus* (Hemiptera: Pyrrhocoridae) midgut cells: purification and properties. Insect Biochem Mol Biol 25:487–494

Silva CP, Terra WR (1997) α-Galactosidase activity in ingested seeds and in the midgut of *Dysdercus peruvianus* (Hemiptera: Pyrrhocoridae). Arch Insect Biochem Physiol 34:443–460
Silva CP, Terra WR, de Sá MFG et al (2001) Induction of digestive α-amylases in larvae of *Zabrotes subfasciatus* (Coleoptera: Bruchidae) in response to ingestion of common bean α-amylase inhibitor 1. J Insect Physiol 47:1283–1290
Silva MC, Terra WR, Ferreira C (2004) The role of carboxyl, guanidine and imidazole groups in catalysis by a midgut trehalase purified from an insect larvae. Insect Biochem Mol Biol 34:1089–1099
Silva MC, Ribeiro AF, Terra WR et al (2009) Sequencing of *Spodoptera frugiperda* midgut trehalases and demonstration of secretion of soluble trehalase by midgut columnar cells. Insect Mol Biol 18:769–784
Silva MC, Terra WR, Ferreira C (2010) The catalytic and other residues essential for the activity of the midgut trehalase from *Spodoptera frugiperda*. Insect Biochem Mol Biol 40:733–741
Silva W, Terra WR, Ferreira C (2015) Conformational changes on ligand binding in wild-type and mutants from *Spodoptera frugiperda* midgut trehalase. Biochem Biophys Rep 4:215–223
Spencer KC (1988) Chemical mediation of coevolution in the *Passiflora–Heliconius* interaction. In: Spencer KC (ed) Chemical mediation of coevolution. Academic, London p167–p240
Stam MR, Danchin EGJ, Rancurel C et al (2006) Dividing the large glycoside hydrolase family 13 into subfamilies: towards improved functional annotations of α-amylase-related proteins. Prot Eng Des Sel 19:555–562
Stanley D (2006) The non-venom insect phospholipases A2. Biochim Biophys Acta Mol Cell Biol Lipids 1761:1383–1390
Strobl S, Maskos K, Betz M et al (1998) Crystal structure of yellow meal worm α-amylase at 1.64 A resolution. J Mol Biol 278:617–628
Takewaki SI, Kimura A, Kubota M et al (1993) Substrate specificity and subsite affinities of honeybee α-glucosidase II. Biosci Biotechnol Biochem 57:1508–1513
Tamaki FK, Pimentel AC, Dias AC et al (2014) Physiology of digestion and the molecular characterization of the major digestive enzymes from *Periplaneta americana*. J Insect Physiol 70:22–35
Tamaki FK, Souza DP, Souza VP et al (2016) Using the amino acid network to modulate the hydrolytic activity of β-glycosidases. PLoS One. https://doi.org/10.1371/journal.pone.0167978
Terra WR, Ferreira C (1994) Insect digestive enzymes: properties, compartmentalization and function. Comp Biochem Physiol B 109:1–62
Terra WR, Ferreira C (2012) Biochemistry and molecular biology of digestion. In: Gilbert LI (ed) Insect molecular biology and biochemistry. Academic/Elsevier, London, pp 365–418
Terra WR, Ferreira C, Bastos F (1985) Phylogenetic considerations of insect digestion. Disaccharidases and the spatial organization of digestion in the *Tenebrio molitor* larvae. Insect Biochem 15:443–449
Terra WR, Barroso IG, Dias RO et al (2019) Molecular physiology of insect midgut. Adv Insect Physiol 56:117–163
Tsuchida K, Wells MA (1988) Digestion, absorption, transport and storage of fat during the last larval stadium of *Manduca sexta*. Changes in the role of lipophorin in the delivery of dietary lipid to the fat body. Insect Biochem 18:263–268
Valaitis AP (1995) Gypsy moth midgut proteinases: Purifi- cation and characterization of luminal trypsin, elastase and the brush-border membrane leucine aminopeptidase. Insect Biochem Mol Biol 25:139–149
Vance JE, Vance DE (2008) Biochemistry of lipids, lipoproteins and membranes Elsevier, Amsterdam
Verhaest M, Van den Ende W, Le Roy K et al (2005) X-ray diffraction structure of a plant glycosyl hydrolase family 32 protein: fructan 1-exohydrolase IIa of Cichorium intybus. Plant J 41:400–401
Whitworth ST, Blum MS, James T (1998) Proteolytic enzymes from larvae of the fire ant, *Solenopsis invicta*. Isolation and characterization of four serine endopeptidases. J Biol Chem 273:14430–14434

Withers SG, Rupitz K (1990) Measurement of active- site homology between potato and rabbit muscle α-glucan phosphorylases through use of a free energy relationship. Biochemistry 29:6405–6409

Zagrobelny M, Bak S, Rasmussen AV et al (2004) Cyanogenic glucosides and plant-insect interactions. Phytochemistry 65:293–306

Zeng F, Zhu Y-C, Allen C, Cohen AC (2002) Molecular cloning and partial characterization of a trypsin-like protein in salivary glands of *Lygus hesperus* (Hemiptera: Miridae). Insect Biochem Mol Biol 32:455–464

Chapter 5
Midgut Cells, Microvillar Membranes, and Secretory Mechanisms

Abstract Most midgut cells are columnar, and in addition to them, there are other cell types, exemplified by the lepidopteran goblet cells and the dipteran oxyntic cells. Hemipterans have a particular type of midgut cell characterized by the presence of membranes associated with their microvilli that vary among the hemipteran infraorders. The best technique for obtaining purified microvillar membranes involves divalent cation precipitation of midgut homogenates to remove non-microvillar membranes, followed by Tris treatment to free microvilli from the cytoskeleton and, finally, isolation of membranes by centrifugation. Microvillar densities are higher for higher holometabolans and have aminopeptidase as an enzyme marker. Microvilli-associated membranes of hemipteran heteropterans, the perimicrovillar membranes, were separated from the microvillar membranes by gradient-density centrifugation. Their enzyme marker is α-glucosidase. Microvillar proteins were identified by proteomics of isolated microvillar membranes. Several mechanisms for protein secretion were identified in insects, but the emphasis was given to microapocrine secretion. In this mechanism, secretory vesicles bud from the microvilli as double-membrane vesicles, which set free their contents by luminal detergents or by solubilization in very alkaline midgut contents. Based on proteomics and other procedures, it was hypothesized that secretory vesicles move inside the microvilli powered by myosin, with gelsolin clearing the way by its actin-filament-severing activity. The proteins of the secretory machinery are apparently recruited by proteins present in detergent-resistant domains of the microvillar membrane.

5.1 Initial Considerations and Midgut Cell Types

The data discussed in Chap. 2 (Table 2.2) showed that enzymes involved in terminal digestion, for example, maltases, are almost completely restricted to midgut cells in Diptera and Lepidoptera. These enzymes could be cytosolic acting after nutrient absorption or could be bound to the microvillar membrane acting at the surface of the midgut cells. In this chapter, we will focus on where those enzymes are in the

W. R. Terra et al., *Molecular Physiology and Evolution of Insect Digestive Systems*, Entomology in Focus 7, https://doi.org/10.1007/978-3-031-39233-7_5

cells, but first, we will review the major types of cells found in the insect midguts (Fig. 5.1). The most abundant cells are the columnar cells, which apical domains are modified into finger-like projections, the microvilli (Figs. 5.1a, e and 5.2) and that have basal plasma membrane infoldings. Sometimes columnar cells are called "enterocytes." However, as called attention by Caccia et al. (2019), this is incorrect, "because enterocytes" are mammalian cells of the small intestine only involved in absorption and not having a role in secretion. The putative function of columnar cells in the absorption or secretion of water may be inferred from certain morphological features. The cells which have basal cell membranes modified into well-developed infoldings, giving origin to many channels with associated mitochondria, are involved in water absorption. Those membrane infoldings form an extracellular compartment with reduced access to hemolymph, because of the small number of openings to the underlying extracellular space. Pumping ions into that compartment results in a gradient of osmotic pressure between the compartment and midgut lumen, thus energizing water absorption. This membrane organization was described in several tissues active in water transport (Berridge 1970; Martoja and Ballan-Dufrançais 1984). In the case of water secretion, the basal infoldings show many openings to the hemolymph, like in the Malpighian tubules (Berridge 1970).

Other cell types found in all insect midguts are the endocrine cells (see Fig. 5.1h and Chap. 8) and the stem cells (Fig. 5.1f), responsible for midgut growth and for replacing cells lost by routine desquamation or injury.

Lepidopteran goblet cells (Figs. 5.1c, d) are involved in eliminating excess K^+ ingested by the larvae by the action of a specialized pump (H^+-V-ATPase) and an antiport (H^+/K^+) and acting in midgut buffering (see Chap. 6). Dipteran oxyntic cells (Fig. 6.1b) have a role in maintaining the acid pH in the middle midgut (see Chap.

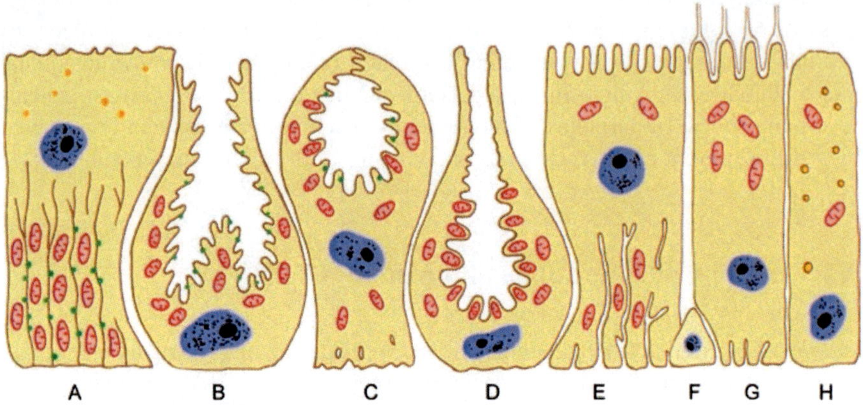

Fig. 5.1 Diagrammatic representation of typical insect midgut cells: (**a**) columnar cell with long and narrow basal membrane infoldings; (**b**) cyclorrhaphan dipteran oxyntic (cuprophilic) cell; (**c**) lepidopteran stalked goblet cell; (**d**) lepidopteran long-neck goblet cell, note mitochondria inside microvilli of goblet cells; (**e**) columnar cell with basal membrane infoldings with numerous openings to underlying space; (**f**) regenerative (stem) cell; (**g**) hemipteran columnar cell; (**h**) endocrine cell

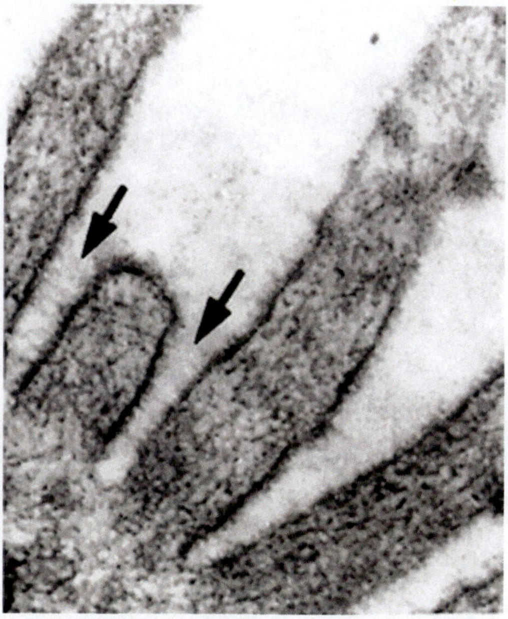

Fig. 5.2 Electron micrograph of a columnar cell of the anterior midgut of *Erinnyis ello*. Detail of microvilli showing glycocalyx (*arrows*). Magnification 52,000 X. (Reproduced with permission from Santos CD, Ribeiro AF, Terra WR (1984). The larval midgut of the cassava hornworm (*Erinnyis ello*). Ultrastructure, fluid fluxes and the secretory activity in relation to the organization of digestion. *Cell Tis Res* 237: 565–574)

6). In addition to the information in Chap. 6, other details on midgut cells, except those of hemipterans, may be found in Caccia et al. (2019). Hemipterans have modified columnar cells, as discussed below.

As discussed in Chap. 1, all hemipterans evolved from a sap-sucking ancestor and only they developed a sap-sucking habit, despite the fact that insects from other orders have suitable piercing-sucking mouthparts. This suggests that the sap-sucking habit depends on modifications in the midgut that enable hemipterans to deal with large amounts of dilute fluid food. Among the modifications, there is the evolution of the hemipteran midgut cell that characteristically has microvilli-associated lipoprotein membranes (Figs. 5.1g and 5.3). These membranes may unsheathe the microvilli as glove fingers (Lane and Harrison 1979) and are now known as perimicrovillar membranes (Terra 1990). They may also form bundles of microvilli (bundle-forming perimicrovillar membranes) (Del Bene et al. 1991; Utiyama et al. 2016). Although the last kind of microvilli-associated membranes (Fig. 5.3d) were named as perimicrovillar membranes (*Mahanarva posticata*, Silva et al. 2004; *Quesada gigas*, Fonseca et al. 2010) or filamentous coats (*Lepyronia coleopterata*, Zhong et al. 2013), these membranes are better described as bundle-forming perimicrovillar membranes.

The apex of the midgut cells of the aphid *Acyrthosiphon pisum* is formed by a network of lamellae (apical lamellae), which apparently lack core filaments and are linked to one another by trabeculae in the same cell and between neighboring cells (O'Loughlin and Chambres 1972; Ponsen 1987; Ponsen 1991; Cristofoletti et al. 2003). Membranes associated with the tips of the apical lamellae are named modified perimicrovillar membranes.

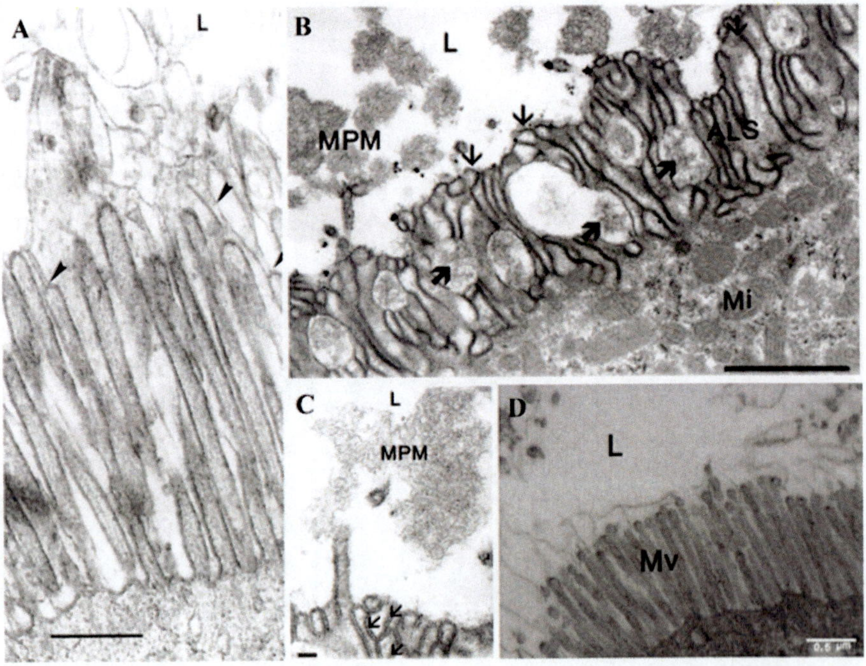

Fig. 5.3 Microvilli-associated membranes: (**a**) Electron micrograph of the apex of the midgut from *Dysdercus peruvianus*, showing microvilli enclosed by perimicrovillar membranes. (**b**) Electron micrograph of the apical surface of anterior midgut cells of *Acyrthosiphon pisum* showing the lamellar system with associated modified perimicrovillar membranes (MPM) projecting into the lumen. Note trabecullae (*small arrows*) between lamellae and MPM masses moving along lamellae. (**c**) Detail of modified perimicrovillar membranes associated with lamellae. Note trabecullae (*arrows*) between lamellae. (**d**) Electron micrograph of the apex of midgut cells of *Mahanarva fimbriolata* with microvilli (Mv) surrounded by microvillar-bundle-forming membranes. Other abbreviations: *L* lumen, *Mi* mitochondria. (Reproduced with permission from Terra WR, Ferreira C (2020) Evolutionary trends of digestion and absorption in the major insect orders. *Arthr. Struc. Dev.* 56: 100931)

The origin of the perimicrovillar membranes of Heteroptera was investigated with the help of electron micrographs and immunocytolocalizations employing antibodies raised against a perimicrovillar membrane marker enzyme (α-glucosidase, see Sect. 5.2 below). The data suggested that the perimicrovillar membranes are originated in double-membrane Golgi cisterna that on budding form double-membrane vesicles, which eventually fuse at the cell apex, the outer vesicle membrane with the microvillar membrane and the inner vesicle membrane with the perimicrovillar one (see Fig. 5.4e) (Silva et al. 1995). With the help of electron micrographs, a model of the origin of the modified perimicrovillar membranes of *Acyrthosiphon pisum* (Sternorrhyncha) was proposed. According to the model, vesicles budded from Golgi cisternae were enclosed by reticulum endoplasmic membranes that form a vesicle which enclose membranes. These vesicles eventually empty their membrane content into the interlamellar space. The membrane mass

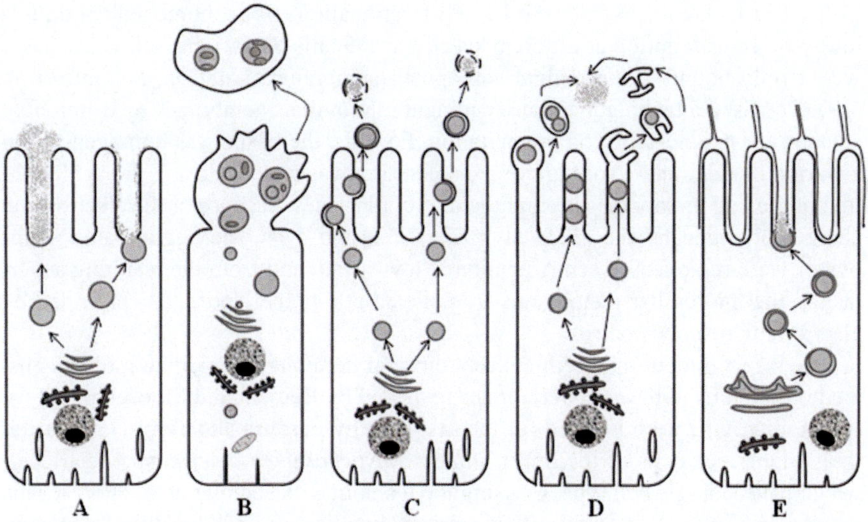

Fig. 5.4 Diagrammatic representation of the secretory processes of insect digestive enzymes: (**a**) Exocytic secretion; (**b**) apocrine secretion; (**c**) microapocrine secretion with budding double-membrane secretory vesicles; (**d**) microapocrine secretion with pinched-off secretory vesicles; (**e**) modified exocytic secretion in hemipteran midgut cell. (Reproduced with permission from Terra WR, Ferreira C (2009) Digestive system. In Resh VH, Cardé RT (Eds) *Encyclopedia of Insects*. Second ed. Academic Press, San Diego, pp. 273–281. © Elsevier)

progresses along the interlamellar space and finally becomes associated with the tips of the lamellae (Cristofoletti et al. 2003). The origin of the bundle-forming perimicrovillar membranes of Auchenorrhyncha is different from the former microvilli membrane-associated membranes. Here, according to electron and scanning electron micrographs, vesicles are formed by apical microvilli constrictions resulting in beads-on-a-string-like structures. The vesicles apparently then flatten and collapse, forming sheets that cover groups of microvilli (Utiyama et al. 2016). In short, perimicrovillar membranes are typical of Heteroptera; modified perimicrovillar membranes of Sternorrhyncha; and bundle-forming perimicrovillar membranes of Auchenorrhyncha.

5.2 Enzymes Associated with Midgut Microvilli, Glycocalyx, and Microvilli-Associated Membranes

The involvement of the midgut cells' apices in the transport of water and organic compounds is known since at least the middle of the last century (Terra et al. 2006). Digestive enzymes were considered to be secreted into midgut contents or acting intracellularly, following absorption, with midgut cell apical membranes playing no role in digestive events. The same was considered to be true for mammals up till

Miller and Crane (1961) showed by cell fractionation (tissue homogenization followed by centrifugation at different speeds, see details below) that disaccharidases were firmly bound to the midgut cell apical membranes. Later on, Schmitz et al. (1973) devised a technique to purify midgut microvillar membranes by differential calcium (or magnesium) ion precipitation. For this, the tissue was homogenized in a waring blender and a solution of a divalent ion was added causing the agglutination of the cell membranes, except the microvillar ones, because of the electrostatic charges provided by the glycocalyx (Figs. 5.2 and 7.1). The agglutinated membranes were removed by centrifugation at low speed, and from the resulting supernatant, the microvillar membranes are pelleted at a higher speed, as confirmed by electron microscope controls.

The insect midgut microvilli, except those of hemipteran modified midgut cells, are homologous to those of vertebrates reviewed by Bement and Mooseker (1996). Insect microvilli were isolated for the first time by Ferreira and Terra (1980) from the midgut ceca cells of the lower Diptera *Rhynchosciara americana* (Sciaridae), which have a single cell type, by using the technique of Schmitz et al. (1973). This paper was also the first to show that enzymes involved in final digestion, at least in Diptera, are microvillar enzymes. Cioffi and Wolfersberger (1983) developed a successful, although time-consuming, method to isolate apical, lateral, and basal plasma membranes of columnar and goblet cells of lepidopterans by the stepwise disruption of the midgut by ultrasound. With this technique, Wieczorek et al. (1984) located the H^+-pump responsible for K^+ elimination, as mentioned above, exclusively in the modified (containing mitochondria) microvilli of the goblet cell. A comparison of the methods for preparing microvilli of midgut columnar cells of lepidopterans, with electron monitoring, including cell fractionation, differential calcium precipitation, and stepwise ultrasound disruption, showed that the differential precipitation method provides the highest yields with rare contaminants (Santos et al. 1986). After complementary work by Eisen et al. (1989), the method of choice for preparing microvillar membranes from insect midgut columnar cells became the differential precipitation method. Thus, besides lower Diptera and Lepidoptera, this method was used successfully for preparing midgut microvillar membranes from columnar cells of, among others, cockroaches (Dictyoptera, Parenti et al. 1986), beetles (Coleoptera, Ferreira et al. 1990), and flies (higher Diptera, Lemos and Terra 1992).

Insect midgut microvilli are substantially free from contaminants, although microvilli still contain some cytoskeleton structures, as observed in the electron microscope (Houk et al. 1986; Santos et al. 1986). The contaminants may be quantified by the use of enzyme markers, taking into account that except for succinate dehydrogenase (mitochondria), lactate dehydrogenase, or glucose 6-phosphate dehydrogenase (cytosol) and aminopeptidase (microvilli), the enzyme markers are not always suitable. For example, the microvillar enzyme marker γ-glutamyl transferase is not suitable for Coleoptera and Hymenoptera, and alkaline phosphatase is useless for Coleoptera and adults of Diptera and Lepidoptera (Terra et al. 2006).

The enrichments of microvillar membrane preparations (ratio of specific activities of the enzyme markers in the preparation and in tissue homogenate) vary widely, depending on the midgut region and on the phylogenetic group of the insect used as a sample. Thus, higher enrichments are observed from small microvilli relative to cell size or microvillar membranes poor in protein components or, finally, from the fact that only parts of the tissue have columnar cells (Terra et al. 2006). Microvillar preparations are frequently known as brush borders, because they are actually tops of cells with microvilli still having cytoskeleton elements. Purified microvillar membranes may be obtained by a two-step procedure. First, the microvillar preparation is treated with hyperosmotic Tris buffer that dissociates the core material from the microvilli. On dilution, the core material is avoided to re-polymerize and remains in the supernatant of a centrifugation to pelleting the purified microvillar membranes, checked by electron microscopy (Jordão et al. 1995; Capella et al. 1997).

Enzymology of purified midgut microvillar membranes showed that, as a rule, aminopeptidase is a microvillar enzyme in the columnar cells of all insects, except Hemiptera, whereas other enzymes involved in the terminal digestion (as maltase) are microvillar enzymes in Diptera, but not in Lepidoptera. If the enzymes active in the terminal digestion in lepidopterans are not microvillar enzymes, are they cytosolic enzymes active only after absorption? To answer this question, lepidopteran midgut cells were submitted to cell fractionation by differential centrifugation of tissue homogenates. For this, midgut tissues were homogenized with the aid of a cylindrical Teflon-made pistil moving in a well-adjusted (gap between pistil and vial between 50 and 100 μm) cylindrical glass vial with a round bottom (Potter-Elvehjem device) in the presence of an isosmotic buffer (e.g., 5 mM Tris, 5 mM EDTA, 215 mM mannitol, pH 7.1). After passing the homogenate through a Nylon net 45 μm pore size to remove non-homogenized material, the resulting filtrate was centrifuged (always at 4 °C) at 600 ×g (the speed in terms of revolutions per min will depend on the diameter of the rotor) for 10 min. The pellet was labeled P1. The obtained supernatant was centrifuged at 3300 ×g for 10 min with the resulting pellet labeled P2. In sequence, the resulting supernatant was centrifuged at 25000 ×g for 10 min forming pellet P3 and a supernatant. The new supernatant was then centrifuged at 100000 ×g (now with the aid of an ultracentrifuge) for 60 min, resulting in pellet P4 and the final supernatant. Inspection of the fractions in the electron microscope showed that P1 contains cell nuclei, large vesicles displaying brush borders, and large mitochondria that are characteristic of insect cells; P2 contains small mitochondria; P3, small vesicles corresponding to fragments of plasma membranes, lysosomes, and secretory vesicles; and P4, very small vesicles formed mostly by the endoplasmic reticulum and some fragments of plasma membranes. The final supernatant contains the soluble material of the cell cytosol.

Assaying all fractions for soluble (not microvillar bound) digestive enzymes and the enzyme markers, Santos and Terra (1984) and Santos et al. (1986) found that major amounts of enzymes involved in the final phases of digestion (like maltase) occurred both in P1 and in the final supernatant. Lactate dehydrogenases (a cytosolic marker) abound only in the final supernatant. Furthermore, they observed that most cytosol leaks from the cells during ultrasound disruption (done as described by Cioffi and Wolfersberger 1983), whereas the soluble digestive enzymes remained mainly in the tissue. These data mean that digestive enzymes are not in the cytosol and they proposed that they are trapped in the glycocalyx. The glycocalyx is the coat of carbohydrate chains of the microvillar proteins and glycolipids facing the lumen (see Figs. 5.2, 7.1). The enzymes in P1 were thought to be trapped in the glycocalyx, as electron microscope controls showed large vesicles with brush borders, obviously corresponding to large fragments on the top of the cells. On freezing and thawing these samples, the vesicles would change in structure freeing the enzymes. Soluble enzymes in the final supernatant were thought to be those loosely bound to the cell glycocalyx that on tissue homogenization are set free.

To test this hypothesis, amaranth, a non-absorbable dye, was fed larvae of *Tenebrio molitor* and after rinsing the dissected midguts, they were homogenized and submitted to differential cell fractionation as before. As expected from the hypothesis, the dye associated with the tissue was recovered after cell fractionation mainly at P1 and in the final supernatant, in agreement with the hypothesis it was associated with the glycocalyx, as thought to occur with the soluble enzymes retained on the cell surface (Ferreira et al. 1990).

After these works, and others compiled by Terra and Ferreira (1994), it became established that, as a rule, the enzymes involved in the intermediary digestion, but mainly those active in the final phase of digestion of carbohydrates are microvillar enzymes in dipterans and glycocalyx associated in lepidopterans. The enzymes performing the digestion of proteins, represented by aminopeptidase, are microvillar enzymes both in dipterans and lepidopterans. It should be noticed, however, that the subcellular distribution of enzyme markers varies among insects, with the exception of succinate dehydrogenase (mitochondria) and lactate dehydrogenase (cytosol). For example, γ-glutamyl transferase, which is a plasma membrane marker for Diptera (Bodnaryk et al. 1974; Espinoza-Fuentes et al. 1987), Lepidoptera (Giordana et al. 1982), and Dictyoptera (Parenti et al. 1986), occurs only in trace amounts in *T. molitor* (Jordão et al. 1995), whereas acid phosphatase, a good marker for lysosomes in some tissues (e.g., mammalian liver; Evans 1978), is found chiefly in the cytosol of midgut cells of Diptera, Lepidoptera (Santos and Terra 1984), and Coleoptera (Jordão et al. 1995).

The study of the properties of hemipteran apical cell membranes is more complicated than that of the other insect groups, because besides the microvillar and basal membranes, hemipterans midgut cells have microvilli-associated membranes. Only the membranes of heteropterans were studied. The first step in this study was to found enzyme markers for the microvillar and perimicrovillar membranes. For this, Ferreira et al. (1988) homogenized the posterior midgut of *R. prolixus* in hypotonic and hypertonic media, followed by differential centrifugation with electron microscopy monitoring. Membrane-bound α-mannosidase predominates in fractions rich in brush-border vesicles (vesicles with microvilli), whereas α-glucosidase is found associated with large membrane sheaths, despite having a subcellular distribution different from α-mannosidase. This distribution suggests that α-mannosidase may be linked to the microvillar membrane and α-glucosidase to the perimicrovillar membrane. Soluble aminopeptidase was thought to be enclosed in the perimicrovillar space to account for the fact it sediments with vesicles having brush-borders frequently still associated with perimicrovillar membranes from where they are set free by freezing and thawing. This agrees with previous cytochemical data (Billingsley and Downe 1985) that showed that aminopeptidase is associated with the cell microvilli of *R. prolixus*.

The perimicrovillar membrane of *R. prolixus* midgut cells is almost devoid of protein particles, thus resembling myelin sheaths (Lane and Harrison 1979) with predicted low densities. This prompted Ferreira et al. (1988) to try to separate the perimicrovillar from the microvillar membrane with the aid of sucrose-density-gradient ultracentrifugation, with the enzyme markers: α-mannosidase (*R. prolixus*) for microvilli and basolateral membranes and α-glucosidase for the perimicrovillar membranes. As expected, α-glucosidase was found to be bound to the lighter membrane (the perimicrovillar membrane) and α-mannosidase to the heavier membrane (microvillar membrane).

A similar procedure was used by Silva et al. (1996) to study the apical midgut cell membranes of *Dysdercus peruvianus* and found that β-glucosidase is a marker of the microvillar membranes and α-glucosidase of the perimicrovillar membranes. They were also successful in separating both membranes by density-gradient centrifugation.

5.3 Chemistry of Microvillar and Microvilli-Associated Membranes

Densities of midgut microvillar membranes were determined by sucrose-density-gradient ultracentrifugation with aminopeptidase as the enzyme marker and showed contamination by a lighter membrane summing up to 5% of total membranes (Terra et al. 2006). The lighter membranes were supposed to be basolateral membranes. The microvillar membrane densities are in the range of 1.116–1.159, depending on the insect order and midgut region: anterior midgut, smaller density; posterior midgut, larger density. Basal plasma membrane densities are in the range of 1.069–1.128 and depend only on the insect order, not on the midgut region (Terra et al. 2006). The density of the purified microvillar membrane linearly increases with the protein-lipid mass ratio that may vary in the range of 1.41–3.13 (Terra et al. 2006), whereas among mammalian enterocytes the range is 1.54–2.44 (Proulx 1991). In insect microvillar membranes, the cholesterol and carbohydrate content vary inversely with the protein-lipid ratio (or membrane density) (Terra et al. 2006). This was confirmed in a detailed chemical study of midgut microvilli (microvillar membranes still contaminated with cytoskeleton) of *Bombyx mori* (Leonardi et al. 2001). The protein-lipid ratio found was smaller in the anterior than in the posterior midgut, with phospholipids amounting to 77% and glycolipids accounting for 8% of total lipids.

The perimicrovillar membranes of both *R. prolixus* and *D. peruvianus* have low densities (1.068 and 1.087, respectively), and the perimicrovillar membranes have a high lipid-protein ratio (1090 μg/mg protein) (Ferreira et al. 1988; Silva et al. 1996).

Microvillar enzymology led to the identification of many membrane-bound digestive enzymes as reviewed in Sect. 5.2 above. Nowadays, the study of microvillar proteins is carried out with microvillar preparations following one of two different approaches. The immunoscreening technique to identify microvillar proteins uses antibodies prepared against purified microvillar membranes (or purified microvilli, brush borders) to found positive clones by screening an expressing midgut-specific cDNA library. The positive clones were sequenced and searched for similarities in international databases (Ferreira et al. 2007; Silva et al. 2013). The proteomics approach to identify microvillar proteins is based on the separation of the microvillar proteins followed by mass spectrometry for identification (McNall and Adang 2003; Krishnamoorthy et al. 2007; Popova-Butler and Dean 2009; Bayyareddy et al. 2009; Javed et al. 2019; Fuzita et al. 2019).

The number of proteins identified in the microvillar membranes is very large, but the major proteins may be grouped into a few classes, although the proteins inside each class vary somewhat among the insect taxa. In Coleoptera (*Tenebrio molitor*), probably because of the small coverage of proteins, only three classes were recognized: digestive enzymes, putative peritrophic membrane ancillary protein (PMAP), and peritrophic membrane proteins. PMAP was later thoroughly studied (Ferreira et al. 2008). The major midgut microvillar proteins in dipterans (Popova-Butler and Dean 2009; Bayyareddy et al. 2009) and lepidopterans (McNall and Adang 2003; Ferreira et al. 2007; Krishnamoorthy et al. 2007; Javed et al. 2019; Fuzita et al. 2019) may be classified in many classes: digestive enzymes; midgut protection (exemplified by thioredoxin peroxidase, protein disulfide isomerase, aldehyde dehydrogenase, serpin); membrane-tight-bound cytoskeleton proteins (fimbrin, actin, cadherin, flotilin); and membrane transporters. In addition, to these classes, lepidopterans have two other classes: peritrophic membrane proteins and proteins associated with microapocrine secretion (see Sect. 5.4 below). The difference between dipterans and lepidopterans is a consequence of dipteran larvae not synthesizing peritrophic membrane along the midgut (Peters 1992) nor having microapocrine secretion.

5.4 Mechanisms of Digestive Enzyme Secretion

Animal proteins, including insect digestive enzymes, are synthesized in the rough endoplasmic reticulum, processed in the Golgi complex, and enclosed in secretory vesicles. Digestive enzymes are set free from the secretory vesicles by four different mechanisms. The most common mechanism is exocytosis, where secretory vesicles coming from the Golgi area fuse with the midgut apical membranes, releasing their contents with no cytoplasm loss (Fig. 5.4a). During holocrine secretion, secretory vesicles stored in the cell cytoplasm are discharged in the midgut contents with the whole secretory cell. Apocrine secretion implies the loss of approximately 10% of the apical cytoplasm accompanying the release of secretory vesicles (Fig. 5.4b). Finally, microapocrine secretion is the secretory mechanism leading to a small loss of cytoplasm (Fig. 5.4c, d).

The digestive secretory routes were usually identified by immunocytolocalization, which is the visualization of the cell sites where antibodies raised against selected proteins are bound. Midgut cell renewal described by histology, mainly in insects other than higher Holometabola, is frequently misidentified as holocrine secretion (Terra and Ferreira 1994). However, trypsin-containing vesicles in the opaque zone cells of adult stable flies were shown by immunocytolocalization to discharge their contents, suggesting holocrine secretion (Jordão et al. 1996a). Exocytic secretion of digestive enzymes was first described by immunocytolocalization in adult mosquitoes (Graf et al. 1986), later also in larval flies (Jordão et al. 1996b), *Tenebrio molitor* (Ferreira et al. 2002), and caterpillars (Bragatto et al. 2010).

Microapocrine secretion is usually referred to two different secretory mechanisms: the release of budding double-membrane secretory vesicles (Fig. 5.4c) and another less common mechanism that consists of pinched-off vesicles that may contain a single or several vesicles (Fig. 5.4d). The last process is closer to the microvilli microvesiculation than to the budding double-membrane mechanism. Microvilli microvesiculation is the widespread release of unilamellar vesicles from the tips of microvilli thought to enable midgut cells (actually described for vertebrate enterocytes) to distribute digestive enzymes (mainly alkaline phosphatase) in the intestinal lumen (McConnell et al. 2009). Thus, from now on we will call microapocrine secretion only the budding double-membrane mechanism of secretion, which is characteristic of lepidopterans (Fig. 5.4c). The contents of the secretory vesicles released by microapocrine secretion are set free by membrane fusion, by luminal detergents, or by solubilization by the high pH of the midgut contents. There is a small loss of cytoplasm in this kind of secretory mechanism that corresponds to the cytoplasm caught between the microvillar and the secretory vesicle membrane.

Digestive enzyme secretory mechanisms depend mainly on midgut regions: Exocytosis is usually observed in the posterior midgut and alternate mechanisms are found in the anterior midgut (Terra and Ferreira 2012). Hemipteran cells have peculiar secretory mechanisms associated with the mechanism of perimicrovillar formation (see Sect. 5.1 and Fig. 5.4, Silva et al. 1995). Digestive enzymes addressed to midgut contents occur inside the double-membrane secretory vesicles; the ones addressed to the microvillar membrane, in the external membrane of the vesicle; those to the perimicrovillar membrane, in the internal membrane; and, finally, the ones to the perimicrovillar space (space between the microvillar and the perimicrovillar membrane), in the space between the membranes of the vesicle.

The mechanisms underlying the insect midgut secretory mechanisms are unknown, except for microapocrine secretion. Candidate proteins to be involved in the machinery of microapocrine secretion in *S. frugiperda* cells were chosen from previous surveys (Ferreira et al. 2007; Silva et al. 2013). These surveys tried to identify microvillar proteins and microapocrine vesicle proteins by immunoscreening a *S. frugiperda* cDNA midgut library with antibodies raised against purified microvillar membranes and microapocrine vesicles. Microvillar membranes were purified by the technique of differential calcium precipitation. The purification of microapocrine vesicles was carried out in three steps. At first, the luminal side of the midgut tissue was rinsed, followed by low-speed centrifugation to remove food residues. The pellet was discarded and the supernatant was centrifuged at high speed to pellet the microapocrine vesicles, confirmed by electron microscopy (Ferreira et al. 2007; Silva et al. 2013).

From the data discussed above, the following mechanism of microapocrine secretion was proposed. Calmodulin, in response to some signaling process, changes Ca^{2+} concentration affecting annexin and gelsolin activities. The movement of the secretory vesicles inside the microvilli along the microfilament track is putatively powered by myosin with gelsolin clearing the way by its actin-filament-severing activity. Finally, the microapocrine vesicle fusion with the microvillar membrane is thought to be promoted by annexin. Suppression of gelsolin synthesis by RNA interference led to the accumulation of secretory vesicles at the base of microvilli, in accordance with its proposed role in freeing the way for the microapocrine vesicles, caused by its actin-filament-severing activity (Silva et al. 2016).

The microapocrine vesicle on budding is enveloped by the microvillar membrane, which may somehow recruit the proteins of the secretory machinery. A proteomic study of the microapocrine vesicles showed that they are rich in proteins anchored to the membranes by glycosylphosphatidylinositol (GPI-anchored proteins) (Fuzita et al. 2019). It is known that there are microdomains in membranes that are resistant to detergent (detergent-resistant membranes, DRM) and rich in GPI-anchored proteins (Braccia et al. 2003; Danielsen and Hansen 2003; Parish et al. 2011) and act as platforms to recruit membrane-bound proteins, playing a variety of roles, like in protein traffic (Rose et al. 2014). As there are some reports about the occurrence of DRMs in insect midgut cells (Bayyareddy et al. 2012; Zhuang et al. 2002), attempts were made to associate DRMs with microapocrine vesicle budding. For this, *S. frugiperda* midgut microvillar membranes were incubated with and without (control) the detergent Triton-X-100 at different times and temperatures, followed by sucrose density-gradient centrifugation to separate membrane fractions from the soluble material. DRMs were isolated and their protein composition was identified by proteomics. The results showed that DRMs are rich in GPI-anchored proteins and that most of them are also found in the microapocrine vesicles, favoring the view that DRMs may recruit proteins of the microapocrine secretory machinery (Fuzita et al. 2019).

Insect midguts secrete digestive enzymes by both the exocytic and microapocrine routes. To identify which enzymes follow which of the routes, a proteomic study was performed with *S. frugiperda* (Fuzita et al. 2022). For this, the following samples were prepared: peritrophic membrane contents (endoperitrophic contents), microapocrine vesicle membranes, microapocrine vesicle soluble fraction, and the washings of the peritrophic membrane: one for freeing loosely bound material and another for tightly bound material. PM washings correspond to proteins isolated from the mucus layer surrounding PM (see details in Chap. 7). Proteins absent from microapocrine vesicles but found in PM washings or PM contents were thought to be released by exocytosis. Proteins occurring in the microapocrine vesicles and in PM washings were considered to be secreted by a microapocrine route. Based on these criteria, most endopeptidases are from the serine type secreted by exocytosis and able to cross the PM. The same is true for triacylglycerol lipases and amylase.

Aminopeptidases are microvillar membrane-bound proteins, with some of them secreted by a microapocrine mechanism. Carboxypeptidase isoforms follow different routes, and some pass and others do not pass through PM. Maltases and β-glycosidases are mostly secreted by exocytosis, although some maltases are secreted membrane-bound to the microapocrine vesicle and some β-glycosidases are released as the soluble contents of microapocrine vesicles (Fuzita et al. 2022). It should be noted, however, that most maltases and β-glycosidases are secreted and remain bound to the midgut cell glycocalyx and hence are not approached by this analysis.

References

Bayyareddy K, Andacht TM, Abdullah MA, Adang MJ (2009) Proteomic identification of Bacillus thuringiensis subsp. israelensis toxin Cry4Ba binding proteins in midgut membranes from Aedes (Stegomyia) aegypti Linnaeus (Diptera, Culicidae) larvae. Insect Biochem Molec Biol 39:279–286

Bayyareddy K, Zhu X, Orlando R, Adang MJ (2012) Proteome analysis of Cry4Ba toxin-interacting *Aedes aegypti* lipid rafts using gel C-MS/MS. J Proteome Res 11:5843–5855

Bement WM, Mooseker MS (1996) The cytoskeleton of the intestinal epithelium: components, assembly, and dynamic rearrangements. In: Hesketh JE, Pryme JF (eds) The cytoskeleton: a multi-volume treatise, vol 3. JAI Press, Greenwich, pp 359–404

Berridge M (1970) Structural analysis of intestinal absorption. In Neville AC (ed) Insect ultrastructure, symposium of the Royal Entomological Society of London, vol 5, pp 135–151

Billingsley PF, Downe AER (1985) Cellular localization in the midgut of Rhodnius prolixus Stal (Hemiptera: Reduviidae) during blood digestion. Cell Tissue Res 241:421–428

Bodnaryk K, Bronskill JF, Fetterly JR (1974) Membrane-bound γ-glutamyl transpeptidase and its role in phenylalanine absorption-reabsorption in the larva of *M. domestica*. J Insect Physiol 20:167–181

Braccia A, Villani M, Immerdal L et al (2003) Microvillar membrane microdomains exist at physiological temperature – role of galectin-4 as lipid raft stabilizer revealed by "superrafts". J Biol Chem 278:15679–15684

Bragatto I, Genta FA, Ribeiro AF et al (2010) Characterization of a β-1,3-glucanase active in the alkaline midgut of *Spodoptera frugiperda* larvae and its relation to β-glucan-binding proteins. Insect Biochem Molec Biol 40:861–872

Caccia S, Casartelli M, Tettamanti G (2019) The amazing complexity of insect midgut cells: types, peculiarities, and functions. Cell Tis Res 377:505–525

Capella AN, Terra WR, Ribeiro AF et al (1997) Cytoskeleton removal and characterization of the microvillar membranes isolated from two midgut regions of *Spodoptera frugiperda* (Lepidoptera). Insect Biochem Mol Biol 27:793–801

Cioffi M, Wolfersberger MG (1983) Isolation of separate apical, lateral and basal plasma membrane from cells of an insect epithelium. A procedure based on tissue organization and ultrastructure. Tissue Cell 15:781–803

Cristofoletti PT, Ribeiro AF, Terra WR (2003) Apocrine secretion of amylase and exocytosis of trypsin along the midgut of *Tenebrio molitor* larvae. J Insect Physiol 47:143–155

Danielsen EM, Hansen GH (2003) Lipids rafts in epitelial brush borders: atypical membrane microdomains with specialized functions. Biochim Biophys Acta 1617:1–9

Del Bene G, Dallai R, Marchini D (1991) Ultrastructure of the midgut and the adhering tubular salivary glands of *Frankliniella occidentalis* (Pergande) (Thysanoptera: Thripidae). Intern. J Insect Morphol Embryol 20:15–24

Eisen NS, Fernandes VF, Harvet WR et al (1989) Comparison of brush border membrane vesicles prepared bt three methods from larval *Manduca sexta* midgut. Insect Biochem 19:337–342

Espinoza-Fuentes FP, Ferreira C, Terra WR (1987) Microvillar and secreted digestive enzymes from *Musca domestica* larvae. Subcellular fractionation of midgut cells with electron microscopy monitoring. Insect Biochem 17:819–827

Evans WH (1978) Preparation and characterization of mammalian plasma membrane. In: Work TS, Work E (eds) Laboratory techniques of biochemistry and molecular biology, part 1. North Holland, Amsterdam, pp 103–121

Ferreira C, Terra WR (1980) Intracellular distribution of hydrolases in midgut caeca cells from na insect with emphasis on plasma membrane-bound enzymes. Comp Biochem Physiol B 66:467–473

Ferreira C, Ribeiro AF, Garcia ES et al (1988) Digestive enzymes trapped between and associated with the double membranes of *Rhodnius prolixus* posterior midgut cells. Insect Biochem 18:521–530

Ferreira C, Bellinello GL, Ribeiro AF et al (1990) Digestive enzymes associated with the glycocalyx microvillar membranes and secretory vesicles from midgut cells of *Tenebrio molitor* larvae. Insect Biochem 20:839–847

Ferreira AHP, Ribeiro AF, Terra WR et al (2002) Secretion of β-glycosidase by middle midgut cells and its recycling in the midgut of *Tenebrio molitor* larvae. J Insect Physiol 48:113–118

Ferreira AHP, Cristofoletti PT, Lorenzini DM et al (2007) Identification of midgut microvillar proteins from *Tenebrio molitor* and *Spodoptera frugiperda* by cDNA library screenings with antibodies. J Insect Physiol 53:1112–1124

Ferreira AH, Cristofoletti PT, Pimenta DC et al (2008) Structure, processing and midgut secretion of putative peritrophic membrane ancillary protein (PMAP) from *Tenebrio molitor* larvae. Insect Biochem Mol Biol 38:233–243

Fonseca FV, Silva JR, Samuels RI et al (2010) Purification and partial characterization of a midgut membrane-bound α-glucosidase from *Quesada gigas* (Hemiptera: Cicadidae). Comp Biochem Physiol B 155:20–25

Fuzita FJ, Pimenta DC, Palmisano G et al (2019) Detergent-resistant domains in *Spodoptera frugiperda* midgut microvillar membranes and their relation to microapocrine secretion. Comp Biochem Physiol B 235:8–18

Fuzita FJ, Palmisano G, Pimenta DC et al (2022) A proteomic approach to identify digestive enzymes, their exocytic and microapocrine secretory routes and their compartmentalization in the midgut of *Spodoptera frugiperda*. Comp Biochem Physiol B 257:110670

Giordana B, Sacchi FV, Hanozet GM (1982) Intestinal amino acid absorption in lepidopteran larvae. Biochim Biophys Acta 692:81–82

Graf R, Raikhel AS, Brown MR et al (1986) Mosquito trypsin: immunocytochemical localization in the midgut of blood-fed *Aedes aegyti* (L.). Cell Tissue Res 245:19–27

Houk EJ, Arcus YM, Hardy JL (1986) Isolation and characterization of brush border fragments from mosquito mesenterons. Archs Insect Biochem Physiol 3:135–146

Javed MA, Coutu C, Theilmann DA et al (2019) Proteomics analysis of *Trichoplusia ni* midgut epithelial cell brush border membrane vesicles. Insect Sci 26:424–440

Jordão BP, Terra WR, Ferreira C (1995) Chemical determinations in microvillar membranes purified from brush borders isolated from the larval midgut from one Coleoptera and two Diptera species. Insect Biochem Mol Biol 25:417–426

Jordão BB, Lehane MJ, Terra WR et al (1996a) An immunocytochemical investigation of trypsin secretion in the midgut of *Stomoxys calcitrans*. Insect Biochem Mol Biol 26:445–453

Jordão BB, Terra WR, Ribeiro AF et al (1996b) Trypsin secretion in *Musca domestica* larval midguts. A biochemical and immunocytochemical study. Insect Biochem Molec Biol 26:337–346

Krishnamoorthy M, Jurat-Fuentes JL, McNall RJ et al (2007) Identification of novel CryIAc binding proteins in midgut membranes from *Heliothis virescens* using proteomic analyses. Insect Biochem Molec Biol 37:189–201

Lane NJ, Harrison JB (1979) An unusual cell surface modification: a double plasma membrane. J Cell Sci 39:355–372

Lemos FJA, Terra WR (1992) A high yield preparation of *Musca domestica* larval midgut microvilli and the subcellular distribution of amylase and trypsin. Insect Biochem Mol Biol 22:433–438

Leonardi MG, Marciani P, Montorfano PG et al (2001) Effect of fenoxycarb on leucine uptake and lipid composition of midgut brush border membrane in the silkworm, *Bombyx mori* (Lepidoptera, Bombycidae). Pest Biochem Physiol 70:42–51

Martoja R, Ballan-Dufrançais C (1984) The ultrastructure of the digestive and excretory organs. In: King RC, Akai H (eds) Insect ultrastructure. Plenum Press, New York, pp p199–p268

McConnell RE, Higginbotham JN, Shifrin DA Jr et al (2009) The enterocyte microvillus is a vesicle-generating organelle. J Cell Biol 185:1285–1298

McNall RJ, Adang MJ (2003) Identification of novel *bacillus thuringiensis* Cry1 ac binding proteins in *Manduca sexta* midgut through proteomic analysis. Insect Biochem Molec Biol 33:999–1010

Miller D, Crane RK (1961) The digestive function of the epithelium of the small intestine. II. Localization of disaccharide hydrolysis in the isolated brush border portion of intestinal epithelial cells. Biochim Biophys Acta 52:293–298

O'Loughlin GT, Chambres TC (1972) Extracellular microtubules in the aphid gut. J Cell Biol 53:575–578

Parenti P, Sacchi FV, Hanozet GM et al (1986) Na-dependent uptake of phenylalanine in the midgut of a cockroach (*Blabera gigantea*). J Comp Physiol B 156:549–556

Parish LA, Colquhoun DR, Mohien CU et al (2011) Ookinete-interacting proteins on the microvillar surface are partioned into detergent-resistant membranes of *Anopheles gambiae* midguts. J Proteome Res 10:5150–5162

Peters W (1992) Peritrophic membranes. Springer-Verlag, Berlin

Ponsen MB (1987) Alimentary tract. In: Mincks AK, Harrewijn P (eds) Aphids: their biology, natural enemies and control, vol 4. Elsevier, Amsterdam, pp 79–97

Ponsen MB (1991) Structure of the digestive system of aphids, in particular *Hyaloptterus* and *Coloradoa*, and its bearing on the evolution of filterchambers in the Aphidoidea. Wageningen Agricultural University Papers 91–95:3–61

Popova-Butler A, Dean DH (2009) Proteomic analysis of the mosquito *Aedes aegypti* midgut brush border membrane vesicles. J Insect Physiol 55:264–272

Proulx P (1991) Structure-function relationships in intestinal brush border membranes. Biochim Biophys Acta 1071:255–271

Rose SL, Fulton JM, Brown CM et al (2014) Isolation and characterization of lipid rafts in *Emiliana huxleyi*: a role for membrane microdomains in host-virus interactions. Env Microbil 16:1150–1166

Santos CD, Terra WR (1984) Plasma membrane-associated amylase and typsin: intracelular distribution of digestive enzymes in the midgut of the cassava hornworm, *Erinnyis ello*. Insect Biochem 14:587–595

Santos CD, Ribeiro AF, Terra WR (1986) Differential centrifugation, calcium precipitation and ultrasonic disruption of midgut cells of *Erinnyis ello* caterpillars. Purification of cell microvilli and inferences concerning secretory mechanisms. Can J Zool 64:490–500

Schmitz J, Preiser H, Maestracci D et al (1973) Purification of the human intestinal brush border membrane. Biochim Biophys Acta 323:98–112

Silva CP, Ribeiro AF, Gulbenkian S et al (1995) Organization, origin and function of the outer microvillar (perimicrovillar) membranes of *Dysdercus peruvianus* (Hemiptera) midgut cells. J Insect Physiol 41:1093–1103

Silva CP, Ribeiro AF, Terra WR (1996) Enzyme markers and isolation of the microvillar and perimicrovillar membranes of dysdercus peruvianus (Hemiptra: Pyrrhocoridae) midgut cells. Insect Biochem Mol Biol 26:1011–1018

Silva CP, Silva J, Vasconcelos FF et al (2004) Occurrence of midgut perimicrovillar membranes in paraneopteran insect orders with comments on their function and evolutionary significance. Arthropod Struct Dev 33:139–148

Silva W, Cardoso C, Ribeiro AF et al (2013) Midgut proteins released by microapocrine secretion in *Spodoptera frugiperda*. J Insect Physiol 59:70–80

Silva W, Ribeiro AF, Silva MCP et al (2016) Gelsolin role in microapocrine secretion. Insect Mol Biol 25:810–820

Terra WR (1990) Evolution of digestive systems of insects. Annu Rev Entomol 35:181–200

Terra WR, Ferreira C (1994) Insect digestive enzymes: properties, compartmentalization and function. Comp Biochem Physiol B 109:1–62

Terra WT, Ferreira C (2012) Biochemistry and molecular biology of digestion. In: Gilbert LI (ed) Insect molecular biology and biochemistry. Academic/Elsevier, London, pp 365–418

Terra WR, Costa RH, Ferreira C (2006) Plasma membranes from insect midgut cells. An Acad Bras Cien 86:1–15

Utiyama AH, Terra WR, Ribeiro AF (2016) The digestive system of the leafhopper *Bucephalogonia xanthophis* (Hemiptera. Cicadellidae): the organization of the luminal membrane complex. J Ent Res 40:339–346

Wieczorek HC, Cioffi M, Harvey WR et al (1984) KCl-stimulated ATPase activity in purified goblet cell apical membrane from *Manduca sexta* larval midgut. Pro. First Intern Congress Comp Physiol Biochem, Liege, Beligium, B-101

Zhong H, Zhang Y, Wei C (2013) Anatomy and fine structure of the alimentary canal of the spittlebug *Lepyronia coleopterata* (l.) (Hemiptera: Cercopoidea). Arthr Strut Dev 42:521–530

Zhuang MB, Oltean DI, Gomez L et al (2002) Heliothis virescens and manduca sexta lipid rafts are involved in Cry1A toxin binding to the midgut epithelium and subsequent pore formation. J Biol Chem 277:13863–13872

Chapter 6
Midgut pH Buffering, Nutrient Absorption, Fluid Fluxes, and Enzyme Recycling

Abstract Transporting proteins may be permanently opened (pores), opened or closed (channels), or permanently closed, transporting solutes by conformational changes (transporters). Ion transporters include pumps that transport ions energized by the hydrolysis of adenosine triphosphate (ATP) and transporters driving solutes by a decreasing gradient of concentration. Nutrient transporters are responsible for the transport of amino acids, monosaccharides, fatty acids, sterols, etc. Nutrients may be transported down a concentration gradient or against a concentration gradient powered by the cotransport with ions that go down a concentration gradient. These ions may be protons and sodium or potassium ions. The proton gradient usually is established by the active transport of a pump associated with the hydrolysis of adenosine triphosphate (ATP). The known luminal midgut pH of representative insects and the early attempts to reveal the existence of midgut water fluxes in insects were reviewed. The best-known transporters responsible for midgut pH buffering and for midgut water fluxes are the ones present in *Musca domestica*. Transcriptomic data of transporters' expression along the midgut were combined with electrochemical determinations to develop models for the underlying molecular processes of midgut pH buffering. Transcriptomic data were also combined to experiments with dyes that show regions where water is being absorbed or secreted along the midgut supporting a model for the origin of the countercurrent fluxes of water that power enzyme recycling, thus avoiding enzyme excretion. Both models were tested with the use of inhibitors selected for critical transporters. Data on the molecular mechanisms underlying midgut pH buffering in lepidopterans, mosquitoes, and beetles were also discussed. The sites of nutrient absorption along the midgut were also identified with the use of transcriptomic data.

6.1 Overview

The lipid bilayer of plasma cell membranes is not permeable to polar (except water) and charged molecules. Most molecules are absorbed with the aid of transmembrane proteins located in the apex of midgut cells. There are three types of these

W. R. Terra et al., *Molecular Physiology and Evolution of Insect Digestive Systems*, Entomology in Focus 7, https://doi.org/10.1007/978-3-031-39233-7_6

transporting proteins: pores, channels, and transporters. Pores are permanently opened, channels may be opened or closed, and transporters are permanently closed, transporting solutes by conformational changes.

One of the largest groups of transporters is the solute carrier (SLC) superfamily, also known as the SLC gene series. These transporters are responsible for the transport of nutrients, ions, and xenobiotics down concentration gradients without directly hydrolyzing ATP. The individual families of the SLC superfamily are monophyletic, have similar substrate specificities (Denecke et al. 2020), and most of them were present before the divergence of bilaterian species (Hoglund et al. 2011).

The transporters may move single ions or molecules (uniporters), several of them in one direction (symporters or cotransporters) or several of them in opposite directions (antiporters or exchangers). Solutes are driven by a decreasing gradient concentration in the case of uniporters. One of the solutes moved by a symporter or exchanged by an antiporter moves down a concentration gradient allowing the other solute to be carried against a concentration gradient (secondary active transport). The decreasing gradient of one of the transported solutes is usually formed by a primary active transport of a pump that moves it against a concentration gradient associated with the hydrolysis of ATP.

Table 6.1 provides a list of the best-known transporters, channels, and pumps. Details of the transmembrane transporter proteins will be given below in the context of physiological processes.

6.2 Midgut Conditions That Affect Digestion: pH Buffering, Redox Media, and Fluid Fluxes

The important internal environmental conditions of the midgut that affect digestive enzymes, and hence digestion, are the presence of allelochemicals, redox conditions, pH values, and fluid fluxes. Several allelochemicals occur in the insect gut lumen, such as alkaloids, hydroxamic acids, phenols, and terpene aldehydes (Appel 1994) but, except for phenols and alkaloids, there are no data regarding their effect on digestion (see Chap. 3). Phenols affect digestive enzymes, but their effect is decreased in reducing conditions (Appel and Martin 1990). Reducing conditions are found in the midguts of clothes and other moths, dermestid beetles (Appel and Martin 1990), and in Hemiptera (Silva and Terra 1994). In clothes moth and dermestid beetles, a reducing medium opens disulfide bonds in dietary keratin (Apple and Martin, 1990), making easier their hydrolysis by endopeptidases, whereas in hemipterans, it maintains the activity of their major cysteine endopeptidases (cathepsin L and cathepsin B) (see item 9).

Another important condition affecting enzyme activities is the pH of the contents of the midgut. Midgut correlates better with insect phylogeny than with insect diet (Terra and Ferreira 1994; Clark 1999). This is probably a result of the adaptation of ancestral insects to a particular diet, followed by diet divergence among their descendants while still retaining the ancestral pH.

Table 6.1 pH of gut contents of representative families of the major insect orders

Order			Ventriculus		
Family (Stage)	Foregut	Caeca	Anterior	Middle	Posterior
Dictyoptera					
Blattidae (Adult)	5.6	5.8	5.8	n.d.	6.0
Orthoptera					
Acrididae (Adult)	5.8	6.2	6.2	n.d	7.3
Gryllidae (Adult)	5.0	6.4	7.0	n.d	7.3
Phasmatodea (Adult)	5.3	–	n.d	6.3/8.0	9.1/8.5
Hemiptera					5.8
Miridae (Adult)	n.d.	–	5.2	5.4	4.8
Pyrrhocoridae (Adult)	n.d.	–	6.4	5.7	5.4
Pentatomidae (Adult)	n.d.	–	5.6	5.7	5.8
Coleoptera					
Carabidae (Adult)	5.9	–	6.4	n.d.	6.6
Scarabaeidae (Adult)	8.2	–	8.2	10.4	10.5
Dermestidae (Larva)	n.d.	–	6.7	6.8	6.9
Curculionidae (Larva)	5.5	–	6.5	n.d.	7.6
Cerambycidae (Larva)	n.d	–	4.8	4.9	6.4
Tenebrionidae (Larva)	6.0	–	5.6	5.6	7.9
Chrysomelidae (Larva)	5.9	–	5.9	6.1	6.6
Megaloptera (Larva)	7.2	7.2	n.d.	7.2	n.d.
Hymenoptera					
Argidae (Larva)	n.d.	–	9.6	9.7	8.0
Formicinae (Larva)	n.d.	–	5.5	5.5	6.7
Apidae (Larva)	n.d.	–	6.0	5.7	5.6
Diptera					
Sciaridae (Larva)	n.d.	7.2	9.5	9.4	8.7
Culicidae (Larva)	n.d.	8.8	9.0	10	8.0
Simuliidae (Larva)	n.d.	n.d.	9.8	n.d.	7.5
Muscidae (Larva)	n.d.	n.d.	6.1	3.1	6.8
Calliphoridae (Larva)	6.8	n.d.	7.0	3.3	7.3
Tephritidae (Larva)	n.d.	n.d.	6.5	3.4	6.6
Lepidoptera					
Several families (Larva)	7.0	–	9.8	10.5	9.5

The data are averaged inside each family. A dash means that the structure does not exist, whereas n.d. means not determined. The pH values for Phasmatodea correspond to the anterior and posterior regions of the middle midgut and the anterior and posterior regions of the posterior midgut, respectively. The data were taken from Terra (1988) and A. B. Dias and W. R. Terra (unpublished results)

The pH of the gut contents is usually in the 6–7.5 range, with the remarkable exception of the very alkaline midgut contents of lower dipterans (mosquitoes and black midges), scarab beetles, and lepidopterans and the very acid middle midgut of cyclorrhaphous dipterans. The pH is not equally buffered along the midgut (Table 6.2). The pH optima of several enzymes that are associated with midgut apex

Table 6.2 Overview of putative ion and nutrient transporters, pumps, channels, and aquaporins

Ion transporters		Nutrient transporters		Pumps	
Protein activity	Name	Protein activity	Name	Protein activity	Name
K^+ : Cl^- symporter Na^+: K^+: $2Cl^-$ symporter	KCC NKCC	H^+: PepT symporter	PepT	3 $Na^+/2K^+$ pump H^+ pump	NKA HVA
Na^+/H^+ antiporter	NHE3	H^+:AA symporter	PAT		
Na^+/H^+ antiporter	NHE8	Na^+: AA symporter	NAT	Aquaporins (water channels)	
Na^+/H^+ antiporter	NHA	K^+ :AA symporter	KAAT	Water	DRIP
Cl^-/HCO_3^- antiporter (Na^+-independent)	AE	AA uniporter NH_3 transporter	CAT Rh	Water, urea	PRIP
Na^+ :HCO_3^- /H^+ :Cl^- (Na+-dependent)	NDAE	Glu uniporter	GLUT	Water, glycerol, urea	EGLP
		Na^+: Glu symporter	SGLT		
		Tre uniporter	TRET	Other channels	
				Cl^- (pH-sensitive)	pHCl
				K^+	Ork

AA Amino Acid, *AE* Anion Exchanger, *CAT* Cationic Amino Acid Transporter, *Glu* Glucose, *HVA* Proton V-ATPase, *KCC* Potassium Chloride Cotransporter, *NDAE* Sodium-Driven Anion Exchanger, *NHA* Sodium/Proton Antiporter, *NHE* Sodium/Proton Exchanger, *NKA* Sodium Potassium ATPase, *NKCC* Sodium Potassium Chloride Cotransporter, *Ork* Open Rectifier Potassium channel, *PAT* Proton-coupled Amino Acid Transporter, *Pep* Peptide, *PepT1* Proton-coupled Peptide Transporter 1, *pHCl-2* pH-sensitive Chloride channel 2, *Rh* Ammonium Transporter Rhesus-type, *Tre* Trehalose

are lower than the pH of midgut contents, especially among lepidopterans (see Chap. 4 and Table 6.2). This is supposed to be a consequence of proton retention in the immediate neighborhood of the negatively charged microvillar glycocalyx (Quina et al. 1980), resulting in a pH there lower than in the bulk solution (Terra and Ferreira 2012).

Midgut buffering and fluid fluxes mechanisms have been described in some insects, but before discussing them, it will be necessary to review the most important proteins associated with those mechanisms. These proteins include pumps, ammonium transporters, bicarbonate transporters, chloride and potassium channels, Na^+/H^+ antiporters, and symporters of nutrients with protons and alkalis.

6.3 Ion and Water Transporters

Two pumps are relevant here. The H^+ V-ATPase (vacuolar proton ATPase) pumps protons energized by the hydrolysis of ATP. It is expressed in plasma and organelle membranes and participates in several physiological processes like acidification of cell compartments and coupled transport of small molecules (Forgac 2007). The Na^+/ K^+- ATPase is a widespread pump in animal cells involved in the Na+ and K+ homeostasis. It is usually basolateral, sometimes expressed at the cell apex but in both positions, it energizes the secondary active transporters (Patrick et al. 2006; Okech et al. 2008; Onken et al. 2009; Barroso et al. 2019).

The ammonium transporter Rhesus-type (Rh) was characterized in *Aedes albopictus* and thought to be responsible for the excretion of ammonia derived from amino acid deamination (Wu et al. 2010). This transporter has more affinity for NH_3 than for NH_4^+.

Insects have two main bicarbonate transporters (Cl^-/HCO_3^-): one Na^+-independent known as Cl^-/HCO_3^- anion exchanger (AE; SLC4) and the other, which is specific to insects, named Na^+-dependent $Na^-:HCO_3^-/H^+:Cl^-$ anion exchanger (NDAE) (Romero et al. 2000). These antiporters are involved in the regulation of pH in cells and luminal contents. AE1 has a role in the alkalization of the mucus layer protecting the apex of the cells in the acidic middle midgut of *M. domestica* (Barroso et al. 2019).

The chloride and potassium channels important in midgut buffering are the pH-sensitive chloride channel 2 (pHCl-2) and the Open Rectifier Potassium Channel 1(Ork1), respectively. pHCl-2 is pH-gated and thought to promote the basolateral transport of chloride into oxyntic cells in the middle midgut of *D. melanogaster* (Remnant et al. 2016) and *M. domestica* (Barroso et al. 2019).

The pattern of expression of Ork1 in *M. domestica* is similar to the potassium channel Slowpoke of *D. melanogaster* related to the acidification of the middle midgut (Overend et al. 2016), and it is probable that they are the same protein (Barroso et al. 2019).

There are two members of the cation/proton antiporter (CPA; SLC9) family involved in pH regulation and transepithelial transport that have been well studied in mosquitoes and *D. melanogaster* (Piermarini et al. 2009): the Na^+/H^+ exchangers (NHE) and the Na^+/H^+ antiporters (NHA), both expressed at the cell apex of midgut cells. The amiloride-sensitive NHE8 promotes the extrusion of Na^+ (or K^+) in exchange for extracellular H^+ generated by a H^+ V-ATPase (Kang'ethe et al. 2007; Barroso et al. 2019). NHA is supposed to act likewise NHE8 (Barroso et al. 2019). Midgut pH buffering may also be affected by the proton-amino acid transporter (PAT) and proton-peptide transporter (PepT), which will be discussed in Sect. 6.4.

Three members of the cation-coupled cotransport family (SLC12) are important for midgut pH buffering and fluid fluxes: potassium: chloride cotransporter (KCC) and sodium: potassium: 2 chloride cotransporter 1 (NKCC1) and 2 (NKCC2). KCC moves ions (and water, see below) outside the cells and is more inhibited by furosemide than bumetanide, whereas NKCC transport ions (and water) into the cells and are more inhibited by bumetanide than by furosemide (Lauf and Adragna 2000; Russell 2000). Associated with ion transport by KCC and NKCC there is a transport of water. The energy necessary to move the solutes through the membrane is supplied by the electrochemical gradient of sodium (NKCC) or potassium (KCC), which is formed by the Na^+/ K^+- ATPase energized by ATP hydrolysis. The transport of water starts as soon as the ion transport begins, without a lag that would be expected if the water transport was based on the osmotic increase of solutes in the medium. Thus, the coupling of water and substrate must occur by a mechanism not entirely known within the transporter protein itself. The molar ratio of water/solutes is 500 for KCC and 590 for NKCC (Zeuthen 2010; MacAulay and Zeuthen 2010).

KCC and NKCC are expressed in tissues that transport ions and water like the gut, Malpighian tubules, and anal pads. Considering the direction of the water fluxes driven by KCC, it favors water secretion, if placed at the apex of the cell and cell water absorption, if located at the cell basal membrane. The contrary is true for NKCC. If it is placed at the apex, it causes water absorption, and if located at the basal membrane, it favors water secretion.

Aquaporins (AQPs) or water channels are responsible for the high permeability of cell membranes to water. The direction of the water flux through aquaporins is uniquely dependent on solute gradients. There are types of aquaporins based on their permeability, from which three are important in insects: DRIP, a water-selective AQP; PRIP, selective to water-urea; and EGPL, which is a water-urea-glycerol AQP (or aquaglyceroporin, specific for holometabolans) (Finn et al. 2015). The fourth type of AQP, identified as BIB, is actually an ion channel, despite its relatedness with aquaporins (Yanochko and Yool 2002). Some aquaporins may be activated/inactivated by phosphorylation changing the water permeability of the membrane to water in some cells (Campbell et al. 2008).

6.4 Nutrient Transporters

Nutrient transporters are the transporters involved in the absorption of amino acids (and small peptides), monosaccharides (and the disaccharide trehalose), and fatty acids (Holtof et al. 2019).

Amino acids are absorbed by the action of transporters, from which the most important in insects are the symporters proton-coupled amino acid transporter (PAT, SLC36) (Evans et al. 2009), sodium-dependent nutrient amino acid transporter (NAT, SLC6), potassium-coupled amino acid transporter (KAAT, SLC7), proton-coupled peptide transporter (PepT, SLC15, Roman et al. 1998), and the uniporter cationic amino acid transporter (CAT, SLC7) (Boudko 2012; Holtof et al. 2019). NATs transport the essential amino acids (Trp, Phe, Tyr, Met) coupled to sodium, and individually they may have a broad or a narrow substrate specificity (Boudko 2012; Meleshkevitch et al. 2013). KAAT 1 has a broad specificity and is characteristic of lepidopterans that have a diet rich in potassium and display high pH midgut contents (Boudko 2012). PATs and PepTs use the proton gradient formed mainly by the H^+ V-ATPase (or NHE) as a driving force, whereas NAT and KAAT use the sodium or potassium gradient generated by the Na^+/K^+ ATPase or other antiporters like NHE. As a rule, symporters are found at the cell apex and the uniporter at the apex or base of the cells. PATs differ in transport capacity, substrate specificity, and how they are affected by external pH changes (Thwaites and Anderson 2011). It is interesting to note that PATs may also transport, in addition to the more common

Gly, Ala, and Pro, D-amino acids, which is advantageous in the case of larvae feeding on decaying food rich in bacteria like houseflies (Barroso et al. 2019).

The absorption of monosaccharides like glucose, fructose, mannose, and galactose is carried out by sugar porters (SPs), which are mainly uniporters, and to a lesser extent by symporters. SPs are usually uniporters close to Class III of the mammalian GLUT family, but they may have differences in their sequence pointing to differences in substrate specificity (Pimentel et al. 2018) or may be selective in relation to glucose as Class I GLUT of mammals (Price et al. 2007; Price et al. 2010; Kikuta et al. 2010; Price and Gatehouse 2014). Although SPs of the GLUT family are as a rule uniporters, there are some exceptions, exemplified by NIST8 of *Nilaparvata lugens* (H: trehalose) (Kikuta et al. 2012) and arguably MdSP8 of *M. domestica* (Pimentel et al. 2018).

Sugar porters which are symporters of the SGLT family were functionally characterized in the midgut of *Aphidius ervi* (Caccia et al. 2007) and in *D. peruvianus* (Bifano et al. 2010). The sugar porter of *A. ervi* transport glucose coupled with sodium and that of D. peruvianus transport glucose coupled with potassium and carry water, like mammalian GLUTs and SGLTs (Zeuthen 2010). In mammals, there are sugar-like transporters that actually are glucose sensors (Sclafani et al. 2016). The same seems to occur in *M. domestica*, based on the finding of SPs which expression is upregulated after starch ingestion and that carry critical sequence mutations hampering their transporting capacity (Pimentel et al. 2018).

Trehalose is the main sugar in insect hemolymph and accordingly there are a large number of trehalose transporters (TRETs) in insect tissues, including the midgut. Most of them are uniporters, but there are also symporters with protons (Kikuta et al. 2012; Pimentel et al., 2012). TRETs are usually expressed at the basolateral membranes in most tissues, including the midgut, thus providing trehalose as an energy source for metabolism. At the cell apex, TRETs are involved in the uptake of trehalose from the diet (Pimentel et al. 2018).

Dietary lipids are mainly triacylglycerols that are used as energy storage and phosphatides and galactolipids that are major cell membrane components. (Vance and Vance 2008; Majerowicz and Gondim 2013). On digestion, those lipids originate fatty acids (FA) and glycerol. Even though FA may diffuse across midgut cell membranes down a concentration gradient, due to metabolic demands, a protein-mediated absorption is necessary.

The proteins involved in enhancing FA uptake and in preventing its exit out of the cell include the acyl-CoA synthetases (ACS) that activate FA, forming acylCoA, and proteins that bind FA (fatty acid–binding proteins, FABP) and its activated form (acylCoA-binding protein, ACBP) (Grevengoed et al. 2014). FABPs have been described in the midguts of both Lepidoptera and Hymenoptera (Smith et al. 1992; Huang et al. 2012; Caccia et al. 2007). ACS sequences have considerable homology with conserved ATP/AMP and FA binding sites and are classified according to the length of their preferred substrate (Grevengoed et al. 2014). Some of them are supposed to be able to transport (fatty acid transport protein, FATP) and activate FA (ACS/FATP). ACS have been described in several insect species (Ohnishi et al.

2009; Dourlen et al. 2015; Alves-Bezerra et al. 2016), but their subcellular location along the midgut is known only for *M. domestica* (Barroso et al. 2021).

Sterols confer membrane fluidity in cell membranes and are precursors of steroid hormones like ecdysone (Zheng et al. 2018). Since insects are unable to synthesize sterols, they must acquire them from the diet. Two families of sterol transporters were found in insects: one responsible for the dietary absorption of sterols and the other for sterol absorption and intracellular transport before pupation (Guo et al. 2009; Holtof et al. 2019).

6.5 Models of Midgut pH Buffering

The high pH found in the midgut lumen was studied for a long time, but the first successful model to explain it was Dow's (1992), as detailed in the legend of Fig. 6.1. Details developed after the paper of Dow (1992) are discussed by Harvey (2009).

The first successful mechanism of pH buffering in the midgut *of M. domestica* was proposed by Terra and Regel (1995), based on chemical determinations in the midgut and in the effect of inhibitors added to the diet in the pH of midgut contents. According to the model, protons are pumped by the middle midgut oxyntic cells by an unidentified pump with the involvement of a Na^+/K^+ ATPase. Protons were followed by chloride ions present in high concentrations in the middle midgut. Finally, posterior midgut contents were neutralized by ammonia. More recently, the underlying molecular processes of midgut buffering were described. For this, in vivo electrochemical determinations at different conditions were combined with data on the expression (by RNA-seq) of selected transporters and proteomic analysis of purified microvillar membranes of cells along the midgut (Barroso et al. 2019). In brief, acidification of the middle midgut of larval *M. domestica* is caused by protons pumped by a H^+ V-ATPase (inhibited by bafilomycin) into the canaliculi of the oxyntic cell. The protons are followed by chloride ions transported by KCC (inhibited by furosemide). The top of the middle midgut cells is protected by a mucus layer (see details in Chap. 7) neutralized by bicarbonate secreted with sodium in exchange for proton and chloride ions. The posterior midgut is neutralized by ammonia secreted by Rh (inhibited by dimethylamonia) and protons are internalized coupled with amino acids by PATs. The model of acidification is reproduced in Fig. 6.2 and that of alkalization in Fig. 6.3.

In the beetle *T. molitor*, transcriptomic data combined with physiological data were used to propose that the acidification of the anterior midgut results from the secretion of NH_4^+ and Cl^- in exchange for NH_3, whereas alkalization of the posterior midgut depends on the secretion of bicarbonate in exchange for protons and chloride ions (Moreira et al. 2017). It is probable that similar mechanisms cause acidification of the anterior midgut and alkalization of the posterior midgut in other Coleoptera and many Dictyoptera and Orthoptera.

Phasmids have a characteristic system of midgut tubules at their alkaline posterior midguts. The tubules resemble Malpighian tubules and open in the midgut

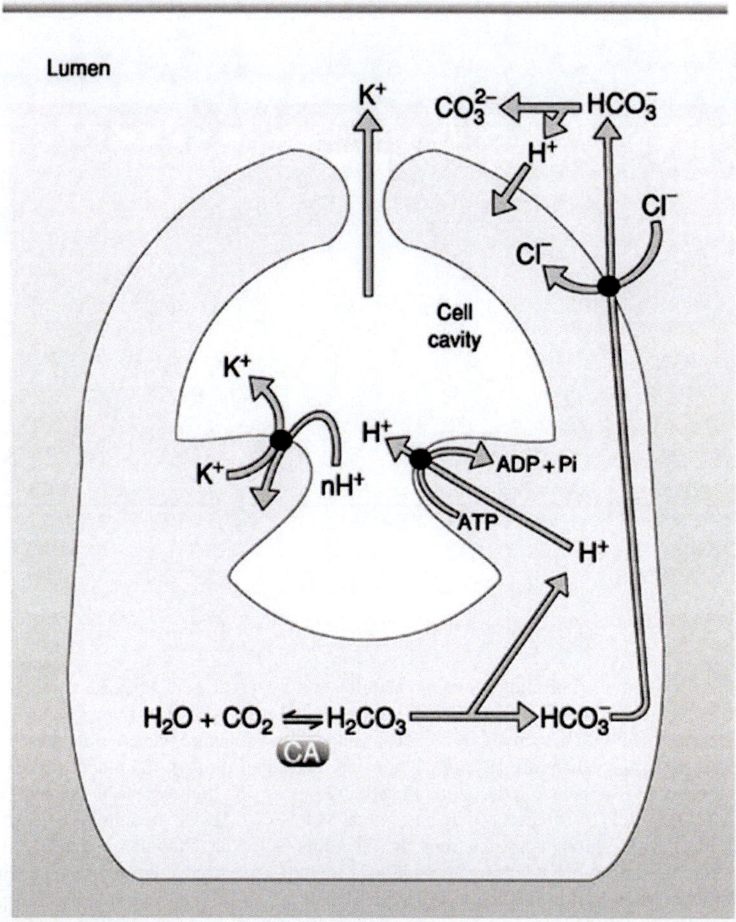

Fig. 6.1 A model for generation of high gut pH by the goblet cells of lepidopteran larvae. Carbonic anhydrase (CA) produces carbonic acid that dissociates into bicarbonate and a proton. The proton is pumped by a V-ATPase into the goblet cell cavity, from where it is removed in exchange with K^+ that eventually diffuses into the lumen. Bicarbonate is secreted in exchange with chloride and loses a proton due to the intense field near the membrane, forming carbonate and raising the gut pH. (Reproduced with permission from Terra WR, Ferreira C, 2012. Biochemistry and molecular biology of digestion. In: Gilbert LI (Ed.), *Insect Molecular Biology and Biochemistry*, Academic Press/Elsevier, London, pp. 365–418)

through midgut protuberances. These protuberances have a cytological acidophilic character and a high carbonic anhydrase activity. Based on this, Monteiro et al. (2014) proposed that the secretion of bicarbonate by the tubules results in the alkalization of the posterior midgut. In a similar way, bicarbonate is transported through a putative bicarbonate-activated ATPase in the mammalian duodenum to neutralize the acid bolus of the stomach.

Scarab beetles have a midgut with three rows of caeca with a ventral groove between the middle and posterior row of ceca. The pH of gut contents raises to

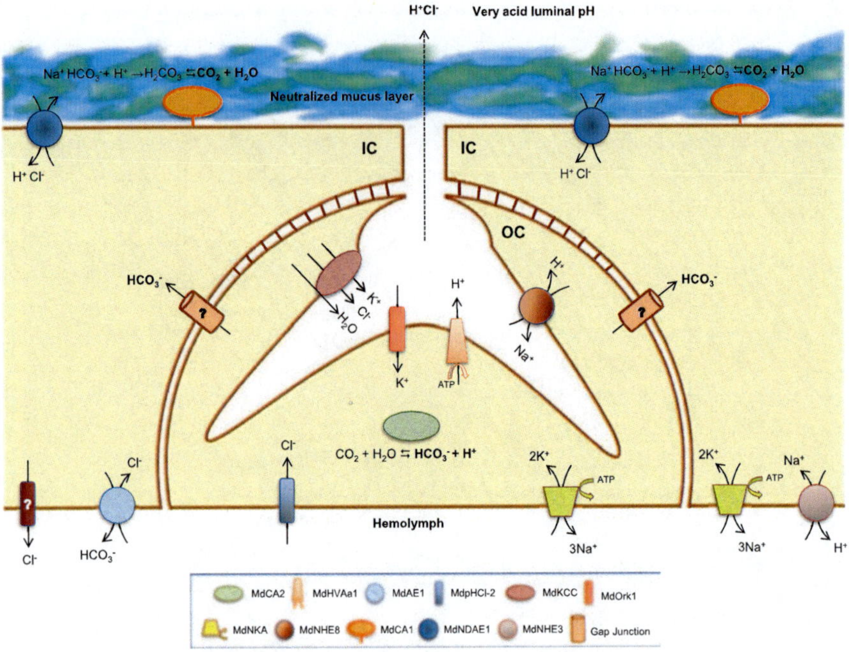

Fig. 6.2 Working model of midgut luminal acidification by oxyntic cells (OC) and mucosal acid protection by interstitial cells (IC). Under the action of the MdHVAa1, H^+ is pumped into the canaliculi, accompanied by Cl^- secreted by MdKCC and water. As a consequence of increased hydrostatic pressure in the canaliculi, H^+ and Cl^- ions are propelled through the mucus layer into the midgut lumen. H^+ resulted from the action of MdCA2 on CO_2. K^+ homeostasis is assured by basolateral MdNKA, and Cl^- is recovered by basolateral MdpHCl-2. HCO_3^- remaining in OC is transferred from OC to IC through gap junctions, thus avoiding OC alkalization. K^+ secreted by MdKCC is recovered by an apical K^+ channel, MdOrk1, in hyperpolarized conditions, enhancing the MdKCC activity. The amiloride-sensitive MdNHE8 controls the H^+ concentration by transporting part of the H^+ extruded by the H^+ pump from the canaliculi to the cell. Interstitial middle midgut cells secrete into the mucus layer HCO_3^- and Na^+ that neutralizes the mucus and are exchanged for H^+ and Cl^- under the action of MdNDAE1. HCO_3^- reacts with H^+ permeated from the lumen into the mucus layer, forming H_2CO_3, which is dehydrated by a GPI-anchored MdCA1, resulting in H_2O and CO_2. Excess acid is eliminated from IC by basolateral amiloride-insensitive MdNHE3 and Cl^- by a putative channel. (Reproduced with permission from Barroso et al. 2019)

almost 12 along the ventral groove (Bayon 1981). Despite the unavailability of data, it is probable that the ventral groove alkalization follows a mechanism like that described for phasmids.

Before blood ingestion, the anterior midgut of mosquitoes is maintained at pH6 by the action of a H^+ V-ATPase. On blood feeding, the posterior midgut is alkalized by the combined action of H^+ V-ATPase inhibition influenced by cAMP and proton removal coupled with amino acid through PATs and bicarbonate transport to midgut contents in exchange for chloride ions (Nepomuceno et al. 2017, 2021).

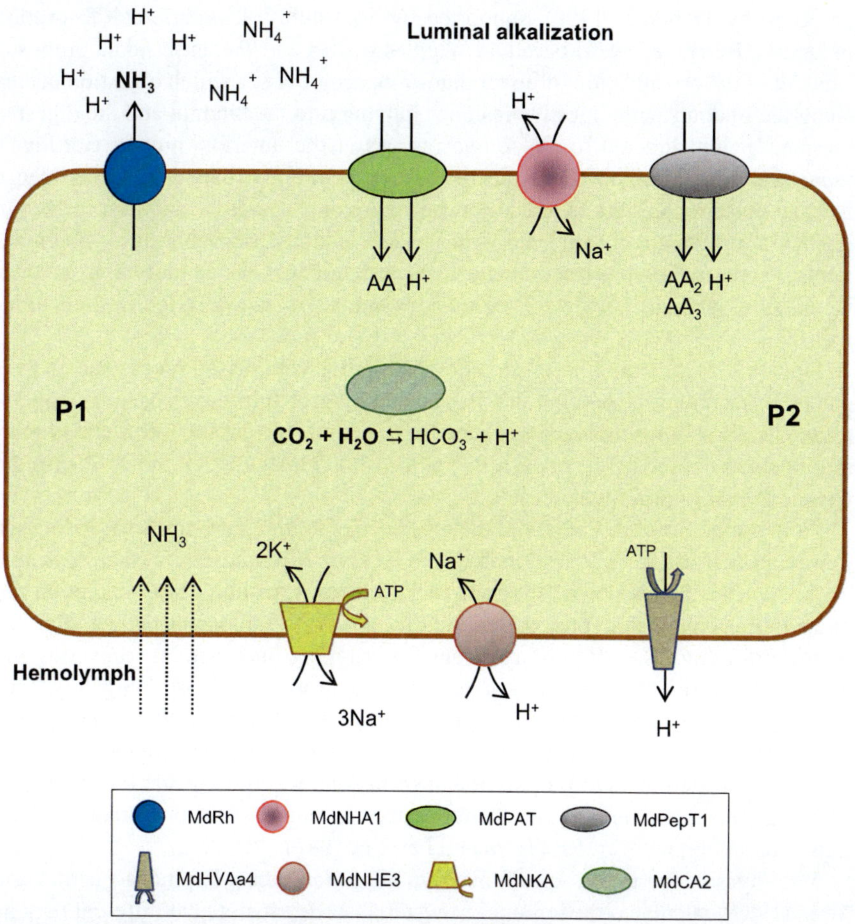

Fig. 6.3 Working model of luminal alkalization at the posterior midgut. Midgut alkalization occurs mainly at the beginning of the posterior region caused by the transport of NH_3 by MdRh. NH_3 reacts with H^+ coming from the middle midgut contents, forming NH_4^+. Cell NH_3 comes from hemolymph by simple diffusion. Alkalization is further increased in a diet containing protein, as a consequence of the symport of H^+ and peptides carried out by MdPept1 and presumably also by H^+ and amino acids by MdPATs. The MdNHA1 maintains mild acid in the neighborhood of MdPATs and MdPepT 1, thus enhancing their activity. Intracellular acidification is circumvented by basolateral MdNHE3, coupled to the Na^+ gradient established by basolateral MdNKA and H^+ pumping by basolateral MdHVAa4. (Reproduced with permission from Barroso et al. 2019)

6.6 Models of Midgut Fluid Fluxes and Enzyme Recycling

Wigglesworth (1933) during his studies on osmoregulation in mosquito larvae was the first to describe midgut fluid fluxes with the aid of dyes. According to him, fluid is absorbed in the midgut ceca at the anterior midgut and secreted by the posterior midgut. A similar role of insect anterior midgut ceca in fluid absorption was proposed by Treherne (1959), supported by his studies of amino acid absorption in locusts. Berridge (1970) based on Wigglesworth's and Treherne's data proposed a model of the organization of insect digestion, according to which digestion occurs along the midgut, with digestive residues moving into the hindgut and the digested nutrients being directed to absorptive regions at the anterior midgut. Berridge's model met wide acceptance, despite the lack of quantitative data and clear evidence that the posterior midgut is actually water secretory. Secretory and absorptive gut regions are qualitatively inferred with the use of non-absorbable dyes, like amaranth. Secretory regions accumulate dyes injected into the hemolymph on their hemal side, whereas absorptive regions accumulate dyes orally fed in their luminal side.

Dow (1981) was the first to provide quantitative data, based on the use of dyes and other techniques, showing that in starving locusts fluid fluxes occur in opposition to the food bolus movement, generating a countercurrent flux. This countercurrent flux was caused by water secretion at the Malpighian tubules and its absorption back in the anterior midgut ceca.

At the same time Dow reported the existence of countercurrent fluxes in locusts, Terra and Ferreira (1981) described in the larva of *Rhynchosciara americana* that countercurrent fluxes power enzymes recycling, thus avoiding their excretion. To come to this conclusion, they determined fecal digestive enzymes delivered at different times onto humid sand layers and, after taking into account recoveries and enzyme stabilities, they calculated enzyme excretory rates. On doing this, they found that the excretory rate of trypsin (an enzyme able to cross PM) was equal to that of trehalase, an enzyme known to be restricted to the ectoperitrophic space (see Chap. 2). This led them to propose that trypsin is moved from inside PM to outside PM at the posterior midgut and then directed forward by the countercurrent flux, thus avoiding being excreted (Terra and Ferreira 1981).

With dyes added to the diet of the larva of *M. domestica,* Espinoza-Fuentes and Terra (1987) calculated the volumes of water absorbed/secreted at different regions of the midgut from the intensities of color and dimensions of each midgut region. From the data, they proposed that a countercurrent flux from the end to the middle midgut is responsible for trypsin recycling and hence for decreasing its excretory rate. Similar findings were reported for Lepidoptera (Bolognesi et al. 2001, 2008), Coleoptera (Ferreira et al. 2002; Caldeira et al. 2007), Orthoptera Ensifera (Biagio et al. 2009), and Phasmida (Monteiro et al. 2014). All these insects, in contrast to Orthoptera Caelifera (Dow 1981; Ferreira et al. 1990), have midgut countercurrent in both fed and starving conditions. The accumulation in the anterior midgut of proteins secreted in the middle (Peterson et al. 1994) or posterior midgut (Borhegyi et al. 1999) suggests the occurrence of midgut countercurrents.

A theoretical model was developed by Bolognesi et al. (2008) to calculate the enzyme distribution along the midgut lumen, given the water absorption and secretion sites. It is interesting to note that the computed enzyme distribution in the midgut of *M. domestica* only agrees with actual experimental data if the chief water absorption site was located at the anterior region of the posterior midgut (as recognized today, see below), instead of in the middle midgut as in the Espinoza-Fuentes and Terra model.

Despite the numerous papers on the subject and the wide acceptance of the midgut countercurrent fluxes and enzyme recycling models, the molecular mechanisms underlying the phenomenon were only recently studied. At first, a transcriptome approach was used. According to this approach, the expression of the genes coding for selected membrane transporter proteins (see Table 6.2) was evaluated at different sites along the gut. The transcriptomic approach assumes that the more a transporter coding gene is expressed, the more the protein coded by it is translated. Although this is not necessary, as a protein may occur in a large amount, if it is stable, even in the event the gene coding for it is lowly expressed and not all transcripts generated will be translated or become functional proteins. In spite of these considerations, the fact is that usually there is a good correlation between gene expression and the amount of the coded protein present. Gene expression is usually evaluated by RNA-seq, which is the determination of how many transcribed RNA sequences correspond to a particular gene in a tissue at a time. In the case of protein-coding genes, for this determination, the usual protocol includes the following steps: total RNA is extracted from the tissue under study, the messenger RNA fraction is enriched, converted into DNA by a reverse transcriptase, the generated cDNA is enzymatically fragmented, and the resulting DNA molecules are sequenced. The obtained sequences (reads, which correspond to fragments of the transcripts) are then aligned against a genome or a "de novo" assembled transcriptome. A "de novo" assembled transcriptome is a series of transcript sequences (contigs) assembled with bioinformatics tools from the original reads. Contigs correspond to transcripts. The number of reads aligned to the exons of a gene or to a contig from a transcriptome is used to measure its expression. Expression values for each gene inside a sample are expressed in a normalized way, frequently as TPM (transcripts per million) values. This approach was used for the first time to propose a putative model of countercurrent fluxes in *T. molitor*, according to which water is absorbed at the anterior midgut with the aid of NKCC and secreted at the posterior midgut favored by KCC (Moreira et al. 2017).

More recently, a detailed investigation was performed to disclose the molecular mechanisms underlying the midgut fluid fluxes in the larva of *M. domestica* (Barroso et al. 2020). With the use of dyes, the secretory and absorptive sites were identified (see Fig. 6.4a, b) and the amount of water moved in each process was quantified. The expression along the midgut of the genes coding for transporters that might be involved in water fluxes was evaluated by RNA-seq. The proteins of the isolated midgut microvillar membranes occurring along the midgut were identified by proteomics. Proteomics in this case consists in the chromatographic separation of the protein fraction, followed by mass spectrometry and matching the resulting peaks with a genome or a "de novo" assembled transcriptome for protein identification. Transcriptomic and proteomic data were combined to identify the sites of occurrence of the transporters along the midgut and whether they are apical (detected by microvillar proteomics) or basal (not detected by microvillar proteomics). All the data were combined in a model that was tested with the use of inhibitors of specific

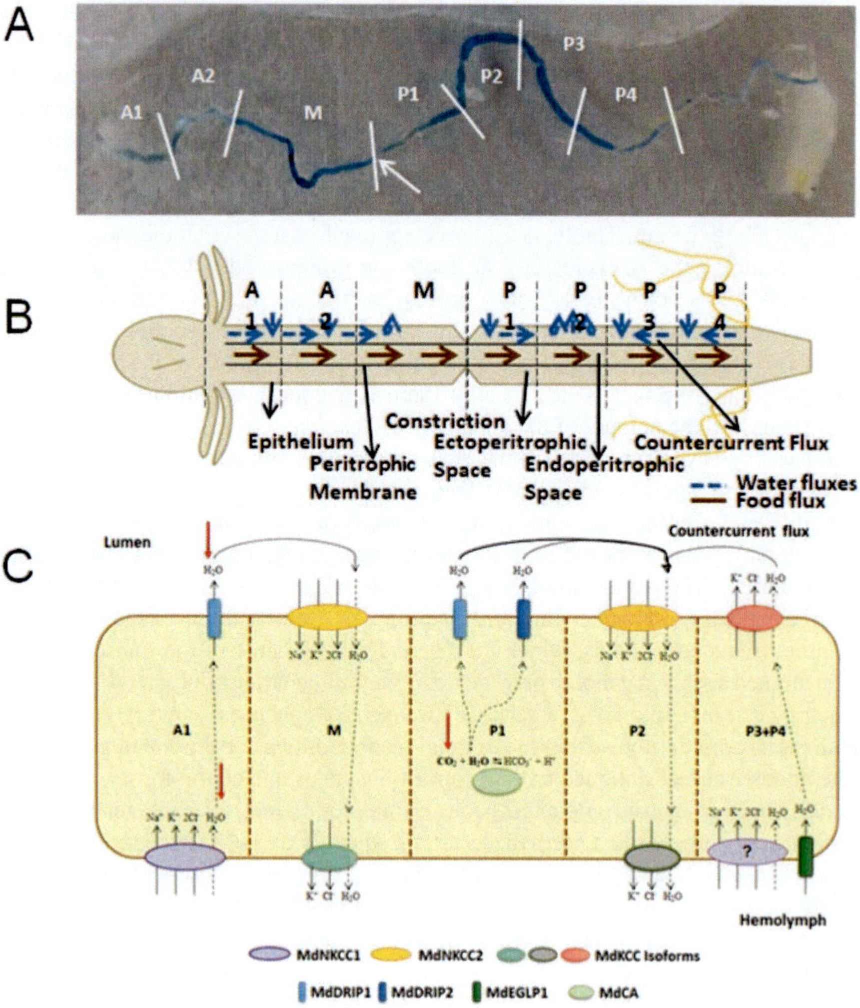

Fig. 6.4 (**a**) Dye distribution along the midgut of *M. domestica* larvae fed on 1% agarose gel, 0.05 % Evans Blue. A secretion of water is suggested at A1 and A2, followed by absorption at M. Posterior sections P1, P3, and P4 apparently secrete water, whereas P2 apparently absorbs water. (**b**) Diagrammatic representation of the midgut water fluxes in *M. domestica* larva. Water is secreted at the anterior midgut (A1, A2) and absorbed at the middle midgut (M). In the posterior midgut, water is secreted at P1, P3, and P4 and absorbed at P2, resulting in a countercurrent flux of water from P3+P4 to P2. Water fluxes from the anterior and middle midgut are probably independent of the posterior midgut because of the existence of a constriction region between the middle and posterior midgut. *Blue dotted arrows* refer to water fluxes, whereas *brown solid arrows* indicate the direction of food movement. (**c**) Simplified working model of molecular mechanisms involved in midgut water fluxes based on transcriptomics, proteomics, and in vivo experiments. Secretion of water at A1 is caused by the action of a basolateral MdNKCC1 and helped by an apical water-selective aquaporin MdDRIP1. Water is absorbed in the middle midgut (M) by the activity of an apical MdNKCC2 and a basolateral MdKCC isoform. In the posterior midgut P1,

(continued)

transporters. According to the model, water is secreted at the posterior region of the posterior midgut with the aid of an apical KCC and a basal NKCC and is absorbed at the middle part of the posterior midgut helped by an apical NKCC and a basal KCC. The use of the inhibitors furosemide (more effective for KCC) and bumetanide (more effective for NKCC) affect both midgut water fluxes and the trypsin activity pattern along the midgut, enhancing its excretion. This is the first direct evidence of the effect of midgut countercurrent fluid fluxes in enzyme recycling. The midgut of *M. domestica* have other sites of water absorption and secretion besides those described, but they do not affect the countercurrent fluxes and hence enzyme recycling (Fig. 6.4a–c).

References

Alves-Bezerra M, Klett EL, De Paula IF et al (2016) Long-chain acyl-CoA synthetase 2 knockdown leads to decreased fatty acid oxidation in fat body and reduced reproductive capacity in the insect *Rhodnius prolixus*. Biochim Biophys Acta Mol Cell Biol Lipids 1861:650–662

Appel HM (1994) The chewing herbivore gut lumen: physicochemical conditions and their impact on plant nutrients, allelochemicals, and insect pathogens. In: Bernays EA (ed) Insect–plant interactions, vol 5. CRC Press, Boca Raton, pp 203–223

Appel HM, Martin MM (1990) Gut redox conditions in herbivorous lepidopteran larvae. J Chem Ecol 16:3277–3290

Barroso IG, Santos CS, Bertotti M et al (2019) Molecular mechanisms associated with acidification and alkalization along the larval midgut of *Musca domestica*. Comp Biochem Physiol A 237:110535

Barroso IG, Fuzita FJ, Ferreira C et al (2020) Midgut fluxes and digestive enzyme recycling in *Musca domestica:* a molecular approach. Comp Biochem Physiol A 241:110627

Barroso IG, Cardoso C, Ferreira C et al (2021) Transcriptomic and proteomic analysis of the underlying mechanisms of digestion of triacylglycerols and phosphatides and absorption and fate of fatty acids along the midgut of *Musca domestica.* Comp Biochem Physiol D 39:100826

Bayon C (1981) Ultrastruture de l'epithelium intestinale et flore parietale chez la larva xylophage d'Oryctes nasicornis larvae (Coleotera: Scarabaidae). Int. J Insect Morphol Embryol 10:359–371

Berridge (1970) A structural analysis of intestinal absorption. Symp Roy Ent Soc 5:135–150

Biagio FP, Tamaki FK, Terra WR et al (2009) Digestive morphophysiology of *Gryllodes sigillatus* (Orthoptera: Gryllidae). J Insect Physiol 55:1125–1133. https://doi.org/10.1016/j.jinsphys.2009.08.015

Fig. 6.4 (continued) water diffuses passively into the lumen through MdDRIP1 and MdDRIP2 located at the cell apex. This is caused by differential osmotic pressure, probably as a consequence of carbonic anhydrase activity and a putative hyperosmotic luminal content discussed in the text. Water absorption at P2 is alike the middle midgut with an apical MdNKCC2 and a basolateral MdKCC isoform. Finally, water secretion at P3+P4 occurs through an apical MdKCC isoform helped by basolateral MdEGLP1 or a putative MdNKCC1, although it was not detected. Thus, water secreted at A1 is absorbed at M, and water secreted at P1 and P3+P4 is absorbed at P2. The water moving from P3+P4 to P2 is the countercurrent flux of water that powers enzyme recycling. (Reproduced with permission from Barroso et al. 2020)

Bifano TD, Alegria TGP, Terra WR (2010) Transporters involved in glucose and water absorption in *Dysdercus peruvianus* (Hemiptera: Pyrrhocoridae) anterior midgut. Comp Biochem Physiol B 157:1–9

Bolognesi R, Ribeiro AF, Terra WR et al (2001) The peritrophic membrane of *Spodoptera frugiperda*: secretion of peritrophins and role in immobilization and recycling digestive enzymes. Arch Insect Biochem Physiol 47:62–75

Bolognesi R, Terra WR, Ferreira C (2008) Peritrophic membrane role in enhancing digestive efficiency Theoretical and experimental models. J Insect Physiol 54:1413–1422

Borhegyi NH, Molnár K, Csikós G et al (1999) Isolation and characterization of an apically sorted 41-kDa protein from the midgut of tobacco hornworm (*Manduca sexta*). Cell Tissue Res 297:513–525

Boudko DY (2012) Molecular basis of essential amino acid transport from studies of insect nutrient amino acid transporters of the SLC6 family (NAT-SLC6). J Insect Physiol 58:433–449

Caccia S, Casartelli M, Grimaldi A et al (2007) Unexpected similarity of intestinal sugar absorption by SGLT1 and apical GLUT2 in an insect *(Aphidius ervi*, Hymenoptera) and mammals. Am J Physiol Regul Integr Comp Physiol 292:2284–2291

Caldeira W, Dias AB, Terra WR et al (2007) Digestive enzyme compartmentalization and recycling and sites of absorption and secretion along the midgut of *Dermestes maculatus* (Coleoptera) larvae. Archs Insect Biochem Physiol 64:1–18

Campbell PM, Cao AT, Hines ER et al (2008) Proteomic analysis of the peritrophic matrix from the gut of the caterpillar, *Helicoverpa armigera*. Insect Biochem Mol Biol 38:950–958

Clark TM (1999) Evolution and adaptive significance of larval midgut alkalinization in the insect superorder Mecopterida. J Chem Ecol 25:1945–1960

Denecke SM, Driva O, Luong HNB et al (2020) The identification and evolutionary trends of the solute carrier superfamily in arthropods. Genome Biol Evol 12:1429–1439

Dourlen P, Sujkowski A, Wessells R et al (2015) Fatty acid transport proteins in disease: New insights from invertebrate models. Prog Lipid Res 60:30–40

Dow JAT (1981) Countercurrent flow, water movements and nutrient absorption in the locust midgut. J Insect Physiol 27:579–585

Dow JAT (1992) pH gradients in lepidopteran midgut. J Exp Biol 172:355–375

Espinoza-Fuentes FP, Terra WR (1987) Physiological adaptations for digesting bacteria - water fluxes and distribution of digestive enzymes in *Musca domestica* larval midgut. Insect Biochem 17:809–817

Evans AM, Aimanova KG, Gill SS (2009) Characterization of a blood-meal-responsive proton-dependent amino acid transporter in the disease vector, Aedes aegypti. J Exp Biol 212:3263–3271

Ferreira C, Oliveira MC, Terra WR (1990) Compartimentalization of the digestive process in *Abracris flavolineata* (Orthoptera:Acrididae) adults. Insect Biochem 20:267–274

Ferreira AHP, Ribeiro AF, Terra WR et al (2002) Secretion of β-glycosidase by middle midgut cells and its recycling in the midgut of *Tenebrio molitor* larvae. J Insect Physiol 48:113–118

Finn RN, Chauvigne F, Stavang A et al (2015) Insect glycerol transportes evoved by functional co-option and gene replacement. Nat Comm. https://doi.org/10.1038/ncomms8814

Forgac M (2007) Vacuolar ATPases: Rotary proton pumps in physiology and pathophysiology. Nat Rev Mol Cell Biol 8:917–929

Grevengoed TJ, Klett EL, Coleman RA (2014) Acyl-CoA metabolism and partitioning. Annu Rev Nutr 34:1–30

Guo XR, Zheng SC, Liu L et al (2009) The sterol carrier protein 2/3-oxoacyl-CoA thiolase (SCPx) is involved in cholesterol uptake in the midgut of Spodoptera litura: Gene cloning, expression, localization and functional analyses. BMC Mol Biol 10:1–18

Harvey WR (2009) Voltage coupling of primary K^+ V-ATPases to secondary Na^+- or K^+-dependent transporters. J Exp Biol 212:1620–1629

Hoglund PJ, Nodstrom KV, Schioth HB et al (2011) The solute carrier families have a remarkable long evolutionary history with the majority of the human families present before divergence of bilaterian species. Mol Biol Evol 28:1531–1541

Holtof M, Lenaert C, Cullen et al (2019) Extracellular nutrient digestion and absorption in the insect gut. Cell Tis Res 377:397-414 https://doi.org/10.1016/j.jinsphys.2017.03.009

Huang Z, Zhou D, Gao G et al (2012) Cloning and characterization of a midgut-specific fatty acid binding protein in *Spodoptera litura*. Arch Insect Biochem Physiol 79:1–17. https://doi.org/10.1002/arch.21001

Kang'ethe W, Aimanova KG, Pullikuth AK et al (eds) (2007) NHE8 mediates amiloride sensitive Na+/H+ exchange across mosquito Malpighian tubules and catalyzes Na+ And K+ transport in reconstituted proteoliposomes. Am J Physiol Ren Physiol 292:F1501–F1512

Kikuta S, Kikawada T, Hagiwara-Komoda Y et al (2010) Sugar transporter genes of the brown planthopper, *Nilaparvata lugens*: A facilitated glucose/fructose transporter. Insect Biochem Mol Biol 40:805–813

Kikuta S, Hagiwara-Komoda Y, Noda H et al (2012) A novel member of the trehalose transporter family functions as an H+-dependent trehalose transporter in the reabsorption of trehalose in Malpighian tubules. Front Physiol 3:1–9

Lauf PK, Adragna NC (2000) K-Cl cotransports: properties and molecular mechanism. Cell Physiol Biochem 10:341–354

MacAulay N, Zeuthen T (2010) Water transport between CNS compartments: contributions of aquaporins and cotransporters. Neuroscience 168:941–956

Majerowicz D, Gondim KC (2013) Insect Lipid Metabolism: Insights into Gene Expression Regulation. In: Mandal SS (ed) Recent trends in gene expression. Nova Science Publishers, Inc, Hauppauge, pp 147–190

Meleshkevitch EA, Voronov DA, Miller MM et al (2013) A novel eukaryotic Na+ Methionine selective symporter is essential for mosquito development. Insect Biochem Mol Biol 43:755–767

Monteiro EC, Tamaki FK, Terra WR et al (2014) The digestive system of the "stick bug" *Cladomorphus phyllinus* (Phasmada, Phasmatidae): a morphological, physiological and biochemical analysis. Arthr Struct Develop 43:123–134

Moreira NR, Cardoso C, Dias RO et al (2017) A physiologically-oriented transcriptomic analysis of the midgut of *Tenebrio molitor*. J Insect Physiol 99:58–66

Nepomuceno DB, Santos VC, Araujo RN et al (2017) pH control in the midgut of *Aedes aegypti* under different nutritional conditions. J Exp Biol 220:3355–3362

Nepomuceno DB, Pessoa GC, Koerich et al (2021) cAMP: asecond messenger involved in the mechanism of midgut alkalinization in *Lutzomia longipalpis* Insect Sci 29: 1059-1070 https://doi.org/10.1111/1744-7917.12980.

Ohnishi A, Hashimoto K, Imai K et al (2009) Functional characterization of the *Bombyx mori* fatty acid transport protein (BmFATP) within the silkmoth pheromone gland. J Biol Chem 284:5128–5136

Okech BA, Boudko DY, Linser PJ et al (2008) Cationic pathway of pH regulation in larvae of *Anopheles gambiae*. J Exp Biol 211:957–968. https://doi.org/10.1242/jeb.012021

Onken H, Patel M, Javoroncov M et al (2009) Strong Alkalinization in the Anterior Midgut of Larval Yellow Fever Mosquitoes (*Aedes aegypti*): Involvement of Luminal Na+/K+ - ATPase. J Exp Zool A Ecol Genet Physiol 311:155–161

Overend G, Luo Y, Henderson L et al (2016) Molecular mechanism and functional significance of acid generation in the *Drosophila* midgut. Sci Rep 6:1–11

Patrick ML, Aimanova K, Sanders HR et al (2006) P-type Na+/K+-ATPase and V-type H+-ATPase expression patterns in the osmoregulatory organs of larval and adult mosquito Aedes aegypti. J Exp Biol 209:4638–4651 https://doi.org/https://doi.org/10.1242/jeb.02551

Peterson AM, Barillas-Mury CV, Wells MA (1994) Sequence of three cDNAs encoding an alkaline midgut trypsin from *Manduca sexta*. Insect Biochem Mol Biol 24:463–471

Piermarini PM, Weihrauch D, Meyer H et al (2009) NHE8 is an intracellular cation/H+ exchanger in renal tubules of the yellow fever mosquito *Aedes aegypti*. AJP Ren Physiol 296:F730–F750

Pimentel AC, Barroso IG, Ferreira JMJ et al (2018) Molecular machinery of starch digestion and glucose absorption along the midgut of *Musca domestica*. J Insect Physiol 109:11–20

Price DRG, Gatehouse JA (2014) Genome-wide annotation and functional identification of aphid GLUT-like sugar transporters. BMC Genomics 15:1–11

Price DRG, Wilikinson HS, Gatehouse JA (2007) Functional expression and characterization of a gut faciltative glucose transporter, NIHT1 from the phloem-feeding insect *Nilapavarta lugens* (rice brown planthopper). Insect Biochem Mol Biol 37:1138–1148

Price DR, Tibbles K, Shigenobu S et al (2010) Sugar transporters of the major facilitator superfamily in aphids; from gene prediction to functional characterization. Insect Mol Biol 19:97–112

Quina FH, Politi MJ, Cuccovia IM et al (1980) Ion exchange in micellar solutions. 4. "Buffered" systems. J Phys Chem 84:361–365

Remnant EJ, Williams A, Lumb C et al (2016) Evolution, expression, and function of nonneuronal ligand-gated chloride channels in Drosophila *melanogaster*. 6:2003–2012. https://doi.org/10.1534/g3.116.029546

Roman G, Meller V, Wu KH et al (1998) The opt1 gene of *Drosophila* melanogaster encodes a proton-dependent dipeptide transporter. Am J Physiol 275:857–869

Romero MF, Henry D, Nelson S et al (2000) Cloning and characterization of a Na+-driven anion exchanger (NDAE1). A new bicarbonate transporter. J Biol Chem 275:24552–24559

Russell JM (2000) Sodium-potassium-chloride cotransport. Physiol Rev 80:211–276

Sclafani A, Koepsell H, Ackroff K (2016) SGLT1 sugar transporter/sensor is required for post-oral glucose appetition. Am J Physiol Integr Comp Physiol 310:R631–R639

Silva CP, Terra WR (1994) Digestive and absorptive sites along the midgut of the cotton seed sucker bug *Dysdercus peruvianus* (Hemiptera: Pyrrhocoridae). Insect Biochem Mol Biol 24:493–505

Smith AF, Tsuchida K, Hanneman E et al (1992) Isolation, characterization, and cDNA sequence of two fatty acid-binding proteins from the midgut of *Manduca sexta* larvae. J Biol Chem 267:380–384

Terra WR, Ferreira C (1981) The physiological role of the peritrophic membrane and trehalase: digestive enzymes in the midgut and excreta of starved larvae of Rhynchosciara. J Insect Physiol 27:325–331

Terra WR, Ferreira C (1994) Insect digestive enzymes: properties, compartmentalization and function. Comp Biochem Physiol B 109:1–62

Terra WT, Ferreira C (2012) Biochemistry and molecular biology of digestion. In: Gilbert LI (ed) Insect molecular biology and biochemistry. Academic/Elsevier, London, pp p365–p418

Terra WR, Regel R (1995) pH buffering in *Musca domestica* midguts. Comp Biochem Physiol 112:559–564. https://doi.org/10.1016/0300-9629(95)02028-4

Thwaites DT, Anderson CMH (2011) The SLC36 family of proton-coupled amino acid transporters and their potential role in drug transport. Br J Pharmacol 164:1802–1816

Treherne JE (1959) Amino acid absorption in the locust (*Schistocerca gregaria* forsk.). J Exp Biol 36:533–545

Vance JE, Vance DE (2008) Biochemistry of lipids, lipoproteins and membranes. Elsevier, Amsterdam

Wigglesworth VB (1933) The function of the anal gills of the mosquito larva. J Exp Biol 10:16–26

Wu Y, Zheng X, Zhang M et al (2010) Cloning and functional expression of Rh50-like glycoprotein, a putative ammonia channel, in *Aedes albopictus* mosquitoes. J Insect Physiol 56:1599–1610

Yanochko GM, Yool AJ (2002) Regulated cationic channel function in *Xenopus* oocytes expressing *Drosophila* big brain. J Neurosci 22:2530–2540

Zeuthen T (2010) Water-transporting proteins. J Membr Biol 234:57–73. https://doi.org/10.1007/s00232-009-9216-y

Zheng W, Rus F, Hernandez A et al (2018) Dehydration triggers ecdysone-mediated recognition-protein priming and elevated anti-bacterial immune responses in *Drosophila* Malpighian tubule renal cells. BMC Biol 16:60

Chapter 7
Midgut Extracellular Layers and Their Function

Abstract The extracellular layers which are not cell extensions are the mucus and the peritrophic membrane (PM). The mucus is a viscous material overlaying the cells made by highly glycosylated proteins known as mucus-forming mucins. The mucus is found widespread in insects, where it plays a role in the protection against coarse food (due to lubrication), microorganism invasion, and chemical injury, as caused by acids in some insect midgut regions. PM is a membranous sheath surrounding the food bolus. The term "peritrophic matrix" proposed to replace peritrophic membrane is based on misconceptions and should be avoided. Most insects have a PM, with the remarkable exception of hemipterans that have instead microvilli-associated membranes. PM is composed of chitin and the integral proteins of PM, the peritrophins. Associated with PM are digestive enzymes which actually occur in a mucus layer surrounding the PM. Peritrophins have chitin-binding domains (CBD) and may also have mucin-like domains. Typical PM peritrophins have more than three CBDs, whereas peritrophins with one or three CBDs are cuticular proteins analogous to peritrophins (CPAPs). PM once formed is modified by chitinases and chitin deacetylases. PM pores are estimated to be 7–8.5 nm from the diameters of enzymes passing or retained by PM. PM peritrophins are thought to originate from mucins, which acquired chitin-binding domains before insect divergence from Nematoda. PM is supposed to result from selective pressures to enhance digestive efficiency, and hence insect performance, by dividing the midgut into two compartments. Models to explain how PM enhances insect digestive efficiency were proposed and tested with experimental models.

7.1 Introduction

There are several structural layers outside the midgut cells: the glycocalyx, the microvilli-associated membranes, the mucus, and the peritrophic membrane (PM). The glycocalyx, the carbohydrate coat of the cells, and the microvillar-associated membranes are actually extensions of the cells and were described in Chap. 5

W. R. Terra et al., *Molecular Physiology and Evolution of Insect Digestive Systems*, Entomology in Focus 7, https://doi.org/10.1007/978-3-031-39233-7_7

(Fig. 5.2) and were represented schematically in Fig. 8.1. The mucus and the peritrophic membrane are truly midgut extracellular layers. The mucous is a viscous material overlaying the midgut cells and the peritrophic membrane, which was briefly described in Chap. 2 is a membrane that separates the food content from the midgut cells (see Fig. 2.1).

7.2 Mucus: Chemical Nature and Function

Mucus is best known from the studies carried out with vertebrates, among which it lubricates the surface of epithelial cells of the respiratory, digestive, and urogenital systems. The mucus layer protects the cells from physical and chemical injury, dehydration, and infection and facilitates the movement of food along the gut (Perez-Villar and Hill 1999). Mucus is made by highly glycosylated proteins known as mucus-forming mucins (Mf-mucins). The membrane-bound mucins, which are attached to the microvillar membranes by a glycosylphosphatidyl inositol (GPI) anchor or a transmembrane helix, are not part of the mucus but are part of the glycocalyx (Gxf-mucins). Mucins form viscous solutions because they are extended molecules, having regions in sequence that are rich in Pro, Thr, and Ser residues (PTS repeats), intercalated by other residues (Jentoff 1990). Mf-mucins may have, in addition to PTS repeats, domains with a few hundred amino acid residues containing 29–33 Cys residues at the carboxyl terminus (CK domains) and domains similar to regions of the Willebrand factor (von Willebrand disease domain, VWD) (Andrianifahanana et al. 2006). Mf-mucins may form gels, resulting from molecular complexes linked by disulfide bonds between residues of the domains rich in them (CK or VWD domains) or viscous solutions, in the absence of those domains (Andrianifahanana et al. 2006).

Gastrointestinal mucus is found in metazoan animals, from Cnidaria (or Coelenterata) to the higher vertebrate groups (Ruppert and Barnes 1994), and is always made up of Mf-mucins, according to phylogenetic tree analyses using their VWD domains (Lang et al. 2007, 2016). Surprisingly, insect Mf-mucins did not appear in those studies, giving apparent support to the hypothesis put forward by Wigglesworth, as early as 1930, according to which the peritrophic membrane replaced the mucus in protecting the insect gut from injuries (Wigglesworth 1972). Despite being unreasonable, from the point of view of the evolutionary theory, to suppose that a more complex structure (PM) replaces a less complex one (the mucus) to play the same role, the hypothesis met wide acceptance (Hegedus et al. 2009). However, an insect mucus was described by Freyvogel and Stäbli (Freyvogel and Stäubli 1965) in *Anopheles maculipennis*, followed later by reports of midgut mucus (at the time named peritrophic gel, see below) in regions of the midgut of several insects (Terra 2001).

Insect Mf-mucins were not found by Lang et al. (2007, 2016), because they lack VWD domains, as shown by Syed et al. (2008) in *Drosophila melanogaster* after a search for proteins containing PTS repeats. Later on, Buchon et al. (2013,

Supplementary Fig. S2) provided histochemical evidence of the presence of a mucus layer on the top of midgut cells of *Drosophila*, which stained dark orange with periodic acid-Schiff and light orange with Alcian blue. More recently, the occurrence of Mf-mucins, and hence a mucus in the insect midgut, was shown to be widespread (Dias et al. 2018). For this, a search was conducted in several transcriptomes and genomes for sequences having a serine and threonine content of more than 20% of total residues and at least three $(P/T/S)_6$ repeats and lacking any other domains. A $(P/T/S)_6$ repeat is a series of six uninterrupted residues of Pro or Thr in any combination. Dias et al. (2018) data also showed that there is a trend from basal to more derived insects of mucins to decrease their Cys content. As a consequence, there are less interchain disulfide bonds in the mucins of more derived insects, resulting in more fluid mucus.

The functions of the midgut mucus layer in insects are supposed to be similar to that in other animals in aspects like protection from mechanical injury by coarse food (consequence of lubrication), microorganism invasion, and chemical injury, as caused by acid (Perez-Villar and Hill 1999; see also Chap. 6), but include particular roles, exemplified by the immobilization of proteins around the peritrophic membrane in *S. frugiperda* (Dias et al. 2018; Terra et al. 2018; Fuzita et al. 2022). Protection against food abrasion and microorganism invasion is more obvious for all insects that do not have a PM in the anterior midgut (see list in Peters 1992) or in which PM is produced only hours after food ingestion, as seen in mosquitoes (Billingsley 1990). The finding that the gene coding for one of the Mf-mucins is overexpressed after blood feeding lends further support for the role of mucins in midgut protection (Allen and Flemström 2005).

7.3 Peritrophic Membrane: Occurrence, Structure, and Formation

In a monograph on the anatomy of a caterpillar, Lyonet (1762) described a membranous sheath surrounding the food bolus. Similar findings were reported for different insects, and in 1890, Balbiani named this anatomical structure peritrophic membrane (Balbiani 1890). More recently, this term has been replaced by some authors for peritrophic matrix on the grounds that PM is not a lipid bilayer, which according to them all membranes should be (Hegedus et al. 2009). This argument is based on several misconceptions. In biology, the morphological term "membrane," according to dictionaries like the *Oxford Dictionary of Biology* (2008), refers to a film that separates compartments. In anatomy, membranes may be a composite structure made of different components, like the tympanic membrane, the interdigital membrane of duck foot, and the nictitating membrane (transparent third eyelid of reptiles and birds). Only in cell biology, membranes are lipid bilayers. Taking into account that PM is an anatomical structure, not a cell part, no one should suppose it is made by lipid bilayers. The use of matrix in this context is also based on misconceptions.

In biology, the term "matrix" does not convey the idea of a film, but of an amorphous material filling a space, exemplified by the nuclear matrix, the mitochondrial matrix, and the extracellular matrix. Hence the term "peritrophic matrix" suggests it is an amorphous material filling the space around the food (the ectoperitrophic space), which obviously was not the intention of the proponents of the name change.

The consistency of the fully formed film surrounding the food bolus is variable. The film is considered to be a peritrophic membrane if it can be picked up with a pair of fine forceps and as peritrophic gel if it cannot (Fig. 7.1). Below we will show that the peritrophic gel actually is a mucus.

Most insects have a PM. Exceptions are found among Hemiptera and Thysanoptera that have instead microvilli-associated membranes (see Chap. 5), lice (Phthiraptera), booklice (Psocoptera), most adult moths and butterflies (Lepidoptera), adult fleas (Siphonaptera), and few others (Peters 1992). The lack of PM in those groups is thought to be a consequence of the adaption of their ancestors to a diet poor in proteins and starch, meaning they do not require luminal digestion or, in the case of tiny hematophagous insects like Siphonaptera and Phthyraptera, because they are able to maintain efficient midgut countercurrent flows (see Chap. 6) allowed

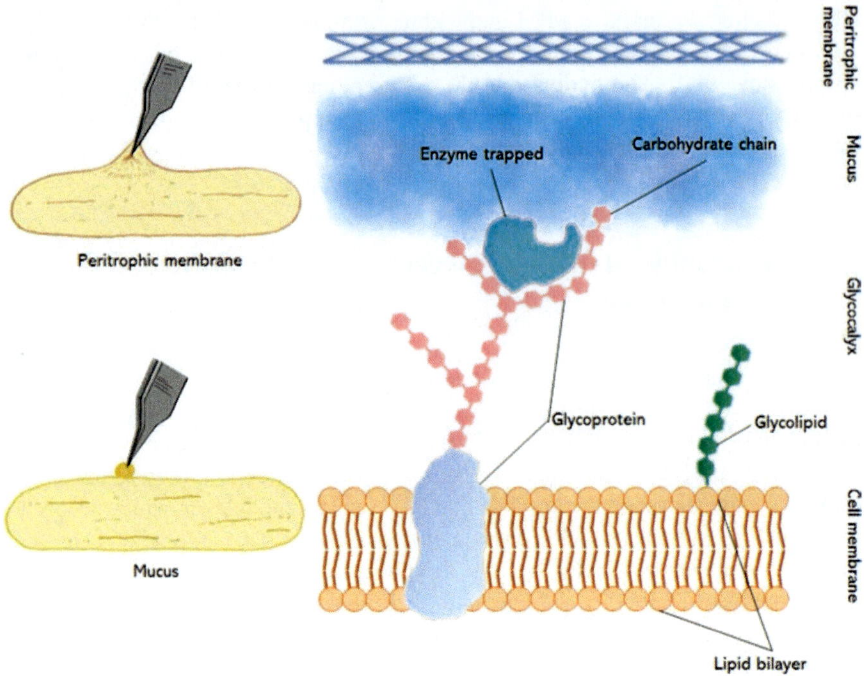

Fig. 7.1 Identification of film types surrounding the food bolus and a diagrammatic representation of glycocalyx, mucus layer, and peritrophic membrane. The kind of film surrounding the food bolus is identified after dissection with the aid of a fine forceps. Contrary to the mucus, the peritrophic membrane can be grabbed with forceps. See micrographs of the peritrophic membrane in Fig. 2.4 and of the glycocalyx in Fig. 5.2

by their blood clot and their small size that facilitate the efficient diffusion of digestion products up to their midgut surface (Terra and Ferreira 2012).

Peritrophic membranes are usually classified into two types (Wigglesworth 1972; Peters 1992). Type I PM is formed either by the whole midgut tissue or by only part of it (usually the posterior region) and is found in cockroaches (Dictyoptera); bees, wasps, and ants (Hymenoptera); moths and butterflies (Lepidoptera); and in hematophagous adult mosquitoes (Diptera). Frequently type I PM is formed after food ingestion. Type II PM is produced by a few rows of cells at the entrance of the midgut (cardia), being molded to the diameter of the anterior midgut and is usually found in insects irrespective of food ingestion. Type II PM is found in larval and adult (except hematophagous ones) mosquitoes and flies (Diptera) and in a few Lepidoptera. Hematophagous mosquitoes that store ingested blood in the dilated posterior midgut have type I PM, despite their larvae possessing type II PM. PMs molded in the narrower anterior midgut (type II PM) cannot expand to encase an ingested food such as that observed in the posterior midgut of a mosquito (Terra 2001). PM production restricted to a belt in the middle third of the middle midgut (the weevil *Cionus*) (Richards and Richards 1977) may be considered a particular case of type I PM.

Peritrophic membrane is formed by proteins and chitin. PM proteins were classified by Tellam et al. (1999) depending on the ease of recovery. Class 1 proteins are removed with saline buffers, class 2 with mild detergents, class 3 with strong denaturants (exemplified by urea), whereas class 4 remains as a residue (Tellam et al. 1999). Class 1 and 2 proteins are mainly digestive enzymes associated with a jelly-like material surrounding the tissue side of PM (Dias et al. 2018; Terra et al. 2018). This material is thought to be a mucus layer partly carried out from the microvilli by the chitin network of the developing PM among the microvilli, as detailed below. This mucus layer contains numerous enzymes that may add up to 13% and 18% of the midgut luminal activity of amylase and trypsin, respectively (Ferreira et al. 1994; Bolognesi et al. 2001) and includes enzymes restricted to the ectoperitrophic fluid and others that eventually pass through the PM into the endoperitrophic space (Fuzita et al. 2022).

Class 3 proteins are the integral proteins of PM and were named by Tellam et al. (1999) as peritrophins, whereas class 4 proteins are likely the same as class 3, because Campbell et al. (2008) were able to completely solubilize PM and the same proteins previously described as class 3 were found. Peritrophins are proteins composed of one to several chitin-binding domains (CBDs), which may include mucin-like domains (sequences having at least three P/T/S repeats) (Tellam et al. 1999). The mucin-like domains may be small or large (Wang and Granados 1997; Shen et al. 1999; Tellam et al. 1999, 2003).

One of the peritrophins of *Trichoplusia ni* was named Insect Intestinal Mucin (IIM) because it has a very large mucin-like domain (Wang and Granados 1997). This proposal led several authors naming of mucin the PM peritrophins with large mucin-like domains (Hegedus et al. 2009). This is misleading and should be avoided, because now it is known that insects have true Mf-mucins that form a vertebrate-like mucus and that IMM is actually a peritrophin. Peritrophins from larval *Ae. aegypti* and *D. melanogaster* (type II PM) have more complex domain structures than those of adult *Ae. aegypti* and *T. castaneum* (type I PM). Furthermore, the mucin-like domains of peritrophins from *T. castaneum* (feeding on rough food) are lengthier than those of adult *Ae. aegypti* (blood feeder). Thus, type I and type II PMs may have variable architectures determined by different peritrophins, which may be partly modulated by diet (Venancio et al. 2009).

Other proteins in addition to peritrophins have CBDs like chitinases and chitin deacetylase. Peritrophin CBDs (pfam ID: PF01607) are formed by 6 Cys residues separated by several amino acids in a consensus distribution described by Tellam et al. (1999) as $CX_{15-17}CX_{5-6}CX_9CX_{12}CX_{6-7}X$ and modified by Jasrapuria et al. (2010) to $CX_{11-24}CX_5CX_{9-14}CX_{12-16}CX_{6-8}X$. Peritrophic membrane peritrophins (PM peritrophins) have usually more than three CBDs and are similar to one kind of cuticle proteins, the cuticular proteins analogous to peritrophins (CPAPs), which have one or three CBDs and are essential for insect development, molting, cuticle integrity, etc. (Jasrapuria et al. 2012; Tetreau et al. 2015). The genes coding for PMPs are more expressed in the midgut, whereas those for CPAPs, in tissues other than the midgut (Jasrapuria et al. 2010, 2012; Tetreau et al. 2015; Dias et al. 2018; Dias et al. 2019). Finally, another kind of peritrophin is the ubiquitous CBD-containing proteins (UCBP) of unclear function that have a varied number of CBDs and which coding genes are expressed at both midgut and carcass (Dias et al. 2019).

During the formation of type I PM, a fibrous material appears at the tips of the microvilli of midgut cells and then is rapidly included in a thin PM surrounding the food bolus (Harper and Hopkins 1997). Thus, most events in PM formation occur among the microvilli. Chitin chains are synthesized by a chitin synthase bound to microvillar membranes employing precursors from inside the cells, once formed the polymer pass to the extracellular space (Zimoch and Merzendorfer 2002; Arakane et al. 2005). Once chitin chains are self-organized in chitin fibers (Hegedus et al. 2009), it interlocks with PM peritrophins that are released by exocytosis (Bolognesi et al. 2001). In the case of type II PM, peritrophins are secreted by exocytosis (Eisemann et al. 2001). The recently formed PM is modified by chitinases and chitin deacetylase (CDAs, E,C. 3.5.1.41). Chitinases cleave glycosyl links of chitin and, at least in mosquitoes, seem to modulate the thickness and permeability of PM (Shen and Jacobs-Lorena 1997). CDAs catalyze the N-deacetylation of chitin to form chitosan. This modification changes the binding properties of chitin. There are several groups of CDAs, from which Group V was found associated with PM and may have a role in the determination of the permeability and other properties of PM (Toprak et al. 2008; Dixit et al. 2008; Arakane et al. 2009; Jakubowska et al. 2010).

The 3D structure of PM is supposed to result from chitin fibrils being interconnected with the chitin-binding domains of peritrophins. Mucin-like domains of peritrophins are proposed to face the ectoperitrophic and endoperitrophic sides of PM. As these domains are highly hydrated, they lubricate both surfaces of PM. In addition, the glucan chains associated with the peritrophin mucin-like domains may provide proteinase resistance to PM (Schorderet et al. 1998; Wang and Granados 2001; Hegedus et al. 2009).

The chitin-peritrophin network is supposed to be responsible for strength, elasticity, and permeability. Thus, inhibition of protein synthesis disturbs the PM formation and inhibition of chitin synthesis causes an increase in permeability (Zimmermann and Peters 1987). The sizes of the physiologically relevant PM pores were determined by comparing the diameters of the enzymes restricted to the ectoperitrophic contents (prepared as described in Chap. 2) with the ones occurring inside PM (Terra and Ferreira 1983; Espinoza-Fuentes et al. 1984; Santos and Terra 1986). Enzyme molecular diameters were determined by interpolation of the molecular weights in a plot of log molecular weight against Stoke's radius for known proteins (La Verde et al. 2017). The Stoke's radius is the effective hydrated radius of a molecule, also known as the hydrodynamic radius. Molecular diameters are the double of the corresponding Stoke's radii. PM pore sizes correspond to the value between the diameter of the enzymes which pass with those of the enzymes that do not pass through PM. The method of choice for molecular weight determination is by density gradient centrifugation, because it preserves complexes of molecules of the same enzyme, if existent, giving true in vivo estimations of molecular weights and, hence, trustful PM pore sizes, which were found to be of 7–8.5 nm (Terra and Ferreira 1983; Espinoza-Fuentes et al. 1984; Santos and Terra 1986).

Other methods of pore size estimations are based on the distribution in and out PM of colloidal gold (Zhuzhikov 1964) or dextrans ingested by the insect (Peters and Wiese 1986; Edwards and Jacobs-Lorena 2000; Barbehen and Martin 1995; Agrawal et al. 2014; Zha et al. 2021) or by using PM mounted as a sac and registering diffusing rates (Miller and Lehane 1990; Wolfersberger et al. 1986). Except for pore sizes determined by the partition of dextrans, which are much variable and may result in artifacts (see below), the other methods agree quite well with pore sizes determined from enzyme distribution. PM pore sizes may differ along the midgut, as will be discussed below.

If PM is derived from mucins, as detailed below, it should have originally been synthesized by all midgut cells and the restriction to midgut sections appearing later in evolution. Therefore, type I PM is the ancestral condition and type II PM the derived one (Terra 2001).

We may return now to the evidence showing that the peritrophic gel is actually the mucus. The midgut may have regions lacking a PM but showing a jelly-like substance surrounding the food named at first as peritrophic gel. As the regions having the so-called peritrophic gel are devoid of chitin, as shown by the anterior midgut of *Dermestes maculatus*, the peritrophic gel was supposed to be made of PM peritrophins in the absence of chitin (Terra 2001). However, ongoing research in our lab (Silva CP, Ferreira C, Terra WR, in preparation) revealed that Mf-mucins coding genes are expressed along the whole midgut of *Zabrotes subfasciatus*. PM peritrophins with more than three CBDs (the typical PM peritrophin) and the proteins regarded as part of the machinery that produces PM (midgut chitinase and chitin deacetylase of group V) are mainly expressed at the posterior midgut, where it is seen a transparent film surrounding the food. These data led to the conclusion that the peritrophic gel is actually a mucus layer considered to be a PM.

7.4 Peritrophic Membrane: Evolution and Function

Peritrophic membrane peritrophins are thought to originate from mucins which acquired chitin-binding domains during insect evolution. Later in evolution, some PM peritrophins lost their mucin-like domains (Terra 2001). PM peritrophin evolution is thought to occur concomitantly with chitin secretion in the insect midgut. Insect genomes usually have two chitin synthase genes: CHS1 (or CSC, cuticular chitin synthase) and CHS2 (or CSM, midgut chitin synthase). The CHS2 gene was proposed to originate from a paraphyletic group and was lost in many Arthropoda lineages (Zakrzewski et al. 2014) and is mostly expressed in midgut cells of different insects; whereas, CHS1 is mainly expressed in cuticle and tracheal cells (Arakane et al. 2005; Dias et al. 2018). The expression sites imply that CHS2 is involved in chitin synthesis during PM production and CHS1 during cuticle formation. The role of CHS2 in PM production is also suggested by its lack in several hemipteran genomes, in which PM is absent (Dias et al. 2019; Wang et al. 2012). If PM peritrophin actually evolved from mucins, PM peritrophin would be originally secreted along the whole midgut, as the ancestral mucus did, and only more recently in evolution it was occasionally restricted to midgut regions or secondarily absent, leaving only the mucus layer.

More recent data suggest that the insect PM origin is somewhat more complex than described above as PM is found in Crustacea and Nematoda (Borgonie et al. 1995; Peters 1992). Indeed, the data showed that PM in Crustacea allows digestive enzyme recycling (endo-ectoperitrophic circulation of digestive enzymes) (Alexandre et al. 2014), as described in insects (see Chap. 2). In addition, genes encoding proteins with several CBDs and some with interspersed mucin domains are present in nematode and crustacean genomes (Tetreau et al. 2015). This, together with detailed phylogenetic studies of peritrophin CBD sequences and peritrophin domain and peritrophin coding gene duplication events, supports the hypothesis of a PM peritrophin originated from mucus only before insect divergence from Nematoda (Tetreau et al. 2015). The selective pressures that led to the insect ancestral PM peritrophin being changed may include food preferences (Tetreau et al. 2015), whereas the understanding of the PM peritrophin changes in insect PM properties can be approached by a detailed comparison of crustacean and insect PMs.

The study of the chemical structure of PM is sometimes used in attempts to reveal PM functions. This addressing is logically inconsistent, because structure refers to the spatial organization of constituent parts, whereas function results from the temporal patterns of activities of those parts. Thus, PM functions are discovered only by studying how PM affects digestion and absorption (Terra 2001).

Peritrophic membrane's physiological roles include protection of the midgut cells against food abrasion, completing the action of mucus, as insects lacking PM may have midgut cells damaged by harsh food (Peters 1992; Tellam 1996; Lehane 1997). PM, however, is not able to replace the mucus in chemically protecting the midgut cells against acid (see Chap. 6). Despite the fact that PM is also usually described as a barrier against invasion by microorganisms in insects that transmit viruses and parasites to human beings (Lehane 1997; Hegedus et al. 2009), this is certainly an opportunistic (not an ancestral) use of PM. PM is supposed to result from selective pressures to enhance digestive efficiency, improving insect performance by dividing the midgut lumen into two compartments (Terra 2001; Bolognesi et al. 2008).

The role of PM in improving insect performance is shown by a decrease in insect growth or fertility caused by impaired nutrient utilization measured by nutritional parameters. This was observed in insects with PM damaged by feeding on plants that have a cysteine endopeptidase with a chitin-binding domain (Pechan et al. 2002; Mohan et al. 2006) and in insects with diets containing chitinase (Rao et al. 2004) or calcofluor that hampers PM production (Bolognesi et al. 2001, 2008). It is interesting to mention here that in mosquitoes without PM, the initial digestion of blood proteins is faster than in controls, but despite the number of eggs laid is the same as in controls, the number of viable eggs seems to be approximately 10% less than in controls (Villalon et al. 2003). Thus, the lack of PM is supposed to affect the intermediate and final stages of digestion, resulting in a decrease in insect performance.

Models to explain how PM enhances insect digestive efficiency were proposed initially by Terra (2001), and the models were later developed and tested with experimental models by Bolognesi et al. (2008). The models were based on the fact that PM compartmentalizes the midgut in an endoperitrophic (inside PM) and an ectoperitrophic (outside PM) space, affecting the distribution of molecules (substrates and enzymes) between the two compartments. According to those models, the most important function of PM is prevention of the excretion of digestive enzymes by allowing enzyme recycling powered by an ectoperitrophic counterflux of fluid that displaces forward enzymes and substrates diffused from the endoperitrophic space. Enzyme recycling and the resulting prevention of the excretion of digestive enzymes have substantial experimental support and were reviewed several times (Bolognesi et al. 2008; Terra and Ferreira 2012; Terra et al. 2019; see also discussion in Chap. 6). However, the best support of the enzyme recycling model is the identification in larval flies of the midgut cell transporters involved in fluid secretion and absorption that propel the ectoperitrophic fluid forward and the demonstration that specifically inhibiting those transporters it is possible to affect the counterflux of fluid and the enzyme excretion rate (Barroso et al. 2020; see also Chap. 6). It should be noticed that the recycling mechanism is functional in the fly larvae, despite the fact that the ectoperitrophic countercurrent flux occurs only between a posterior and an anterior region of the posterior midgut. The same is true for *Zabrotes subfasciatus* larvae (Silva, Ferreira and Terra, manuscript in preparation).

The functions proposed for PM and tested with experimental models include (Bolognesi et al. 2008): (1) the prevention of non-specific binding of undigested food onto the midgut cell surface that would hamper absorption, because it compartmentalizes the midgut contents. To test this, purified microvillar membranes were combined with PM contents. The activities of the microvillar enzymes decreased as expected in these conditions. (2) PM allowance of the diffusion of oligomeric molecules resulting from polymeric molecules from the endoperitrophic space, thus avoiding possible inhibition of polymer hydrolases in that space. This was verified by maintaining dialysis bags containing midgut contents suspended in stirred media. Trypsin activities in stirred bags were 210% over the activities in test tubes. (3) Restriction of the enzymes involved in the intermediate phase of digestion (oligomers hydrolases) in the ectoperitrophic space, thus avoiding that they are partially inhibited by polymeric food molecules (due to non-productive binding). This hypothesis was tested by combining isolated ectoperitrophic fluid with PM contents. As expected, the activity of enzymes restricted to the ectoperitrophic fluid was decreased in these conditions. This enzyme compartmentalization was described in several panorpoid insects (lepidopterans and dipterans) based on a limited number of enzymes (Bolognesi et al. 2008; Terra and Ferreira 2012; Terra et al. 2019). Recently, a proteomic study of several midgut fractions of *Spodoptera frugiperda* confirmed those findings identifying a large number of enzyme molecules restricted to the ectoperitrophic fluid that are involved in intermediate digestion (Fuzita et al. 2022).

The countercurrent model of digestive enzyme recycling has as one of its postulates the recovery of enzymes from the PM contents at the end of the midgut. Thus, a further discussion on PM permeability is necessary here.

The best PM pore determinations, those based on enzyme distribution or on the diffusion rate with PM mounted as a bag, resulted in diameters of 7–8.5 nm (Terra 2001, see also discussion above). However, these are average sizes, meaning that the sizes may differ along the midgut. PM increases in thickness from the anterior to the posterior end of the midgut both in beetles (Agrawal et al. 2014) and in lepidopterans (Zha et al. 2021), probably a consequence of the continuous production of peritrophins along the midgut, for example, in *T. molitor* and *S. frugiperda* (Dias et al. 2018). As expected, PM pores determined with fluorescent dextran particles decrease likewise, attaining at the posterior midgut less than 1.4 nm in beetles (Agrawal et al. 2014) and less than 6 nm in lepidopterans (Zha et al. 2021). Although the decrease of PM pore size is true, the determined pore sizes seem to be artifacts, as the estimated diameter of trypsin molecules is 6.5 nm (Terra 2001) and they are found in midgut contents as described below. *T. molitor* midgut secretes most trypsin molecules at the posterior midgut, as shown by immunocytolocalization (Cristofoletti et al. 2001) and theoretical models based on enzyme distribution (Bolognesi et al. 2008). In *S. frugiperda*, many trypsins are mostly expressed at the posterior midgut and are recovered in midgut contents based on proteomics (Lima et al. 2020). Partial dextran particle adsorption at PM may cause an underestimation of pore sizes, explaining the discrepancy in determined PM pore diameters and the diameters of the enzymes passing through PM.

The peritrophic membrane is also described as important in toxin binding and enzyme immobilization. Toxins may be tannins (Bernays and Chamberlain 1980), several different noxious substances (Barbehenn 1999), or hemin, a product of the metabolism of hemoglobin in blood-feeding insects (Pascoa et al. 2002; Graça-Souza et al. 2006). Toxin binding is a consequence of the chemical properties of PM, which may include several hemin-binding sites (Devenport et al. 2006), and not one of the selective pressures that led to the emergence of PM. A similar affirmation may be advanced regarding enzyme immobilization. After the first demonstrations of the existence of luminal digestive enzymes associated with PM (Terra et al. 1979; Walker et al. 1980; Peters and Kalnins 1985) several ones followed (Terra 2001). Later on, it was found in lepidopterans a large number of digestive enzymes associated with a jelly-like material surrounding the tissue side of PM, adding up to 13% and 18% of the midgut luminal activity of amylase and trypsin, respectively (Ferreira et al. 1994; Bolognesi et al. 2001). More recently, a proteomic study confirmed those findings identifying a large number of digestive enzymes associated with the jelly-like material, including enzymes restricted to the ectoperitrophic fluid and others that eventually pass through PM into the endoperitrophic space (Fuzita et al. 2022). The jelly-like material associated with PM is now thought to be a mucus layer partly carried from the microvilli by the chitin-peritrophin network of the forming PM that occurs among the microvilli (Dias et al. 2018; Terra et al. 2018).

References

Agrawal S, Kelkenberg M, Begum K et al (2014) Two essential peritrophic matrix proteins mediate matrix barrier functions in insect midgut. Insect Biochem Mol Biol 49:24–34

Alexandre D, Ozório RA, Derner RB et al (2014) Spatial distribution of digestive enzymes in the midgut of the Pacific white shrimp (*Litopenaeus vannamei*) indicates the existence of endo-ectoperitrophic circulation in Crustacea. Comp Biochem Physiol B 172-173:90–95

Allen A, Flemström G (2005) Gastroduodenal mucus bicarbonate barrier: protection against acid and pepsin. Am J Cell Physiol 288:C1–C19

Andrianifahanana M, Moniaux N, Batra SK (2006) Regulation of mucin expression: mechanistic aspects and implications for cancer and inflammatory diseases. Biochim Biophys Acta 176:189–222

Arakane Y, Muthukrishnan S, Kramer KJ et al (2005) The Tribolium chitin synthase genes TcCHS1 and TcCHS2 are specialized for synthesis of epidermal cuticle and midgut peritrophic matrix. Insect Mol Biol 14:453–463

Arakane Y, Dixit R, Begum K et al (2009) Analysis of functions of the chitin deacetylase family in Tribolium castaneum. Insect Biochem Mol Biol 39:355–365

Balbiani EG (1890) Études anatomiques et histologiques sur le tube digestif des Crytops. Arch Zool Exp Gen 8:1–82

Barbehen RV, Martin MM (1995) Peritrophic envelope permeability in herbivorous insects. J Insect Physiol 4:303–311

Barbehenn RV (1999) Non-absorption of ingested lipophilic and amphiphilic allelochemicals by generalist grasshoppers: the role of extractive ultrafiltration by the peritrophic envelope. Arch Insect Biochem Physiol 42:130–137

Barroso IG, Fuzita FJ, Ferreira C et al (2020) Midgut fluxes and digestive enzyme recycling in *Musca domestica:* a molecular approach. Comp Biochem Physiol A 241:110627

Bernays EA, Chamberlain DJ (1980) A study of tolerance of ingested tannin in *Schistocerca gregaria*. J Insect Physiol 26:415–420

Billingsley PF (1990) The midgut ultrastructure of hematophagous insects. Annu Rev Entomol 35:219–248

Bolognesi R, Ribeiro AF, Terra WR (2001) The peritrophic membrane of *Spodoptera frugiperda*: secretion of peritrophins and role in immobilization and recycling digestive enzymes. Arch Insect Biochem Physiol 47:62–75

Bolognesi R, Terra WR, Ferreira C (2008) Peritrophic membrane role in enhancing digestive efficiency. Theoretical and experimental models. J Insect Physiol 54:1413–1422

Borgonie G, Claeys M, Vanfleteren J (1995) Presence of peritrophic-like membranes in the intestine of three bacteriophagous nematodes (Nematoda: Rhabdita). Fundam Appl Nematol 18:227–233

Buchon N, Osman D, David FA (2013) Morphological and molecular characterization of adult midgut compartmentalization in Drosophila. Cell Rep 3:1725–1738

Campbell PM, Cao AT, Hines ER et al (2008) Proteomic analysis of the peritrophic matrix from the gut of the caterpillar, *Helicoverpa armigera*. Insect Biochem Mol Biol 38:950–958

Cristofoletti PT, Ribeiro AF, Terra WR (2001) Apocrine secretion of amylase and exocytosis of trypsin along the midgut of *Tenebrio molitor* larvae. J. Insect Physiol 47:143–155

Devenport M, Alvarenga PH, Shao L et al (2006) Identification of the *Aedes aegypti* peritrophic matrix protein AeIMUCI as a heme-bindingprotein. Biochemistry 45:9540–9549

Dias RO, Cardoso C, Pimentel AC et al (2018) The roles of mucus-forming mucins, peritrophins and peritrophins with mucin domains in the insect midgut. Insect Mol Biol 27:46–60

Dias RO, Cardoso C, Leal CS et al (2019) Domain structure and expression along the midgut and carcass of peritrophin and cuticle proteins analogous to peritrophins in insects with and without peritrophic membrane. J Insect Physiol 114:1–9

Dixit R, Arakane Y, Specht CA et al (2008) Domain organization and phylogenetic analysis of proteins from chitin deacetylase gene family of *Tribolium castaneum* and three other species of insects. Insect Biochem Mol Biol 38:440–451

Edwards MJ, Jacobs-Lorena M (2000) Permeability and disruption of the peritrophic matrix and caecal membrane from *Aedes aegypti* and *Anopheles gambiae* mosquito larvae. J Insect Physiol 46:1313–1320

Eisemann C, Wijffels G, Tellam RL (2001) Secretion of type 2 pritrophic matrix protein, peritrophin-15 from the cardia. Arch Insect Biochem Physiol 47:76–85

Espinoza-Fuentes F, Ferreira C, Terra WR (1984) Spatial organization of digestion in the larval and imaginal stages of the sciarid fly *Trichosia pubescens*. Insect Biochem 14:631–638

Ferreira C, Capella AN, Sitnik RE et al (1994) Digestive enzymes in midgut cells, endo- and ectoperitrophic contents, and peritrophic membranes of Spodoptera frugiperda (Lepidoptera) larvae. Archs Insect Biochem Physiol 26:299–313

Freyvogel TA, Stäubli W (1965) The formation of peritrophic membrane in culicidae. Acta Trop 22:118–147

Fuzita FJ, Palmisano G, Pimenta DC et al (2022) A proteomic approach to identify digestive enzymes, their exocytic and microapocrine secretory routes and their compartmentalization in the midgut of *Spodoptera frugiperda*. Comp Biochem Physiol B 257:110670

Graça-Souza AV, Maya-Monteiro C, Paiva-Silva GO et al (2006) Adaptations against heme toxicity in blood-feeding arthropods. Insect Biochem Mol Biol 36:322–335

Harper MS, Hopkins TL (1997) Peritrophic membrane structure and secretion in European corn borer larvae (*Ostrinia nubilalis*). Tissue Cell 29:461–475

Hegedus D, Erlandson M, Gillott C et al (2009) New insights into peritrophic matrix synthesis, architecture, and function. Annu Rev Entomol 54:285–302

Jakubowska AK, Caccia S, Gordon KH et al (2010) Down regulation of chitin deacetylase –like response to baculovirus infection. J Virol 84:254–255

Jasrapuria S, Arakane Y, Osman G et al (2010) Genes encoding proteins with peritrophin A-type chitin-binding domains in *Tribolium castaneum* are grouped into three distinct families based on phylogeny, expression and function. Insect Biochem Mol Biol 40:214–227

Jasrapuria S, Specht CA, Kramer KJ et al (2012) Gene families of Cuticular proteins analogous to Peritrophins (CPAPs) in *Tribolium castaneum* have diverse functions. PLoS One 7:e49844

Jentoff N (1990) Why are protein o-glycosylated? Trends Biochem Sci 15:291–294

La Verde V, Dominici P, Astegno A (2017) Determination of hydrodynamic radius of proteins by size exclusion chromatography. Bio-protocol 7(8):e2230

Lang T, Hansson GC, Samuelsson T (2007) Gel-forming mucins appeared early in metazoan evolution. Proc Natl Acad Sci U S A 104:16209–16214

Lang T, Klasson S, Larsson E et al (2016) Searching the evolutionary origin of epithelial mucus protein components-mucins and FCGBP. Mol Biol Evol 33:1921–1936

Lehane MJ (1997) Peritrophic matrix structure and function. Annu Rev Entomol 42:525–550

Lima LR, Dias RO, Fuzita FJ et al (2020) The evolution, gene expression profile, and secretion of digestive peptidases in Lepidoptera species. Catalysts 10:217

Lyonet P (1762) Traite Anatomique de la Chenille. La Haye, Grosse Pinet

Miller N, Lehane MJ (1990) *In vitro* perfusion studies on the peritrophic membrane of the tsetse fly *Glossina morsitans morsitans* (Diptera, Glossinidae). J Insect Physiol 36:813–818

Mohan S, Ma PWK, Pechan T et al (2006) Degradation of the *S. frugiperda* peritrophic matrix by an inducible maize cysteine protease. J Insect Physiol 52:21–28

Oxford Dictionary of Biology, 6th edn (2008) Oxford University Press, Oxford

Pascoa V, Oliveira PL, Dansa-Petretski M et al (2002) *Aedes aegypti* peritrophic matrix and its interaction with heme during blood digestion. Insect Biochem Mol Biol 32:517–523

Pechan T, Cohen A, Williams WP et al (2002) Insect feeding mobilizes a unique defense protease that disrupts the peritrophic matrix of caterpillars. Proc Natl Acad Sc USA 99:13319–13323

Perez-Villar J, Hill RL (1999) The structure and assembly of secreted mucins. J Biol Chem 274:31751–31754

Peters W (1992) Peritrophic membranes. Springer-Verlag, Berlin

Peters W, Kalnins M (1985) Aminopeptidases as immobilized enzymes on the peritrophic membranes of insects. Entomol Gener 11:25–32

Peters W, Wiese B (1986) Permeability of the peritrophic membranes of some Diptera to labelled dextrans. J Insect Physiol 32:43–49

Rao R, Fiandra L, Giordana B et al (2004) AcMNPV ChiA protein disrupts the peritrophic membrane and alters midgut physiology of *Bombyx mori* larvae. Insect Biochem Mol Biol 34:1205–1213

Richards AG, Richards PA (1977) The peritrophic membranes of insects. Annu Rev Entomol 22:219–240

Ruppert EE, Barnes RD (1994) Invertebrate zoology, 6th edn. Saunders, Philadelphia

Santos CD, Terra WR (1986) Distribution and characterization of oligomeric digestive enzymes from *Erinnyis ello* caterpillars and inferences concerning secretory mechanisms and the permeability of the peritrophic membrane. Insect Biochem 16:691–700

Schorderet S, Pearson RD, Vuocolo T et al (1998) cDNA and deduced amino acid sequences of a peritrophic membrane glycoprotein, "Peritrophin-48", from the larvae of *Lucilia cuprina*. Insect Biochem Mol Biol 28:99–111

Shen Z, Jacobs-Lorena M (1997) Characterization of a novel gut-specific chitinase gene from the human malaria vector *Anopheles gambiae*. J Biol Chem 272:28895–28900

Shen Z, Dimopoulos G, Kafatos FC et al (1999) A cell surface mucin specifically expressed in the midgut of the malaria mosquito Anopheles gambiae. Proc Natl Acad Sci U S A 96:5610–5615

Syed ZA, Härd T, Uv A et al (2008) A potential role for drosophila mucins in development and physiology. PLoS One 3:e3041

Tellam RL (1996) The peritrophic matrix. In: Lehane MJ, Billingsley PF (eds) Biology of the insect midgut. Chapman and Hall, London, pp 86–114

Tellam RL, Wijffels G, Willadsen P (1999) Peritrophic matrix proteins. Insect Biochem Mol Biol 29:87–101

Tellam RL, Vuocolo T, Eisemann C et al (2003) Identification of an immuno-protective mucin-like protein, peritrophin-55, from the peritrophic matrix of *Lucilia cuprina* larvae. Insect Biochem Mol Biol 33:239–252

Terra WR (2001) The origin and functions of the insect peritrophic membrane and peritrophic gel. Arch Insect Biochem Physiol 47:47–61

Terra WR, Ferreira C (1983) Further evidence that enzymes involved in the final stages of digestion by *Rhychosciara americana* do not enter the endoperitrophic space. Insect Biochem 13:143–150

Terra WT, Ferreira C (2012) Biochemistry and molecular biology of digestion. In: Gilbert LI (ed) Insect molecular biology and biochemistry. Academic/Elsevier, London, pp 365–418

Terra WR, Ferreira C, De Bianchi AG (1979) Distribution of digestive enzymes among the endo- and ectoperitrophic spaces and midgut cells of *Rhynchosciara* and its physiological significance. J Insect Physiol 25:487–494

Terra WR, Dias RO, Oliveira PL et al (2018) Transcriptomic analyses uncover emerging roles of mucins, lysosome/secretory addressing and detoxification pathways in insect midguts. Curr Opin Insect Sci 29:34–40

Terra WR, Barroso IG, Dias RO et al (2019) Molecular physiology of insect midgut. Adv Insect Physiol 56:117–163

Tetreau G, Dittmer NT, Cao X et al (2015) Analysis of chitin-binding proteins from Manduca sexta provides new insights into evolution of peritrophin A-type chitin-binding domains in insects. Insect Biochem Mol Biol 62:127–141

Toprak U, Baldwin D, Erlandson M et al (2008) A chitin deacetylase and putative insect intestinal lipases are components of the *Mamestra configurata* (Lepidoptera: Noctuidae) peritrophic matrix. Insect Biochem Mol Biol 17:573–585

Venancio TM, Cristofoletti PT, Ferreira C et al (2009) The *Aedes aegypti* larval transcriptome: a comparative perspective with emphasis on trypsins and the domain structure of peritrophins. Insect Mol Biol 18:33–44

Villalon JM, Ghosh A, Jacobs-Lorena M (2003) The peritrophic matrix limits the rate of digestion in adult *Anopheles stephensi* and *Aedes aegypti* mosquitoes. J Insect Physiol 49:891–895

Walker VK, Geer W, Williamson JH (1980) Dietary modulation and histochemical localisation of leucine aminopeptidase activity in *Drosophila melanogaster*. Insect Biochem 10:543–548

Wang P, Granados RR (1997) Molecular cloning and sequencing of a novel invertebrate intestinal mucin. J Biol Chem 272:16663–16669

Wang P, Granados RR (2001) Molecular structure of the peritrophic membrane (PM): identification of potential PM target sites for insect control. Arch Insect Biochem Physiol 47:110–118

Wang Y, Fan HW, Huang HJ et al (2012) Chitin synthase 1 gene and its two alternative splicing variants from two sap-sucking insects, *Nilaparvata lugens* and *Laodelphax striatellus* (Hemiptera: Delphacidae). Insect Biochem Mol Biol 42:637–646

Wigglesworth VB (1972) The principles of insect physiology, 7th edn. Chapman and Hall, London

Wolfersberger MG, Spaeth DD, Dow JT (1986) Permeability of the peritrophic membrane of tobacco hornworm larval midgut. Am Zool 26:356

Zakrzewski AC, Weigert A, Helm C et al (2014) Early divergence, broad distribution, and high diversity of animal chitin synthases. Genome Biol Evol 6:316–325

Zha X-L, Wang H, Sun W et al (2021) Characteristics of the peritrophic matrix proteins of the silkworm, *Bombyx mori* and factors influencing its formation. Insects 12:516

Zhuzhikov DP (1964) Function of the peritrophic membrane in *Musca domestica* L. and *Calliphora erythrocephala* Meig. J Insect Physiol 10:273–278

Zimmermann D, Peters W (1987) Fine structure and permeability of peritrophic membrane of Calliphora erytrocephala (Meigen) (Insecta: Diptera) after inhibition of chitin and protein synthesis. Com Biochem Physiol 86:353–360

Zimoch L, Merzendorfer H (2002) Immunolocalization of chitin synthase in the tobacco hornworm. Cell Tissue Res 308:287–297

Chapter 8
Endocrine Regulation of Insect Digestion

Abstract Insect digestion is regulated mainly by peptides that affect food intake, gut motility, digestive enzyme release, and nutrient absorption. The regulatory peptides are produced by the central nervous system, the stomatogastric nervous system, and midgut endocrine cells. There are peptides with stimulatory (allatotropin, proctolin, sulfakinins, tachyinins, kinins, etc.) or inhibitory (allatostatin, myoinhibitory peptides, myosuppressins, etc.) effects. In this chapter, we will review the chemical identity, site of production, and function of those peptides that differ somewhat among different insects.

8.1 Initial Considerations

The feeding behavior of insects is the result of several factors that occur in a coordinated way and that involve hormonal regulation. The act of feeding is associated with physiological processes such as: the visual or olfactory perception that leads to the recognition of the food source; the nutritional status of the insect, whether it is satiated or fasting; coordination between feeding and ecdysis is finely regulated; oogenesis influences feeding; the effects of parasites are decisive in the feeding behavior of insects; the action of pathogens can influence the insect search for food, as well as the axis between the intestinal microbiota; and the insect's immune response has an effect on feeding behavior. Once food is recognized, mechanical and sensory activities continue under hormonal regulation resulting in food entering the oral cavity of the alimentary canal (foregut). These actions include ingestion and swallowing of food, intestinal peristalsis, release of digestive enzymes, absorption of nutrients, and evacuation of undigested material and waste at the end of the hindgut. In this chapter, details of hormone regulation from a selection of the most important and recognized neuropeptides involved in the regulation of intestinal activities are discussed.

W. R. Terra et al., *Molecular Physiology and Evolution of Insect Digestive Systems*, Entomology in Focus 7, https://doi.org/10.1007/978-3-031-39233-7_8

8.2 The Nervous System and the Regulation of Digestion in Insects

8.2.1 The Stomatogastric Nervous System of Insects

The functioning of the alimentary tract of insects is under neuroendocrine regulation by the central nervous system (CNS), by the stomodaeal or stomatogastric nervous system (SNS), and by the activity of midgut neurosecretory cells (enteroendocrine cells, EEs). The coordinated actions of CNS and SNS in addition to peripheral organs such as fat body and Malpighian tubules regulate food intake, gut peristalsis, digestive enzyme secretion, nutrient absorption, endocrine activity, diuresis, and defecation. Despite an enormous variety of morphological types of alimentary tracts of insects adapted to different types of food, a common organizational pattern of the SNS can be discerned (Fig. 8.1). The principal and most common components of the insect SNS are as follows: the frontal ganglion (FG), the hypocerebral ganglion (HCG), and the ventricular or ingluvial ganglia (VG). The SNS lays on the dorsal surface of the foregut and it innervates the anterior parts of the alimentary tract (the foregut and anterior midgut). The SNS is present in most known insect species. However, one very important experimental model the fruit fly *Drosophila melanogaster*, lacks the FG and HCG. In most insect species, the FG is coupled to the brain via the frontal ganglion connectives and the recurrent nerve links it to the HCG. The HCG is connected to the corpora cardiaca (CC)-corpora

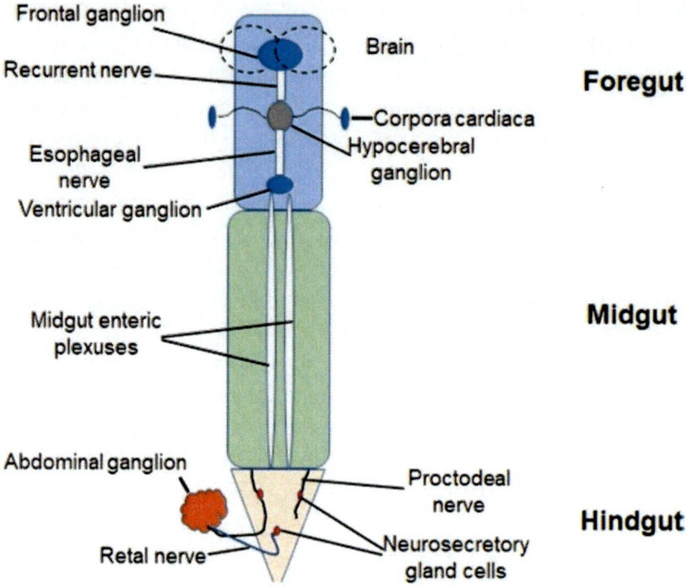

Fig. 8.1 Schematic representation showing the most important connectors, ganglia, and nerves of the enteric nervous system that are essential for the innervation of the intestinal tract in insects

allata (CA) complex, which are paired neurosecretory organs. A pair of esophageal nerves links the HCG to VG. From the VG diffuse nerve plexuses extend across the foregut surface and a pair of midgut enteric plexuses extends along the superficial midgut musculature. The terminal abdominal ganglion (TAG) of the CNS innervates the hindgut via the proctodeal and rectal nerves, which are connected to neurosecretory gland cells present in the hindgut tissue (Fig. 8.1) (Copenhaver 2007).

8.2.2 The Frontal Ganglion

The frontal ganglion (FG) is a component of the SNS that plays an important role in the digestive process of insects by regulating muscle contractions of the foregut. The co-expression of different neuropeptides in endocrine cells of FG has been described in several insect species (Abou El Asrar et al. 2020; Copenhaver 2007; Duve and Thorpe 2003). The FG is connected to the brain (tritocerebrum) by the frontal ganglion connective, while the recurrent nerve connects it to the HCG. In some insect species, the recurrent nerve innervates the muscles of some foregut parts, such as crop, proventriculus, and stomodaeal valve, reaching the anterior midgut region. The stomodaeal valve is innervated by the posterior neurons of the FG in different insect species. The FG may contain about 35 neuron cells, including motor neurons that control the movements of the foregut. Several studies have already demonstrated that the removal of the FG or the rupture of the connection through the recurrent nerve causes an impairment in the emptying of the crop, reinforcing the role of the FG in the regulation of the entry and transit of food along the foregut and its entry into the midgut. Despite the generalizations made here, there is indeed a wide variation in these themes depending on the insect group.

8.2.3 The Hypocerebral Ganglion

The hypocerebral ganglion (HCG), which is linked to the corpora cardiaca-corpora allata complex, is also connected to FG via the recurrent nerve and to the VG via the esophageal nerve. The HCG gives rise to nerve plexuses that innervate the foregut musculature. In chewing insects, the actions of the adductor and abductor muscles of the mandibles are regulated by HCG and by the subesophageal ganglion (SOG), which allows food fragmentation. But beyond the simple act of cutting the food, the mechanoreceptors located on the mandible are also connected to HCG, showing that this ganglion is very important in the ingestion of food. Recent studies have shown that HCG is crucial for the regulation of the muscles that operate the crop, an essential organ in controlling the passage of food from the foregut to the midgut (Abou El Asrar et al. 2020; Rand and Ayali 2010). In sucking insects, the cibarial and pharyngeal muscles are innervated by both FG and HCG. In many insects, the coordinated action of FG and HCG controls the passage of food unidirectionally from the oral cavity to the midgut after passing through different parts of the foregut. The interplay activities of FG, HCG, and the SOG are necessary for normal food ingestion in insects.

8.2.4 The Ventricular Ganglion

The ventricular ganglion (VG), or the ingluvial ganglion in some insects, is the most distal of the three ganglia of the SNS. From the VG nerves extend over the surface of the foregut, and it is also from where the midgut enteric plexus extends along the superficial musculature of the midgut, suggesting that all SNS components are involved in the innervation of the foregut and anterior midgut musculatures.

8.2.5 The Terminal Abdominal Ganglion

The terminal abdominal ganglion (TAG) is the most distal ganglion of the CNS and it innervates the hindgut musculature through the proctodeal and rectal nerves. Some branches of the proctodeal nerve also extend onto the posterior midgut surface. Several neurosecretory cells have been described in the TAG. These cells are responsible for the release of neuropeptides involved in muscular contractions of the hindgut.

8.2.6 The Midgut Endocrine Cells (Enteroendocrine Cells)

The midgut tissue of several insect species has cells which express different neuropeptides. These cells are called enteroendocrine cells (EEs). Enteroendocrine cells secrete different neuropeptides along the midgut, implying that there are different populations of these cells. To date, more than ten different types of hormonal peptides secreted by endocrine cells of the midgut have been described (Wu et al. 2020; Abou El Asrar et al. 2020). Some of these neuropeptides are synthesized by both Ees and the CNS, being considered brain-gut peptides. Some of the neuropeptides produced act on columnar cells of the midgut involved in the secretion of digestive enzymes or the absorption of nutrients, while others act on different organs, including nerve tissues like the brain ganglia (Wu et al. 2020).

8.2.7 Brain and the Subesophageal Ganglion

The insect CNS is composed by the brain with a dorsal location in the head capsule and the ventral nerve cord (VNC), which is ventrally located and extends along the thorax and the abdomen (Copenhaver 2007). The insect brain is divided into three regions: the protocerebrum, the deutocerebrum, and the tritocerebrum (Fig. 8.1). Each of these parts innervates certain head appendages, such as the eyes (protocerebrum), antennas (deutocerebrum), and mouthparts (tritocerebrum). The tritocerebrum is the smallest subdivision of the brain, but it is very important In terms of neuronal control of the digestive system. It is linked to the FG via the frontal ganglion connectives and to the SOG through the periesophageal connectives, which bypass the esophagus linking the brain to the VNC. The SOG is the first ganglion of the VNC and it contains important mechanosensory neurons involved in the control of the antenna, neck, and mouthparts movements. The SOG is also involved in sensing the chemical composition and texture of dietary material (Copenhaver 2007).

8.3 Types of Hormones That Regulate Insect Gut Physiology

Insect hormones are chemically diverse and can be exemplified by steroids (ecdysone), sesquiterpenes (juvenile hormone), polypeptides (insulin-like peptides), and neuropeptides. By far, the neuropeptides represent the largest and most diverse class of insect hormones and they are the main regulators of digestion in insects (Abou El Asrar et al. 2020). Approximately 40 families of insect neuropeptides are now recognized as functioning as paracrine and/or circulating hormones released by the CNS, SNS, and eEs of the midgut. Some neuropeptides act at a short distance after release from peptidergic neurons or enteroendocrine cells, while others are released into the hemolymph from neurosecretory cells or enteroendocrine cells, which are major loci of peptide hormones secretion (the so-called enteroendocrine peptides). Some enteroendocrine peptides can act on different organs other than the alimentary tract and they can also be secreted by the CNS. Some of the neuropeptides that act in the alimentary tract can also operate in other organs, such as nervous ganglia, and help to regulate diverse aspects of insect physiology, such as oogenesis, immune activity, lifespan, myomodulatory activities, reproduction, or metamorphosis and are called pleiotropic hormones. The neuropeptides already described as regulators of different stages of insect digestion have from 5 to 85 amino acid residues and are produced after post-translational processing of larger protein precursors. Post-translational modifications include the formation of internal disulfide bonds and modifications at the amino or carboxy-terminal ends, which increases the stability of mature neuropeptides. Traditionally, these peptides have been studied through their action on metabolism after isolation or purification of extracts from parts of the CNS, SNS, or intestinal epithelium, through analytical techniques by liquid chromatography coupled with mass spectrometry, immunolocalization or in situ hybridization, etc. Today, omics techniques are bringing a flood of new discoveries, including neuropeptides and their receptors which were discovered through in silico–based techniques.

The vast majority of neuropeptides involved in the regulation of digestion in insects act by coupling to receptors belonging to the family of G-protein coupled receptors (GPCRs). Some of these neuropeptides are so important in regulating digestion and other metabolic processes that they are considered promising new tools in insect and vector control, as will be discussed in Chap. 15 of this book. In addition to the possibility of assisting in the control or integrated management of pests and disease vectors, greater knowledge about the control exerted by these hormones can also help in the efficient creation of edible insects or insect species as sources of bioactive compounds.

Due to the inherent complexity of the hormonal regulation of digestion in insects, despite the large number of articles in the literature on this subject, knowledge is still fragmented, as much research has focused on a few model insects, mainly on *D. melanogaster*. There is also a wide variety of experimental approaches employed in this area of research. In order to have a greater focus on the regulation of insect digestion, some more common neuropeptides for which there is a greater amount of data that point to their direct participation in the control of digestion were compiled in this chapter and are shown in Table 8.1.

8.3.1 Allatoregulatory Peptides

There are four different and unrelated types of allatoregulatory peptides according to their amino acid sequences: allatotropin (AT) and three types of allatostatins: allatostatin type A (AST-A), type B (AST-B), and type C (AST-C) (Table 8.1). The ASTs are a family of pleiotropic neuropeptides that were first identified as inhibitors of juvenile hormone (JH) synthesis and secretion by the corpora allata (CA) of different insect species, while AT was first recognized for its stimulatory activity of JH secretion by CA. The allatoregulatory peptides are produced in the brain, SNS, and enteroendocrine cells. In some insects, ASTs and AT are co-expressed in components of the SNS, such as the FG, and in some enteroendocrine cells in the midgut (Nagata and Zhou 2019; Rand and Ayali 2010; Stay et al., 1994). It has been demonstrated that the expression of ASTs in the brain and the enteroendocrine cells is sufficient to suppress feeding (Wu et al. 2020).

Allatoregulatory peptides control the movement of ingested food through their myomodulatory actions on gut musculature. The ASTs have myoinhibitory activity decreasing the peristaltic movements of the foregut, midgut, and hindgut, while the function of Ats usually is opposite of ASTs being generally myostimulatory peptides (Elekonich and Horodyski 2003). In Lepidoptera, it has been reported that AST-A can also regulate the secretion of digestive enzymes. Interestingly, species in the same genus, such as *Spodoptera littoralis* and *S. frugiperda*, may present opposing responses to AST-A. In *S. littoralis* this neuropeptide stimulates the secretion of α-amylases and trypsin, while in *S. frugiperda* there is an inhibition in the levels of these enzymes in the midgut (Abou El Asrar et al. 2020; Bahrami et al. 2018; Lwalaba et al. 2010).

Table 8.1 Neuropeptides involved in endocrine regulation of insect digestion

Name	Abbreviation	Sites of expression	Function[a]	InterPro domains (IPR)	Sequence of mature peptide	Insect species and reference
Allatotropin	AT	Frontal ganglion (FG), terminal abdominal ganglion (TAG)	Stimulate spontaneous contractions in foregut and hindgut; increase in the expression levels of amylase and trypsin	EF-hand 1, calcium-binding site (IPR018247)	GFKNVEMMTARGF-NH_2	*Manduca sexta* (Kataoka et al. 1989)
Allatostatin A	AST-A	Brain, FG, enteroendocrine cells (EE) mainly in the posterior midgut	Inhibit contractions of foregut, midgut, and hindgut; regulation of digestive enzyme secretion in the midgut; K^+ absorption	Allatostatin (IPR010276)	SPSGMQRLYGFGL-NH_2	*Periplaneta americana* (Weaver et al. 1994)
Allatostatin B (Myosuppressin)	AST-B	EE in the midgut	Inhibition of foregut and hindgut contractions; inhibit food intake	C2 domain superfamily (IPR035892)	PDVDHVFLRF-NH_2	*Schistocerca gregaria* (Ragionieri et al. 2022)
Allatostatin C	AST-C	EE in the midgut	Inhibit contractions of foregut and hindgut	Allatostatin (IPR020161)	pEVRFRQCYFNPISCF	*M. sexta* (Kramer et al. 1991)

(continued)

Table 8.1 (continued)

Name	Abbreviation	Sites of expression	Function[a]	InterPro domains (IPR)	Sequence of mature peptide	Insect species and reference
CCHamide 1	CCHa1	EE in the posterior midgut	Inhibition of gut muscle contractions	None predicted IPR	SCLSYGHSCWGAH-NH_2	*Cataglyphis nodus* (Habenstein et al. 2021)
CCHamide 2	CCHa2	Brain, EE in the entire midgut	Increase feeding motivation	None predicted IPR	GGCASFGHSCFGGH-NH_2	*C. nodus* (Habenstein et al. 2021)
Crustacean cardioactive peptide	CCAP	Brain, ventral nerve cord (VNC), midgut	Myotropic actions on midgut and hindgut; regulation of the release of digestive enzymes	Crustacean cardioactive peptide (IPR024276)	AAAPPTPVQGMKPW	*S. gregaria* (Ragionieri et al. 2022)
Kinin	–	Central nervous system (CNS), foregut, midgut, hindgut	Myotropic activity on foregut, anti-feedant activity, control of digestive enzyme release	None predicted IPR	RPSFNSWG-NH_2	*P. americana* (Predel et al. 1997)
Neuropeptide F	NPF	Brain, CNS, subesophageal ganglion (SOG), EE in the midgut	Involved in the regulation of food intake; increase sugar sensitivity	IPR020392	LPAGADAGQQRPERPPMFTSPEELRNYLTQL	*S. gregaria* (Ragionieri et al. 2022)
Proctolin	–	All ganglia of the stomatogastric nervous system (SNC), TAG, hindgut	Stimulate muscle contraction in foregut, midgut, and hindgut	None predicted IPR	RYLPT	*P. americana* (Starratt and Brown 1975)

Name	Abbreviation	Sites of expression	Function[a]	InterPro domains (IPR)	Sequence of mature peptide	Insect species and reference
RYamide	RYa	Brain, TAG, EE cells of the anterior midgut	Regulation of feeding and digestion	None predicted IPR	SEDRSSGNSLKESSFFSPGRY-NH_2	*Nasonia vitripennis* (Hauser et al. 2010)
Short neuropeptide F	sNPF	Brain and rest of the CNS	Adjust insect feeding behavior, negatively or positively, dependent on insect species	None predicted IPR	SQRSPSLRLRF-NH_2	*C. nodus* (Habenstein et al. 2021)
Sulfakinin	SK	Brain, foregut and hindgut	SKs have anorexic effect and a stimulatory effect on the release of amylase in the midgut	IPR013259	pQQLDDYGHMRF-NH_2	*C. nodus* (Habenstein et al. 2021)
Tachykinin-related peptide	TRP	CNS, SNS, EE of the entire midgut	Regulate food intake; stimulate contractions of the foregut and midgut; stimulate diuresis	*Leucophaea maderae* tachykinin-related peptide (IPR013206)	TIMGFQGMR-NH_2	*C. nodus* (Habenstein et al. 2021)
Trissin	–	Brain, FG, TAG	Stimulate contractions of the foregut and midgut	None predicted IPR	IKCDTCGKECASACGTKHFRTCCFNYL	*Drosophila melanogaster* (Ida et al. 2011)

[a] Most of these neuropeptides are pleiotropic, that is, they are involved in multiple functions. Only the functions directly involved in digestion are mentioned in this table

8.3.2 CCHamides 1 and 2 Peptides

CCHamide 1 (CCHa1) and 2 (CCHa2) neuropeptides are expressed in both midgut endocrine cells and in brain nerves as typical brain-gut peptides. It has been demonstrated that these peptides regulate gut muscle contractions. The disruption of CCHa1 and CCHa2 expression by the use of CRISPR/Cas9 gene editing system led to a significant reduction in the food intake of CCHa2 mutants, suggesting that these neuropeptides are involved in feeding regulation (Wu et al. 2020).

8.3.3 Neuropeptide F and Short Neuropeptide F

Insect neuropeptide F has two orthologs known as the long form (NPF) and the short form (sNPF). Analysis of the NPF gene sequences suggests that sNPF is a truncated variant of the complete NPF. Both peptides seem to exert a positive influence on appetite. Overexpression of NPF or its injection in the hemolymph increased the amount of food ingestion, whereas the knockdown of NPF caused inhibition of food intake (Nagata and Zhou 2019; Tan et al. 2019; Li et al. 2018). The expression of NPF and sNPF precursors in the CNS and in enteroendocrine cells reinforces the role of these peptides in the regulation of feeding behavior and digestion. The effects of sNPF are not the same on different insect species. In blood-feeding mosquitos, sNPF does not appear to have phagostimulatory activity, but rather it acts as a satiety factor (Abou El Asrar et al. 2020; Wu et al. 2020; Nagata and Zhou 2019).

8.3.4 Tachykinin-Related Peptides and Insect Kinins

Insect members of the vertebrate conserved "tachykinin-related peptide" family are known as "insectatachykinins" and here abbreviated as TRPs. The TRPs are characterized by a C-terminal FXG/AXRamide motif and by their receptors, which are members of the GPCR class of receptors. TRPs are pleiotropic brain-gut neuropeptides expressed in the CNS, SNS, in enteroendocrine cells, and in Malpighian tubule ampullae (Table 8.1). Their expression is consistent with their regulatory role in stimulating muscle contractions, food transit, secretion of digestive enzymes, lipid metabolism in midgut cells, and diuresis (Abou El Asrar et al. 2020; Bhatt et al. 2014; Holman and Cook 1983). The TRPs were identified based on their myotropic activity (stimulatory) in all three regions of the insect gut tract. As TRPs are released in response to starvation, they might stimulate food search, intake, and secretion of digestive enzymes (Lee et al. 2021; Song et al. 2014; Pascual et al. 2008).

8.3.5 *FMRFamide-Related Peptides and Myosuppressins*

In insects and in other animals the FMRFamide-related peptides (FaRPs) comprise a diverse range of neuropeptides, which share the common C-terminal (H/FL/M) RFamide consensus sequence (Abou El Asrar et al. 2020). Three families of FaRPs have been recognized: the FMRFamides, the FLRFamides, and the HMRFamides. Due to a multitude of N-terminally extended RFamides, there is a considerable variety of these neuropeptides in insects that includes products of different genes. Many of these neuropeptides are produced from precursors that undergo limited proteolysis and can produce multiple copies of these hormones. The FaRPs are multifunctional neuropeptides found in nervous and non-nervous tissues acting as neurotransmitters, neuromodulators, and neurohormones. In terms of insect digestion, they have been described as capable of inhibiting gut muscle contractions and are able to increase the secretion of digestive enzymes such as α-amylase, α-glucosidase, and endopeptidases. Therefore, due to their ability to influence both peristaltic movements and the secretion of digestive enzymes, the FaRPs play a significant role in regulating the feeding behavior of insects including the uptake and digestion of food (Wu et al. 2020; Nagata and Zhou 2019; Down et al. 2011; Orchard et al. 2001).

FMRFamide-expressing neurons innervate the midgut and hindgut, and these peptides were detected in the SNS associated with the gut tract of different insect species. The FMRFamide-related peptides are involved in increasing the frequency of spontaneous muscle contraction of the foregut affecting the transit of food in the crop in some insect species. The detection of FMRFamide-related peptides in enteroendocrine cells suggests a role for these peptides in the regulation of midgut and hindgut peristalsis (Godoy et al. 2021).

The insect myosuppressins (MS) are members of a family of N-extended decapeptide–FLRFamide neuropeptides that has the general conserved sequence of XDXXHXFLRFamide. They influence both gut tract muscle contraction and digestive enzyme secretion. Insect MS are multifunctional peptides found in several different tissues, including digestion-related organs such as the foregut, SOG, hindgut, and abdominal ganglia. There is evidence that myosuppressins can be secreted by the brain and be released to the hemolymph to act in the digestive system inhibiting gut motility. Despite some conflicting results, in most insect species investigated so far, the expression, injection, or oral administration of MS results in the inhibition of food consumption.

The HMRFamides from insects are also called sulfakinins (SKs) due to the presence of a sulfated tyrosine residue which is essential for their biological activity (Abou El Asrar et al. 2020; Orchard et al. 2001). The C-terminal active site of SKs is made up of the conserved hexapeptide YGHMRFamide, where the tyrosine residue (Y) is sulfated. As typical brain-gut peptides, SKs are detected in the brain, foregut, and hindgut innervating neurons. Insect SKs are homologous to the vertebrate neuropeptides gastrin and cholecystokinin, which have been linked to satiety and anorexia induction. Like their vertebrate counterparts, SKs act as antifeedant hormones and can also stimulate the release of digestive enzymes into the midgut in different insect species.

8.3.6 Insect Kinins

Insect kinins are pleiotropic neuropeptides characterized by the FxxWG C-terminal sequence (Abou El Asrar et al. 2020). They are expressed in the CNS and in entero-endocrine cells throughout the alimentary tract where they are known for simulating muscle contractions, food transit, secretion of digestive enzymes, and diuresis (Bhatt et al. 2014; Holman and Cook 1983). A striking feature of kinins observed in several insect species is their antifeedant activity. Some analogous and stable molecules of kinins have shown potential as possible agents to be used in integrated pest and vector management programs (Lange et al. 2016; Nachman et al. 2003).

8.3.7 Proctolin

Proctolin was the first bioactive neuropeptide to be isolated and sequenced as a potent myotropic factor with activity on the hindgut of *Periplaneta americana* (Brown and Starratt 1975). Later it was demonstrated that proctolin displays potent myotropic activity in the visceral, heart, and skeletal muscles from a wide range of insect species and other arthropods. As a multifunctional hormone, it also acts as a releasing factor for adipokinetic hormones and JH in different insect species (Bläser and Predel 2020). Proctolin sequences are almost invariably present in the form of the pentapeptide RYLPT-OH, which is regarded as ancestral for Hexapoda (Abou El Asrar et al. 2020; Bläser and Predel 2020). Proctolin and its receptor genes appear to be unique to arthropods with no homologs in vertebrates. Searches for the proctolin precursor gene in the available insect genomes suggest that it is absent in many insect species and absent in the complete order of Lepidoptera.

In several insect species, proctolin immunoreactivity reveals its presence in all ganglia of the SNS and in some thoracico-abdominal ganglia from which neurons extend to innervate the hindgut musculature, supporting its role in the regulation of feeding and gut motility (Abou El Asrar et al. 2020).

8.3.8 Insect RYamides

RYamides represent a group of neuropeptides characterized by a C-terminal FFxxxRY-amide sequence and belonging to the luqin/RYamide-type hormonal signaling system identified in various invertebrates including echinoderms, mollusks, nematodes, and crustaceans, but absent in vertebrates. Orthologs of RYamides have been found in most deposited insect genomes (Guo et al. 2021; Bläser and Predel 2020). Since the report on the occurrence of RYamides in the parasitic wasp *Nasonia vitripennis* and in the mosquito *Aedes aegpti* (Hauser et al. 2010), the expression of these neuropeptides has been documented in different insect species. Their function, however, has not yet been fully elucidated. RYamides are expressed in various neurons in the brain and in the TAG, as well as in the enteroendocrine cells of the anterior midgut and in the hindgut of larvae, pupae, and adults. The cognate RYamide receptor is also expressed mainly in the midgut and hindgut of insects, reinforcing the evidence for the participation of RYamides in the regulation of digestion and water reabsorption (Guo et al. 2021; Abou El Asrar et al. 2020; Veenstra and Khammassi 2017; Roller et al. 2016).

8.3.9 Trissin

The trisin neuropeptide was originally discovered through in silico research. Molecular cloning of the gene that codes for the expression of trissin allowed the identification of this neuropeptide as the ligand of the orphan G protein-coupled receptor CG34381 of *D. melanogaster* (Ida et al. 2011). Trissin sequences are highly conserved and contain a CxxCxxxCxxxCxxxxxxCC consensus structure in all insects studied. Mature peptides are generally composed of 27 amino acid residues and contain 3 disulfide bonds. In situ hybridization showed that trissin precursor expression is restricted to only two pairs of small interneurons in the protocerebrum and four to five large neurons in the FG of *B. mori* (Roller et al. 2016). These neurons innervate the tritocerebrum and the anterior midgut musculature. Little is known about the details of trissin's action in regulating the digestive process in insects. The high expression of trissin in the FG reinforces the evidence of its possible role in the regulation of foregut and midgut contractions and food intake. The expression of trissin receptors has been reported in the midgut and Malpighian tubules, further suggesting a role for trissin in intestinal physiology and water balance of insects (Roller et al. 2016; Yamanaka et al. 2008).

8.4 Peptidergic Regulation of Insect Digestion

8.4.1 Endocrine Regulation of Mouthparts and Foregut Functioning

Insects seem to be constitutively ready to initiate the behavior of seeking their food and ingesting it as soon as possible. This innate tendency is counterbalanced by hormonal satiety factors that lead them to stop eating and eventually refuse to eat more food. Food ingestion is orchestrated by the mouthpart movements that permit food handling and the rhythmic muscle contractions along the foregut, midgut, and hindgut. Endocrine factors and visceral muscle contractions control the transit of the food bolus throughout the alimentary tract from deglutition to defecation.

In the mouthparts of chewing insects, the closing and opening movement of the mandible is performed by the adductor and abductor muscles, respectively. The rhythmic movements of the mouthparts, mainly of the mandibles, allow the manipulation and fragmentation of the food so that adequate pieces are ingested that will compose the food bolus. In many insect species, these muscles are innervated by neurons from the SOG and FG (Abou El Asrar et al. 2020; Wu et al. 2020; Matsumoto et al. 2019; Audsley and Weaver 2009). These two ganglia are connected to the brain by neuronal axons generating a feeding central circuit in several insects. The SOG is a very important ganglion in the innervation of other organs of the head involved in the search and recognition of food, such as the innervation of mandibles, maxillae, labium, hypopharynx, salivary glands, and the muscles responsible for controlling neck movements (Nagata and Zhou 2019; Audsley and Weaver 2009; Copenhaver 2007). Mechanoreceptors located on the oral cavity and mouthparts are also innervated by the SOG and FG and are responsible for perceiving the size, texture, and chemical composition of dietary material before deglutition. In chewing insects, once the dietary components are adequately broken down by chewing, the insects swallow smaller pieces which then pass to the other compartments of the foregut. The contractions of the longitudinal and circular muscles that cover the foregut allow the movements that lead to the swallowing of food. In sucking (fluid-feeding or liquivorous) insects, the cibarial and pharyngeal muscles form a suction pump that brings the dietary liquid into the food canal. Like in chewing insects, these muscles are mainly innervated by the SOG and FG. Therefore, in both chewing and sucking insects, the SOG and FG are crucial ganglia to generate normal feeding behavior.

Among the neuropeptides involved in the control of digestion in the foregut, those that inhibit or stimulate muscle contraction are highlighted below. As expected, these neuropeptides are mainly expressed in the brain, SOG, FG, and enteroendocrine cells, and their receptors are expressed in the gut muscles and in the mechanoreceptors present in the mouthparts. Allatoregulatory peptides, CCHamides, NPF, SK, sNPF, myosuppressin, proctolin, and RYamide among other peptides have been described as significant actors for regulation of food intake. Injection of these neuropeptides in some insect species resulted in a significant reduction of food intake and prolonged duration in seeking diet, which can lead to compromised insect development, such as weight loss and increased mortality. In *B. mori* larvae, injection of AT and RYamide decreased esophageal contraction frequencies, while exposure of pharynx and ileum to AT and RYamide inhibited spontaneous contraction in fed larvae (Matsumoto et al. 2019). Curiously, exposure of pharynx to RYamide did not inhibit muscle contraction in fasting *B. mori* larvae. These results were correlated with the feeding state of the animals and the expression of RYamide receptors in the pharynx. In food-deprived animals, the expression of RYamide receptors is low and the larvae would be more likely to initiate feeding, while in animals that are feeding, the higher expression of the receptors leads to feeding inhibitory effects and cessation of food ingestion. In another lepidopteran species, the tomato moth *Lacanobia oleracea*, injection of AT in larvae did not influence the amount of food intake, demonstrating that the effects of a given hormone on digestion events can be very different depending on the insect species (Duve and Thorpe 2003). These authors also demonstrated the co-expression of AT and AST in the FG of the noctuids *L. oleracea* and *Heliothis virescens*. The co-localization of these physiologically antagonistic peptides within the same neurosecretory cells suggests that the control of the foregut contractile movements is complex and depends on the ratio between myo-inhibitory and myo-excitatory neuropeptides. An inhibitory effect in feeding behavior of RYamide was also reported in the dipteran species *Phormia regina* (Maeda et al. 2015). In this case, the injection of the neuropeptide resulted in suppression of the early feeding behavior called proboscis extension reflex, an accepted indicator of feeding motivation in sucking insects. In the polyneopteran species *Blattella germanica* and *Gryllus bimaculatus*, injection of AST-A reduced the food intake (Meyering-Vos and Woodring 2008; Aguilar et al. 2003). Reinforcing the importance of AST in food ingestion, silencing of AST-A receptor expression in *D. melanogaster* resulted in impairment of sensing dietary nutrients and an increase in food intake (Hentze et al. 2015). Similarly, the silencing of CCHamide-2 in *D. melanogaster* resulted in significant reduction of food intake, suggesting that CCHamide-2 is also involved in regulating the food-seeking behavior in this dipteran species (Ren et al. 2015).

NPF is also another neuropeptide involved in regulation of food intake in insects. The overexpression of NPF in *D. melanogaster* caused prolonged feeding period (Chung et al. 2017), while silencing of NPF precursor genes in *Locusta migratoria* (Tan et al. 2019) and in *Acyrthosiphon pisum* (Li et al. 2018) resulted in inhibition of food intake. These results suggest that NPF acts as an orexigenic neuropeptide. The neuropeptide sNPF (which is analogous to the NPF peptide) was reported to reduce host seeking behavior in the blood-sucking mosquito *A. aegypti* (Christ et al. 2017). The peptide SK was also found to be a satiety factor inhibiting food intake in the following insect species: *B. germanica*, *G. bimaculatus*, *Rhodnius prolixus*, and *Tribolium castaneum* (Abou El Asrar et al. 2020 and references therein). In *B. germanica*, SK induced foregut contractions inhibiting food intake, whereas a myosuppressin called leucomyosuppressin decreases the motility of the foregut muscles (Aguilar et al. 2006). In another cockroach, *Periplaneta americana*, CCAP increased the amplitude of foregut muscle contractions (Maestro et al. 2001), while proctolin accelerated foregut contraction in the kissing bug *R. prolixus* (Orchard et al. 2011).

8.4.2 Endocrine Regulation of Midgut Functioning

The midgut is a key organ for enzymatic digestion of food, absorption of nutrients, detoxification of potentially toxic compounds, shelter of microbiota, sensing of the nutritional value of the lumen content, storage of fuels such as glycogen and triglycerides, and distribution of absorbed compounds to the hemolymph, including the elimination of excess dietary water, among other functions. In general, the midgut of insects is innervated by plexuses that extend from the VG and by neurons that project from peripheral ganglia connected to the TAG. In addition to neuronal control over muscle contraction in the midgut, this organ also has enteroendocrine cells scattered throughout the intestinal epithelium. The main peptides already described in insects as directly involved in the control of peristaltic movements, secretion of digestive enzymes, and nutrient storage in midgut cells are the following: AT, ASTs, CCAP, TRP, sNPF, RYamides, and proctolin (Guo et al. 2021; Abou El Asrar et al. 2020; Bahrami et al. 2018; Audsley and Weaver 2009).

The expression and release of neuropeptides in enteroendocrine peptides are primarily regulated by the quality of nutrients present in the midgut lumen, especially the protein/carbohydrate ratio of the diet (Abou El Asrar et al. 2020). Transcriptional levels of some neuropeptide precursors and their receptors are altered between starvation and feeding states. In different insect species, the following peptides have been shown to be involved in the sensing of nutrients and secretion of digestive enzymes: AST-A, CCAP, and TRP. For example, in *Anopheles* mosquitoes the expression of AST-A receptors in enteroendocrine cells is significantly higher in blood-fed animals than in mosquitoes that ingested glucose instead (Felix et al. 2015). In starved *Spodoptera littoralis* (Lepidoptera: Noctuidae) larvae, the expression of AST-A by enteroendocrine cells in the midgut is relatively low, but it increases following refeeding (Bahrami et al. 2018). In response to AST-A injection in larval *S. littoralis*, the levels of α-amylase and peptidase secretion into the midgut lumen also increase, suggesting that AST-A is crucial for carbohydrate and protein digestion in this noctuid (Bahrami et al. 2018). Curiously, in another noctuid species, *S. frugiperda*, AST-A reduced the midgut levels of α-amylase and trypsin, while AT stimulates α-amylase and trypsin release (Lwalaba et al. 2010), illustrating how different species respond to different neuropeptides, sometimes with opposite effects.

In the American cockroach *P. americana* AST, CCAP, and TRP from enteroendocrine cells increase the secretion of α-amylases, lipases, and peptidases (Mikani 2016; Matsui et al. 2013; Sakai et al. 2004). In this case, a brain structure, the pars intercerebralis (PI) acts as an inhibitory center for the secretion of CCAP and AST by enteroendocrine cells, probably through intervention by sNPF. Removal of this brain structure causes hyperphagia and increased secretion of digestive enzymes in cockroaches under dark conditions, whereas the injection of sNPF into the hemocoel led to a decrease in α-amylase, peptidase, and lipase activities in the midgut (Mikani 2016; Matsui et al. 2013).

In addition to the important role in enzymatic degradation of dietary material, peristaltic movements, and maintenance of fluid flow in the lumen of the midgut are also critical aspects of the digestive process in insects. The regulation of midgut muscle contractions influences food seeking and food intake. In most insects, the allatoregulatory peptides (AST and AT) are the main players exhibiting either stimulatory (AT) or inhibitory (AST) effects on gut contraction and motility, but they are not the unique myoactive peptides (Abou El Asrar et al. 2020; Wu et al. 2020). The midgut contractions are also stimulated by TRPs in *L. migratoria* and in the cockroach *Leucophaea maderae* (Winther and Nässel 2001). Proctolin activated midgut contraction in the hematophagous bug *R. prolixus* (Orchard et al. 2011), while in another blood-feeding heteropteran *T. infestans*, AT increases midgut muscle contraction (Lange et al. 2016). CCAP also has myoregulatory activity in *P. americana*, since this peptide is also capable of increasing the amplitude of midgut contraction (Sakai et al. 2004).

8.4.3 Endocrine Regulation of Hindgut Functioning

Evacuation by producing feces to remove the resulting non-digested dietary material, wastes (such as cells and peritrophic membrane debris), and excreta from both the midgut lumen and the Malpighian tubules is supported by hindgut peristalsis. The muscle contractions that permit the hindgut peristalsis are regulated by neuropeptides secreted mainly by the TAG and nerves connected to it. It is important to note that proper defecation is directly related to other aspects of digestion, such as feeding behavior and food seeking. The most common neuropeptides involved in controlling hindgut physiology are the following: AT, AST, FaRP, TRP, proctolin, and RYamide (Table 8.1). Some representative examples in the literature on hormonal and neuronal regulation of the hindgut are presented in the following paragraphs.

In the cockroach *Leucophaea maderae*, AT has a prominent role in the regulation of hindgut peristalsis. It has been demonstrated that AT stimulates an increase in frequency of spontaneous contractions of the longitudinal muscle in the rectal pads at the distal portion of the cockroach hindgut (Rudwall et al. 2000). In this same cockroach species, exogenous ASTs presented myoinhibitory effects on the hindgut muscle contractions (Rudwall et al. 2000). In the hematophagous bug, *Triatoma infestans*, AT secreted by the Malpighian tubules stimulates hindgut contractions to facilitate the mixing of urine and feces for evacuation during post-prandial diuresis (Sterkel et al. 2010; Santini and Ronderos 2009). AT is also able to increase the frequency and tonus of contractions in the hindgut of the coleopteran *T. castaneum* (Vuerinckx et al. 2011). In the close relatives of *T. castaneum*, the tenebrionids *T. molitor* and *Zophobas atratus*, the application of a synthetic bioanalogue FaLP (NSNFLRFamide) to hindgut preparations increased the muscle contraction frequency in this organ (Marciniak et al. 2020).

In the kissing bug *R. prolixus*, TRPs have been described as potent inhibitors of hindgut muscle contraction following induction by an insect kinin called Rhopr-kinin-2 (Haddad et al. 2018; Zandawala and Orchard 2013). ASTs have also been described as hindgut myoinhibitory neuropeptides in several insect species such as *R. prolixus* (Villalobos-Sambucaro et al. 2016), the locust *L. migratoria* (Robertson and Lange 2010), the tenebrionids *T. molitor* and *Z. atratus* (Lubawy et al. 2018), and in *D. melanogaster* (Chen et al. 2016). Another neuropeptide with myotropic action on the hindgut is proctolin, already documented in insects of different orders, except for the order Lepidoptera (Abou El Asrar et al. 2020; Wu et al. 2020; Nagata and Zhou 2019). Proctolin was originally isolated and sequenced based on its myostimulatory activity on the hindgut of the cockroach *P. americana* (Brown 1975; Starratt and Brown 1975). Proctolin myostimulatory activity was also found in the ringlegged earwig *Euborellia annulipes* (Phitayakorn et al. 2008), in the cockroach *Diploptera punctata* (Fusé and Orchard 1998), in *R. prolixus* (Orchard et al. 2011), in the locust *S. gregaria* (Gray et al. 2000), in the burying beetle *Nicrophorus vespilloides* (Urbanski et al. 2019), and in the fruit fly *D. melanogaster* (Nässel and Zandawala 2020).

The examples above represent some of the classic and other more recent studies on the main neuropeptides directly involved in the hormonal regulation of hindgut functioning. With recent advances in omics approaches, a more detailed understanding of the interconnections between myoexcitatory and myoinhibitory neuropeptides in the regulation of the final stages of digestion in insects is taking shape.

References

Abou El Asrar R, Cools D, Broeck JV (2020) Role of peptide hormones in insect gut physiology. Curr Opin Insect Sci 41:71–78. https://doi.org/10.1016/j.cois.2020.07.004

Aguilar R, Maestro JL, Vilaplana L et al (2003) Allatostatin gene expression in brain and midgut, and activity of synthetic allatostatins on feeding-related processes in the cockroach *Blattella germanica*. Regul Pept 115:171–177. https://doi.org/10.1016/S0167-0115(03)00165-4

Aguilar R, Maestro JL, Belles X (2006) Effects of myoinhibitory peptides on food intake in the German cockroach. Physiol Entomol 31:257–261. https://doi.org/10.1111/j.1365-3032.2006.00513.x

Audsley N, Weaver RJ (2009) Neuropeptides associated with the regulation of feeding in insects. Gen Comp Endocrinol 162:93–104. https://doi.org/10.1016/j.ygcen.2008.08.003

Bahrami MN, Mikani A, Moharramipour S (2018) Effect of caffeic acid on feeding, α-amylase and protease activities and allatostatin-A content of Egyptian cotton leafworm, *Spodoptera littoralis* (Lepidoptera: Noctuidae). J Pestic Sci 43:73–88. https://doi.org/10.1584/jpestics.D17-086

Bhatt G, da Silva R, Nachman RJ, Orchard I (2014) The molecular characterization of the kinin transcript and the physiological effects of kinins in the blood-gorging insect, *Rhodnius prolixus*. Peptides 53:148–158. https://doi.org/10.1016/j.peptides.2013.04.009

Bläser M, Predel R (2020) Evolution of neuropeptide precursors in Polyneoptera (Insecta). Front Endocrinol 11:197. https://doi.org/10.3389/fendo.2020.00197

Brown BE (1975) Proctolin: a peptide transmitter candidate in insects. Life Sci 17:1241–1252

Brown BE, Starratt AN (1975) Isolation of proctolin, a myotropic peptide, from *Periplaneta americana*. J Insect Physiol 21:1879–1881

Chen J, Reiher W, Hermann-Luibl C et al (2016) Allatostatin a signalling in *Drosophila* regulates feeding and sleep and is modulated by PDF. PLoS Genet 12:e1006346. https://doi.org/10.1371/journal.pgen.1006346

Christ P, Reifenrath A, Kahnt J et al (2017) Feeding-induced changes in allatostatin-A and short neuropeptide F in the antennal lobes affect odor-mediated host seeking in the yellow fever mosquito, *Aedes aegypti*. PloS one:e0188243. https://doi.org/10.1371/journal.pone.0188243

Chung BY, Ro J, Hutter SA, Miller KM et al (2017) *Drosophila neuropeptide* F signaling independently regulates feeding and sleep-wake behavior. Cell Rep 19:2441–2450. https://doi.org/10.1016/j.celrep.2017.05.085

Copenhaver PF (2007) How to innervate a simple gut: familiar themes and unique aspects in the formation of the insect enteric nervous system. Dev Dyn 236:1841–1864. https://doi.org/10.1002/dvdy.21138

Down RE, Matthews HJ, Audsley N (2011) Oral activity of FMRFamide-related peptides on the pea aphid *Acyrthosiphon pisum* (Hemiptera: Aphididae) and degradation by enzymes from the aphid gut. Regulatory Pept 171:11–18. https://doi.org/10.1016/j.regpep.2011.05.013

Duve H, Thorpe A (2003) Neuropeptide co-localisation in the lepidopteran frontal ganglion studied by confocal laser scanning microscopy. Cell Tissue Res 311:79–89. https://doi.org/10.1007/s00441-002-0648-2

Elekonich MM, Horodyski FM (2003) Insect allatotropins belong to a family of structurally-related myoactive peptides present in several invertebrate phyla. Peptides 24:1623–1632. https://doi.org/10.1016/j.peptides.2003.08.011

Felix RC, Trindade M, Pires IR et al (2015) Unravelling the evolution of the allatostatin-type A, KISS and galanin peptide-receptor gene families in bilaterians: insights from *Anopheles* mosquitoes. PLoS One 10(7):e0130347. https://doi.org/10.1371/journal.pone.0130347

Fusé M, Orchard I (1998) The muscular contractions of the midgut of the cockroach, *Diploptera punctata*: effects of the insect neuropeptides proctolin and leucomyosuppressin. Regul Pept 77:163–168

Godoy RS, Barbosa RC, Procópio TF et al (2021) FMRF-related peptides in *Aedes aegypti* midgut: neuromuscular connections and enteric nervous system. Cell Tissue Res 7:1–8. https://doi.org/10.1007/s00441-021-03462-3

Gray AS, Hancock JT, Osborne RH (2000) Proctolin antagonists bind to [^{3}H]proctolin binding sites in the locust hindgut. Peptides 21:189–196. https://doi.org/10.1016/S0196-9781(99)00199-0

Guo Z, He X, Jiang C, Shi Y, Zhou N (2021) Activation of *Bombyx mori* neuropeptide G protein–coupled receptor A19 by neuropeptide RYamides couples to Gq protein-dependent signaling pathways. J Cell Biochem 122:456–471. https://doi.org/10.1002/jcb.29874

Habenstein J, Schmitt F, Liessem S et al (2021) Transcriptomic, peptidomic, and mass spectrometry imaging analysis of the brain in the ant *Cataglyphis nodus*. J Neurochem 158:391–412. https://doi.org/10.1111/jnc.15346

Haddad AN, Defferrari MS, Hana S et al (2018) Expression and functional characterization of tachykinin-related peptides in the blood-feeding bug, *Rhodnius prolixus*. Peptides 99:247–254. https://doi.org/10.1016/j.peptides.2017.11.006

Hauser F, Neupert S, Williamson M et al (2010) Genomics and peptidomics of neuropeptides and protein hormones present in the parasitic wasp *Nasonia vitripennis*. J Proteome Res 9:5296–5310. https://doi.org/10.1021/pr100570j

Hentze JL, Carlsson MA, Kondo S et al (2015) The neuropeptide allatostatin a regulates metabolism and feeding decisions in *drosophila*. Sci Rep 5(1):1–16. https://doi.org/10.1038/srep11680

Holman GM, Cook BJ (1983) Isolation and partial characterization of a second myotropic peptide from the hindgut of the cockroach, *Leucophaea maderae*. Comp Biochem Physiol 76C:39–43

Ida T, Takahashi T, Tominaga H et al (2011) Identification of the endogenous cysteine-rich peptide trissin, a ligand for an orphan G protein-coupled receptor in *drosophila*. Biochem Biophys Res Commun 414:44–48. https://doi.org/10.1016/j.bbrc.2011.09.018

Kataoka H, Toschi A, Li JP et al (1989) Identification of an allatotropin from adult *Manduca sexta*. Science 243:1481–1483

Kramer SJ, Toschi A, Miller CA et al (1991) Identification of an allatostatin from the tobacco hornworm *Manduca sexta*. Proc Natl Acad Sci U S A 88:9458–9462

Lange AB, Nachman RJ, Kaczmarek K, Zabrocki J (2016) Biostable insect kinin analogs reduce blood meal and disrupt ecdysis in the blood-gorging Chagas' disease vector, *Rhodnius prolixus*. Peptides 80:108–113. https://doi.org/10.1016/j.peptides.2016.01.012

Lee SH, Cho E, Yoon SE et al (2021) Metabolic control of daily locomotor activity mediated by tachykinin in *Drosophila*. Commun Biol 4:693. https://doi.org/10.1038/s42003-021-02219-6

Li X, Qu MJ, Zhang Y et al (2018) Expression of neuropeptide F gene and its regulation of feeding behavior in the pea aphid, *Acyrthosiphon pisum*. Front Physiol 9:87. https://doi.org/10.3389/fphys.2018.00087

Lubawy J, Marciniak P, Kuczer M, Rosiński G (2018) Myotropic activity of allatostatins in tenebrionid beetles. Neuropeptides 70:26–36. https://doi.org/10.1016/j.npep.2018.05.003

Lwalaba D, Hoffmann KH, Woodring J (2010) Control of the release of digestive enzymes in the larvae of the fall armyworm, *Spodoptera frugiperda*. Arch Insect Biochem Physiol 73:14–29. https://doi.org/10.1002/arch.20332

Maeda T, Nakamura Y, Shiotani H et al (2015) Suppressive effects of dRYamides on feeding behavior of the blowfly, *Phormia regina*. Zoological Lett 1:35. https://doi.org/10.1186/s40851-015-0034-z

Maestro JL, Aguilar R, Pascual N et al (2001) Screening of antifeedant activity in brain extracts led to the identification of sulfakinin as a satiety promoter in the German cockroach—are arthropod sulfakinins homologous to vertebrate gastrinscholecystokinins? Eur J Biochem 268:5824–5830. https://doi.org/10.1046/j.0014-2956.2001.02527.x

Marciniak P, Witek W, Szymczak M et al (2020) FMRFamide-related peptides signaling is involved in the regulation of muscle contractions in two tenebrionid beetles. Front Physiol 11:456. https://doi.org/10.3389/fphys.2020.00456

Matsui T, Sakai T, Satake H, Takeda M (2013) The pars intercerebralis affects digestive activities of the American cockroach, *Periplaneta americana*, via crustacean cardioactive peptide and allatostatin-A. J Insect Physiol 59:33–37. https://doi.org/10.1016/j.jinsphys.2012.06.010

Matsumoto S, Kutsuna N, Daubnerova I et al (2019) Enteroendocrine peptides regulate feeding behavior via controlling intestinal contraction of the silkworm *Bombyx mori*. PLoS One 14:e0219050. https://doi.org/10.1371/journal.pone.0219050

Meyering-Vos M, Woodring J (2008) A-type allatostatins and sulfakinins as satiety effectors in the Mediterranean field cricket *Gryllus bimaculatus*. M. D. Gesell. Allg Ange Entomol 16:409–412

Mikani A (2016) Tachykinin stimulation effects on α-amylase, protease and lipase activities in midgut of American cockroach, *Periplaneta americana* (Blattodea: Blattidae). J Entomol Soc Iran 36:81–88

Nachman RJ, Coast GM, Douat C et al (2003) A C-terminal aldehyde insect kinin analog enhances inhibition of weight gain and induces significant mortality in *Helicoverpa zea* larvae. Peptides 24:1615–1621. https://doi.org/10.1016/j.peptides.2003.06.008

Nagata S, Zhou YJ (2019) Chapter four – feeding-modulating neuropeptides and peptide hormones in insects. In: Jurenka R (ed) Advances in insect physiology. Academic, pp 137–172. https://doi.org/10.1016/bs.aiip.2019.10.002

Nässel DR, Zandawala M (2020) Hormonal axes in *drosophila*: regulation of hormone release and multiplicity of actions. Cell Tissue Res 382:233–266. https://doi.org/10.1007/s00441-020-03264-z

Orchard I, Lange AB, Bendena WG (2001) FMRFamide-related peptides: a multifunctional family of structurally related neuropeptides in insects. Adv Insect Physiol 28:267–329. https://doi.org/10.1016/S0065-2806(01)28012-6

Orchard I, Lee DH, Da Silva R, Lange AB (2011) The proctolin gene and biological effects of proctolin in the blood-feeding bug, *Rhodnius prolixus*. Front Endocrinol 2:59. https://doi.org/10.3389/fendo.2011.00059

Pascual N, Maestro JL, Chiva C et al (2008) Identification of a tachykinin-related peptide with orexigenic properties in the German cockroach. Peptides 29:386–392. https://doi.org/10.1016/j.peptides.2007.11.010

Phitayakorn R, Bennett GP, Rankin SM (2008) Quantification of cockroach allatostatin-like peptide and its myotropic effects in males of the earwig *Euborellia annulipes*. Physiol Entomol 26:11–17. https://doi.org/10.1111/j.1365-3032.2001.00210.x

Predel R, Kellner R, Rapus J et al (1997) Isolation and structural elucidation of eight kinins from the retrocerebral complex of the American cockroach, *Periplaneta americana*. Regul Pept 71:199–205

Ragionieri L, Verdonck R, Verlinden H et al (2022) *Schistocerca* neuropeptides – An update. J Insect Physiol 136:104326. https://doi.org/10.1016/j.jinsphys.2021.104326

Rand D, Ayali A (2010) Neuroanatomy and neurophysiology of the locust hypocerebral ganglion. J Insect Physiol 56:884–892. https://doi.org/10.1016/j.jinsphys.2010.04.001

Ren GR, Hauser F, Rewitz KF et al (2015) CCHamide-2 is an orexigenic brain-gut peptide in *Drosophila*. PLoS One 10:e0133017. https://doi.org/10.1371/journal.pone.0133017

Robertson L, Lange AB (2010) Neural substrate and allatostatin-like innervation of the gut of *Locusta migratoria*. J Insect Physiol 56:893–901. https://doi.org/10.1016/j.jinsphys.2010.05.003

Roller L, Cizmar D, Bednar B, Zitnan D (2016) Expression of RYamide in the nervous and endocrine system of *Bombyx mori*. Peptides 80:72–79. https://doi.org/10.1016/j.peptides.2016.02.003

Rudwall AJ, Sliwowska J, Nässel DR (2000) Allatotropin-like neuropeptide in the cockroach abdominal nervous system: Myotropic actions, sexually dimorphic distribution and colocalization with serotonin. J Comp Neurol 428:159–173. https://doi.org/10.1002/1096-9861(20001204)428:1<159::AID-CNE11>3.0.CO;2-Y

Sakai T, Satake H, Minakata H, Takeda M (2004) Characterization of crustacean cardioactive peptide as a novel insect midgut factor: isolation, localization, and stimulation of alpha-amylase activity and gut contraction. Endocrinology 145:5671–5678. https://doi.org/10.1210/en.2004-0722

Santini MS, Ronderos JR (2009) Allatotropin-like peptide in Malpighian tubules: insect renal tubules as an autonomous endocrine organ. Gen Comp Endocrinol 160:243–249. https://doi.org/10.1016/j.ygcen.2008.12.002

Song W, Veenstra JA, Perrimon N (2014) Control of lipid metabolism by Tachykinin in *Drosophila*. Cell Rep 9:40–47. https://doi.org/10.1016/j.celrep.2014.08.060

Starratt AN, Brown BE (1975) Structure of the pentapeptide proctolin, a proposed neurotransmitter in insects. Life Sci 17:1253–1256

Sterkel M, Riccillo FL, Ronderos JR (2010) Cardioacceleratory and myostimulatory activity of allatotropin in *Triatoma infestans*. Comp Biochem Physiol 155A:371–377. https://doi.org/10.1016/j.cbpa.2009.12.002

Tan SQ, Li AM, Wang Y, Shi WP (2019) Role of the neuropeptide F 1 in regulating the appetite for food in *Locusta migratoria*. Pest Manag Sci 75:1304–1309. https://doi.org/10.1002/ps.5244

Urbanski A, Lubawy J, Marciniak P, Rosinski G (2019) Myotropic activity and immunolocalization of selected neuropeptides of the burying beetle *Nicrophorus vespilloides* (Coleoptera: Silphidae). Insect Sci 26:656–670. https://doi.org/10.1111/1744-7917.12569

Veenstra JA, Khammassi H (2017) Rudimentary expression of RYamide in *Drosophila melanogaster* relative to other *drosophila* species points to a functional decline of this neuropeptide gene. Insect Biochem Molec Biol 83:68–79. https://doi.org/10.1016/j.ibmb.2017.03.001

Villalobos-Sambucaro MJ, Diambra LA et al (2016) Allatostatin-C antagonizes the synergistic myostimulatory effect of allatotropin and serotonin in *Rhodnius prolixus* (Stal). Gen Comp Endocrinol 233:1–7. https://doi.org/10.1016/j.ygcen.2016.05.009

Vuerinckx K, Verlinden H, Lindemans M (2011) Characterization of an allatotropin-like peptide receptor in the red flour beetle, *Tribolium castaneum*. Insect Biochem Molec Biol 41:815–822. https://doi.org/10.1016/j.ibmb.2011.06.003

Weaver RJ, Freeman ZA, Pickering MG, Edwards JP (1994) Identification of two allatostatins from the CNS of the cockroach *Periplaneta americana*: novel members of a family of neuropeptide inhibitors of insect juvenile hormone biosynthesis. Comp Biochem Physiol 107C:119–127

Winther AME, Nässel DR (2001) Intestinal peptides as circulating hormones: release of tachykinin-related peptide from the locust and cockroach midgut. J Exp Biol 204:1269–1280. https://doi.org/10.1242/jeb.204.7.1269

Wu K, Li S, Wang J et al (2020) Peptide hormones in the insect midgut. Front Physiol 11:191. https://doi.org/10.3389/fphys.2020.00191

Yamanaka N, Yamamoto S, Žitňan D et al (2008) Neuropeptide receptor transcriptome reveals unidentified neuroendocrine pathways. PLoS One 3:e3048. https://doi.org/10.1371/journal.pone.0003048

Zandawala M, Orchard I (2013) Post-feeding physiology in *Rhodnius prolixus*: the possible role of FGLamide-related allatostatins. Gen Comp Endocrinol 194:311–317. https://doi.org/10.1016/j.ygcen.2013.10.005

Chapter 9
Recruitment of Lysosomal Cysteine and Aspartic Endopeptidases as Digestive Enzymes

Abstract Protein digestion is carried out by serine endopeptidases in most insects. However, some insects recruit lysosomal proteins as digestive enzymes to help or replace their serine endopeptidases, under different selective pressures. All recruited endopeptidases originate from duplicated genes and may be aspartic endopeptidases (cathepsin D, CAD), active in acidic medium, or cysteine endopeptidases that include cathepsin L (CAL) and cathepsin B (CAB) and which are active in reducing conditions in mildly acidic medium. The CAL 3D structural changes on adapting to a digestive function were investigated. *T. molitor* CALs were produced in recombinant form, crystallized, and had their 3D structures solved. In the case of *Z. subfasciatus*, digestive and lysosomal CALs were compared by a structure alignment-based clustering with an appropriate algorithm with recombinant *T. molitor* CAL. Major differences were found in P2 subsites of enzymes of both insects. Kinetical studies suggested that the structural differences led to less substrate-specific enzymes able to digest prolamins, occurring in seeds. Combining transcriptomic and genomic data for different beetles revealed that the expansion of CAL and CAB coding genes occurred only in the infraorder Cucujiformia, as a putative adaptation to pollen feeding by Cucujiformia ancestors. Hemipterans lost their serine endopeptidases on adapting to feeding protein-poor plant sap. On returning to protein-rich diets, hemipterans recruited lysosomal CAL and CAB as replacement for their lost serine endopeptidases. Larval flies digest bacteria from their bacteria-rich food with the aid of CADs, usually with a proline loop, in their acid middle midguts. Digestive CADs in hemipterans are modulated by feeding and hydrolyzing CAL inhibitors before the food bolus meet the compartment rich in CALs. The sorting mechanism involved in targeting lysosomal proteins to midgut remains under discussion.

W. R. Terra et al., *Molecular Physiology and Evolution of Insect Digestive Systems*, Entomology in Focus 7, https://doi.org/10.1007/978-3-031-39233-7_9

9.1 Introduction

Protein digestion starts under the action of endopeptidases that break internal peptide bonds in proteins. As previously discussed in Chap. 4, there are three major groups of endopeptidases, classified according to their active site groups: serine, aspartic, and cysteine endopeptidases. The major digestive endopeptidases in animals are the serine endopeptidases because they are active in oxidizing and mildly alkaline media, usually found outside the cells of their guts. In contrast, cysteine endopeptidases need a reducing media, and aspartic endopeptidases, a very acidic environment. Thus, cysteine endopeptidases, exemplified by cathepsins L (CAL) and B (CAB), and aspartic endopeptidases, exemplified by cathepsin D (CAD), are typical of lysosomes. Lysosomes are cell organelles involved in intracellular protein turnover in an acidic and reducing media.

Despite the fact that serine endopeptidases are the major digestive endopeptidases in insects, some of them recruit lysosomal CADs, CALs, and CABs to midgut contents to aid or replace their serine endopeptidases, under different selective pressures (Terra et al. 2019).

Amino acid residues of the endopeptidase substrates interact with different enzyme subsites (Fig. 9.1). According to Schechter and Berger (1967), amino acid residues in substrates are numbered P1, P2, P3, ...Pn from the hydrolyzed peptide bond in the direction of the N-terminus and P1′, P2′, P3′, ...Pn’ in the direction of the C-terminus. The enzyme subsites corresponding to the substrate positions are numbered, respectively, S1, S2, S3, ...Sn and S1′, S2′, S3′, ...Sn′.

CADs are soluble lysosomal aspartic endopeptidases, which hydrolyze preferentially at hydrophobic amino acids (mainly Phe residues) at the scissile bond (between P1 and P1′). CADs are synthesized as proenzymes and are self-activated and active at low pH with an optima between 3.5 and 5.0 and are characteristically inhibited by pepstatin (Benes et al. 2008; Barrett et al. 2013). CADs are usually assayed with synthetic substrates that may be chromophoric (Dunn et al. 1986) or internally quenched fluorescent peptides, also named fluorescent resonance energy transfer (FRET) substrates (Pimenta et al. 2001). This kind of substrate is a polypeptide containing a fluorescent group at one extremity and a quenching group at the other extremity, so that the polypeptide is not fluorescent, but the products of hydrolysis fluoresce.

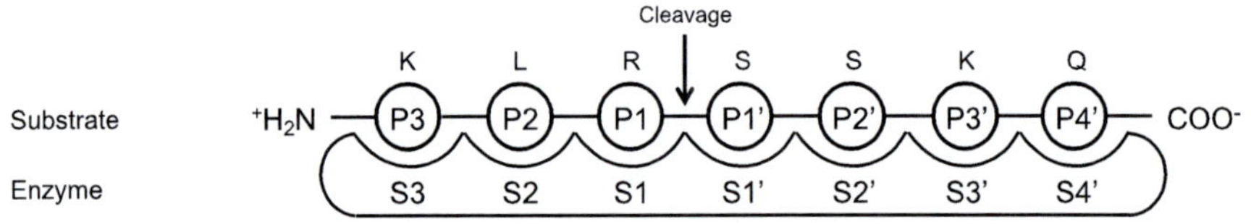

Fig. 9.1 Diagrammatic representation of endopeptidase subsites (S3-S3′) bound to substrate amino acid residues (P3-P3′) and an indication of the peptide bond to be cleaved

CALs are cysteine endopeptidases with a preference toward substrates with aromatic residues in P2 that are usually assayed at pH 5–6 in the presence of activating sulfhydryl agents (dithiothreitol, DTT, or cysteine) with compounds like carbobenzoxy-phenylalanine-arginine (Z-FR)-p-nitroanilide or Z-FR-4-methyl coumarin 7 amide (MCA). CAL sequences have the features characteristic of family C1 of cysteine endopeptidases: N-terminal propeptide, which is autocatalytically (or by the action of CAD, Nishimura et al. 1989) removed to activate the enzyme, the catalytic triad (Cys 25, H169, Asn 175, papain numbering), the ERFNIN motif in the propeptide, and inhibition by trans-epoxysuccinyl-L-leucyl-amide (4-guanidino butane) (E-64).

CABs are cysteine endopeptidase like CALs but differ from them in the fact that they prefer Arg residues in P2 (a good substrate is Z-RR-MCA) and have an occluding loop occupying their S′ subsites. This occlusion leads to a dipeptidyl carboxypeptidase activity and lowers endopeptidase activity and lower affinity for cysteine endopeptidase inhibitors (Musil et al. 1991). The removal of the occluding loop from CABs increases their endopeptidase activity and abolishes their exopeptidase activity (Nagler et al. 1997; Illy et al. 1997).

9.2 Digestive Cathepsins L and B in Coleoptera

Luminal digestive cysteine endopeptidases were first reported in seed beetles (Chrysomelidae: Bruchinae) (Gatehouse et al. 1985; Kitch and Murdock 1986; Wieman and Nielsen 1988; Campos et al. 1989; Lemos et al. 1990; Silva and Xavier-Filho 1991) and were at first considered to occur in all Coleoptera. However, Terra and Cristofoletti (1996) using species of different Infraorders proposed that the occurrence of digestive cysteine endopeptidases in Coleoptera is actually restricted to the Infraorder Cucujiformia. Furthermore, they showed that cysteine endopeptidases occur in the anterior and middle midgut, whereas the serine endopeptidases occur in the posterior midgut of the larvae. The occurrence and distribution of cysteine endopeptidases were interpreted as an adaptation to circumvent the inhibition of serine endopeptidases by protein inhibitors present in the seed flour ingested by the larvae. Some coleopteran CALs were characterized in detail as recombinant proteins, exemplified by those of *Sphenophorus levis* (Fonseca et al. 2012), *Diaprepes abbreviatus* (Ben-Mahmoud et al. 2015), *Tribolium castaneum* (Dvoryakova et al. 2022), and *T. molitor* (see below).

Structural changes are expected to have occurred with CALs on adapting to the digestive role. Detailed studies were done with the CALs of the beetle *T. molitor*. This insect has three major CALs: one lysosomal (TmCAL1) and two digestive (TmCAL2 and TmCAL3) enzymes, as confirmed by enzyme assays in isolated midgut contents and immunocytolocalization (Cristofoletti et al. 2005; Beton et al. 2012). Although both TmCAL2 and TmCAL3 are found in the midgut of the Brazilian strain of *T. molitor* (Cristofoletti et al. 2005), only TmCAL3 is found in the USDA (Manhatan) strain (Prabhakar et al. 2007). N-terminal sequencing of in vitro autoactivated recombinant TmCAL3 resulted in a sequence with two residues lengthier than the in vivo mature enzyme. These two residues are probably removed in vivo by dipeptidyl peptidase (Damasceno et al. 2017).

The 3D structures of procathepsins TmCAL2 and TmCAL3 were solved by crystallography and showed remarkable differences, mainly at their S2 subsites and in kinetic properties. Whereas TmCAL3 hydrolyzes both Z-FR-MCA and Z-RR-MCA, TmCAL2 acts only on Z-FR-MCA (Beton et al. 2012). The specificity of TmCAL3 was also investigated using 64 FRET substrates with different replacements in the positions P2, P1, P1′, and P2′. As the human lysosomal cathepsin L (HsCAL1) (Puzer et al. 2004) was assayed with the same substrates as TMCAL3, it was possible to compare their subsites with free energy relationships (Withers and Rupitz 1990). These are plots of log (k_{cat}/ K_m) for substrates varying in a position (P1 as an example) of TmCAL3 against the same parameters for HsCAL1. A rectilinear line indicates that the subsites are similar in both enzymes, whereas the absence of a straight line implies that the subsites differ. These plots showed that only the S1 subsite of TmCAL3 is similar to the corresponding one in HsCAL1. Nevertheless, even in this case, several substrates with different amino acid replacements did not fit the straight line indicating they interact differently with TmCAL3 and HsCAL1 (Damasceno et al. 2017).

Another way of comparing subsite substrate specificity is measuring the relative k_{cat}/K_m obtained for each amino acid replacement in a position in TmCAL3 and HsCAL1. The results showed that the preferential amino acid on substrate position P1 (primary specificity) is Arg for both enzymes. However, when the substrate has Gln in P1, the k_{cat}/K_m value is 5.2% for HsCAL1 (Puzer et al. 2004) and 47% for TmCAL3 (Terra et al. 2019), relative to Arg in P1 taken as 100%. Similar differences were obtained with other amino acid replacements. This means that TmCAL 3 is less substrate selective than HsCAL1, which may be an adaptation to digest prolamins, in accordance with the finding of Goptar et al. (2012) that *T. molitor* cysteine endopeptidases and those of *T. castaneum* (Dvoryakova et al. 2022) hydrolyze prolamins better than the serine endopeptidases. Prolamins are storage proteins of seed cereals that are rich in Gln and Pro, exemplified by wheat (gliadin), barley (hordein), and corn (zein) (Shewry and Halford (2002).

Taking into account the 3D structures and the peptides used in the kinetics studies, the identification of the subsites of the lysosomal human CAL (HsCAL1), TmCAL3, and TmCAL2 was done by molecular docking (Damasceno et al. 2017). In establishing a subsite residue composition, the frequency of predicted interactions was taken into account, because the amino acid side chains are flexible and interactions may differ depending on the complexed protein. Some of the amino acid positions in the different subsites were previously described in other cysteine endopeptidases (Turk et al. 1998).

The ratio of the transition state binding energy (calculated from k_{cat} values and substrate binding energy (calculated from K_m values) permits to identify the roles of the four TmCAL3 subsites. The results of Damasceno et al. (2017) showed that the ratio for S2 was larger than one, indicating it participates mainly in catalysis, as expected. The ratios for S1 and S2′ were smaller than one, implying they are mainly involved in substrate binding, and, finally, the ratio for S1′ was one, meaning it participates both in catalysis and substrate binding (Damasceno et al. 2017). The mechanism of hydrolysis of cysteine endopeptidases includes the formation of an acyl-enzyme intermediate, followed by a deacetylation step. According to data, the rate-limiting step in TmCAL3 activity is the acylation step (Damasceno et al. 2017).

Another interesting property of the subsites is their hydrophobicities calculated from kinetic data. The subsite hydrophobicity (H) is an index that evaluates the hydrophobic character of the subsites calculated from the efficiency of hydrolysis (k_{cat}/K_m) with which the enzyme act on substrates with different amino acid replacements in the peptide (Lopes et al. 2006). The binding subsite hydrophobicity (Hs) is similar to H, but calculated from K_m, instead from k_{cat}/K_m (Damasceno et al. 2017). H represents mainly the residues involved in enzyme-transition state complexes, whereas Hs, the residues participating in interactions with the substrate. Hydrophobicities may also be calculated from the hydropathic values of the interacting amino acids identified by docking and taking into account the hydropathic indexes for the amino acids (Kyte and Doolittle 1982). The results obtained from kinetics (mainly Hs) and docking were closer for S1 and S2′, than for S1′ and S2, indicating that the residues that were more involved in substrate binding are different from those participating in transition state binding (Damasceno et al. 2017).

There are many more genes coding for CALs in *T. molitor* than TmCAL1, TmCAL2, and TmCAL3. Martynov et al. (2015) described a large number of genes coding for CALs and CABs, some of the latter with shortened occluding loops. These findings were confirmed and extended for other Coleoptera groups in a study combining transcriptomic data of isolated tissues of *Dermestes maculatus*, *Tenebrio molitor*, and *Zabrotes subfasciatus* and genomic data from 13 different beetles (Silva et al. 2022). As every single cell has lysosomal enzymes (Barrett et al. 2013), these enzymes are expected to be similarly or more expressed in the extraintestinal tissues than in the midgut. In contrast, digestive enzymes are expected to predominate tissues than in the midgut. Lysosomal and luminal digestive cathepsins were recognized by the grouping of their sequences with sequences known to be secreted to the midgut lumen or to be lysosomal. The data (Silva et al. 2022) showed that the expansion of CAL and CAB genes, including genes coding for luminal digestive enzymes, were found only among the Infraorder Cucujiformia. CAL genes varied from 12 to 38 (9–37 digestive) and CAB genes from 4 to 12 (2–11, digestive), many of the latter coding CABs without complete occluding loops.

The 3D structural differences observed in *T. molitor* CALs on adapting to a digestive function were investigated among other Coleoptera. Thus, the 3D structures of digestive and lysosomal CALs of *Z. subfasciatus* (identified as described above) were compared by a structure alignment-based clustering of those proteins based on the overall structures using the STRALCP web server (Zemia et al. 2007). The results showed that digestive ZsCALs differ widely from the lysosomal ZsCAL, although less than in the case of *T. molitor*. This finding suggests that on adapting to luminal digestion, *Z. subfasciatus* CALs diverged less in their 3D structures from the ancestral lysosomal CAL than those of *T. molitor* (Silva et al. 2022). The technique of structure alignment-based clustering of proteins is an easy method to evaluate differences in the overall structures of proteins, in spite of not permitting to compare isolated subsites. For this, it is necessary the use of techniques of molecular dynamics.

Selective pressures that resulted in the recruitment of lysosomal CALs as digestive enzymes were at first thought to be an adaptation to ingest seeds rich in serine endopeptidase inhibitors (Terra and Critofoletti, 1996). However, none of the seed (or seed flour) feeding Cucujiformia (Chrysomelidae, Curculionidae, and Tenebrionidae) are basal Cucujiformia taxa (McKenna et al. 2019). Thus, a new proposal was advanced by Silva et al. (2022). According to them, CAL and CAB recruitment occurred as an adaptation to ingesting pollen at the beginning of Cucujiformia diversification as pollinators (McKenna et al. 2019). Pollen, as seeds, are rich in serine endopeptidase inhibitors (Höllbacher et al. 2017).

9.3 Digestive Cathepsins L and B in Hemiptera

Cysteine endopeptidases as digestive enzymes in Hemiptera were at first described in true bugs (Heteroptera), including blood feeder *Rhodnius prolixus* (Houseman and Downe 1980; Terra et al. 1988) and seed feeders *Dysdercus peruvianus* (Silva and Terra 1994) and *Pyrrhocoris apterus* (Kodrick et al. 2012). According to Houseman and Downe (1980), the Hemiptera common ancestral species lost its digestive serine endopeptidases on adapting to a low-protein plant sap; a characteristic maintained in the order. On returning to a high protein diet, the hemipterans recruited lysosomal cysteine endopeptidases as digestive enzymes. The expansion of CAB genes is found in all hemipterans, attaining within Sternorrhyncha up to 15 genes in a single species genome, whereas CAL genes expansion is characteristic of Heteroptera (Rispe et al. 2008; Pimentel et al. 2020; Sparks et al. 2020). The best-known hemipteran CAB and CAL are those from *Diaphorini citri* (Ferrara et al. 2015) and *D. peruvianus* (Pimentel et al. 2020), respectively.

The identification of hemipteran digestive and lysosomal CALs and CABs and the comparison of their overall and subsite structures were performed as described previously for the enzymes of Coleoptera (see 9.2). *D. peruvianus* have one lysosomal and several digestive CALs that are more expressed in the middle and posterior midgut and CABs more expressed in the anterior midgut. *R. prolixus* data are similar to *D. peruvianus* if we take into account that the anterior midgut in this insect is only a storage region. Thus, the middle midgut in *R. prolixus* corresponds to the anterior midgut in *D. peruvianus*, characterized by a higher expression of CAB, whereas its posterior midgut is alike that region in *D peruvianus*. *Mahanarva fimbriolata* lacks digestive CALs and their CABs are expressed mainly in the anterior midgut (Pimentel et al. 2020).

Expression, overall structure, and subsite structure data led to the proposal that in Heteroptera (exemplified by *R. prolixus* and *D. peruvianus*) CABs are necessary in the first steps of protein digestion, followed by the action of CALs with substrate less specific than the corresponding lysosomal CAL. In Auchenorrhyncha (exemplified by *M. fimbriolata*), there are no digestive CALs, and the occurring CABs are supposed to be associated only with the inactivation of sap noxious proteins (Pimentel et al. 2020). The *Acyrthosiphon pisum* (Sternorrhyncha) genome has a single gene coding for CAL. In spite of that, a digestive CAL has been described in this insect (Cristofoletti et al. 2003; Deraison et al. 2004), putatively digesting noxious proteins.

The apparent advantage of CABs over CALs on starting the digestion of native proteins is supposed to rely on their dipeptidyl carboxypeptidase activity, through which CABs progress along the substrate polypeptide chain, causing the stepwise unfolding of the protein. As a consequence, the hindered substrate peptide bonds become available, turning the substrate more susceptible to the endopeptidase activity of CAB itself and to other endopeptidases (Pimentel et al. 2020)

9.4 Digestive Cathepsin D in Diptera Cyclorrhapha, Hemiptera, and Coleoptera

Greenberg and Paretsky (1955) found a high proteolytic activity at pH 2.5–3.0 in whole body homogenates of *Musca domestica*, which they identified as "fly pepsin." Later on, Lemos and Terra (1991) showed that the activity was by midgut-originated CAD. Padilha et al. (2009) identified three CADs expressed in *M. domestica* (MdCAD1, MdCAD2, and MdCAD3), all having the catalytic Asp33 and Asp 229 (bovine cathepsin D numbering) and the conserved substrate-binding pockets. MdCAD1 is expressed in all tissues and has the proline loop with the motif DxPxPx(G/A)P), as the vertebrate lysosomal cathepsin D. MdCAD 2 and MdCAD3 are expressed only in midguts and lack the proline loop, thus resembling vertebrate pepsin. MdCAD3 was produced in recombinant form, which was autocatalytically activated and used to produce antibodies in a rabbit. With these antibodies, MdCAD3 was shown to be secreted from midgut cells by exocytosis and was immunolocalized in midgut contents (Padilha et al. 2009).

The recruitment of CADs for midgut luminal digestion in *M. domestica* is thought to be an adaption to deal with bacteria-rich food (dung in nature, fermented meal in laboratory cultures) in an acid region of the midgut, in parallel to what happened with vertebrates. The loss of the proline loop may be associated with the extracellular role of both pepsin and digestive MdCAD3.

Among hemipterans, CADs were at first described in *R. prolixus* midgut contents (Houseman and Downe 1982), where they account for around 15% of the total proteolytic activity (Terra et al. 1988). Digestive CADs were also found in other hemipterans like *Oncopeltus fasciatus* (Defferrari et al. 2011) and *Triatoma infestans* (Balczun et al. 2012), which have one digestive CAD with and another without a proline loop. In *D. peruvianus*, there are 10 genes coding for CADs, from which 9 express digestive CADs detected in midgut contents by proteomics. CADs amount to 40% of the total proteolytic activity in the anterior midgut, which coding genes have their expression modulated by feeding (Pimentel et al. 2017). CADs together with CABs are supposed to hydrolyze cysteine endopeptidase inhibitors known to be present in the cotton seeds ingested by the insect, in the anterior midgut, before the food bolus meets CALs in the middle midgut (Pimentel et al. 2017). This proposal is supported by the finding that seed beetles (Bruchinae) overexpress CAB (Koo et al. 2008) and CAD (Ahn and Zhu-Salzman 2009) on ingesting cystatin, a CAL inhibitor.

The expansion of genes coding for CADs in Coleoptera is much more discreet than that found for CALs and CABs. There are, as a rule, two genes coding for CADs. In the case there are more than two CAD genes, some of them code CADs without a proline loop (Silva et al. 2022).

9.5 Targeting Recruited Lysosomal Enzymes to Midgut Lumen

The mechanism of recruitment of lysosomal proteins to midgut lumen is thought to be a modification of the mechanism of sorting those proteins to lysosomes. In mammals, CALs are sorted to lysosomes using the mannose 6-phosphate (M6P) pathway. In this pathway, M6P are added to the enzyme surface and were subsequently recognized by specific receptors in the Golgi apparatus, which finally direct the enzymes to lysosomes. Mammalian M6P machinery is complete in the genome of some primitive Metazoa but is almost entirely lost in Annelida and Insecta (Kumar and Bhamidimarri 2015; Damasceno et al. 2017; Terra et al. 2018). Apparently, there is an evolutionary trend resulting in the substitution of the M6P ancestral sorting mechanism by another under investigation (Dennes et al. 2005; Hasanagic et al. 2015; Terra et al. 2018). It has been hypothesized that overexpression would saturate specific cell receptors acting on sorting and direction of intracellular proteins, resulting in moving them to midgut lumen (Kane et al. 1988; Brömme and Wilson 2011; Damasceno et al. 2017). However, without a known mechanism for lysosomal targeting of the lysosomal enzymes, the question of how those enzymes may be re-directed to midgut lumen remains without a clear answer.

References

Ahn JE, Zhu-Salzman K (2009) CmCatD, a cathepsin D-like protease has a potential role in insect defense against a phytocystatin. J Insect Physiol 55:678–685

Balczun C, Siemanowski J, Pausch JK et al (2012) Intestinal aspartate proteases TiCatD and TiCatD2 of the haematophagous bug *Triatoma infestans* (Reduviidae): sequence characterisation, expression pattern and characterisation of proteolytic activity. Insect Biochem Mol Biol 42:240–250

Barrett AJ, Rawlings ND, Woessner JF (2013) Handbook of proteolytic enzymes, 3rd edn. Academic, New York

Benes P, Vetvicka V, Fusek M (2008) Cathepsin D--many functions of one aspartic protease. Crit Rev Oncol Hematol 68:12–28

Ben-Mahmoud S, Ramos JE, Shatters RG Jr et al (2015) Cloning and characterization of a basic cysteine-like protease (Cathepsin L1) expressed in the gut of larval *Diaprepes abbreviatus* L. (Coleoptera: Curculionidae). J Insect Physiol 72:1–13

Beton D, Guzzo CR, Ribeiro AF et al (2012) The 3D structure and function of digestive cathepsin L-like proteinases of *Tenebrio molitor* larval midgut. Insect Biochem Mol Biol 42:655–664

Brömme D, Wilson S (2011) Role of cysteine cathepsins in extracellular proteolysis. In: Parks WC, Mecham RP (eds) Extracellular matrix degradation, biology of extracellular matrix 2. Springer, Berlin, pp 23–51

Campos FAP, Xavier-Filho J, Silva CP et al (1989) Resolution and partial characterization of proteinases and α-amylases from midguts of larvae of the bruchid beetle *Callosobruchus maculatus* (F.). Comp Biochem Physiol B 92:51–57

Cristofoletti PT, Ribeiro AF, Deraison C et al (2003) Midgut adaptation and digestive enzyme distribution in a phloem feeding insect, the pea aphid *Acyrthosiphon pisum*. J Insect Physiol 49:11–24

Cristofoletti PT, Ribeiro AF, Terra WR (2005) The cathepsin L-like proteinases from the midgut of *Tenebrio molitor* larvae: sequence, properties, immunocytochemical localization and function. Insect Biochem Mol Biol 35:883–901

Damasceno TF, Dias RO, de Oliveira JR et al (2017) Active subsite properties, subsite residues and targeting to lysosomes or midgut lumen of cathepsins L from the beetle *Tenebrio molitor*. Insect Biochem Mol Biol 89:17–30

Defferrari MS, Demartini DR, Marcelino TB et al (2011) Insecticidal effect of *Canavalia ensiformis* major urease on nymphs of the milkweed bug *Oncopeltus fasciatus* and characterization of digestive peptidases. Insect Biochem Mol Biol 41:388–399

Dennes A, Cromme C, Suresh K et al (2005) The novel drosophila lysosomal enzyme receptor protein mediates lysosomal sorting in mammalian cells and binds mammalian and drosophila GGA adaptors. J Biol Chem 280:12849–12857

Deraison C, Darboux I, Duportets L et al (2004) Cloning and characterization of a gut-specific cathepsin L from the aphid Aphis gossypii. Insect Mol Biol 13:165–177

Dunn BM, Jimenez M, Parten BF et al (1986) A systematic series of synthetic chromophoric substrates for aspartic proteinases. Biochem J 237:899–906

Dvoryakova EA, Klimova MA, SimonyanTR et al. (2022) Recombinant cathepsin L of *Tribolium castaneum* and its potential in the hydrolysis of immunogenic gliadin peptides. Int J Mol Sci 23, 7001. doi:https://doi.org/10.3390/ijms231237001

Ferrara TF, Schneider VK, Kishi LT et al (2015) Characterization of a recombinant Cathepsin B-like cysteine peptidase from *Diaphorina citri* Kuwayama (Hemiptera: Liviidae): a putative target for control of citrus Huanglongbing. PLoS One 10:e0145132

Fonseca FPP, Soares-Costa A, Ribeiro AF et al (2012) Recombinant expression, localization and in vitro inhibition of midgut cysteine peptidase (SL-CathL) from sugar cane weevil, *Sphenophorus levis*. Insect Biochem Mol Biol 42:58–69

Gatehouse AMR, Butler KJ, Fenton KA et al (1985) Presence and partial characterization of a major proteolytic enzyme in the larval gut of *Callosobruchus maculatus*. Entomol Exp Appl 39:279–286

Goptar IA, Semashko TA, Danilenko SA et al (2012) Cysteine digestive peptidases function as post-glutamine cleaving enzymes in tenebrionid stored-product pests. Comp Biochem Physiol B Biochem Mol Biol 161:148–154

Greenberg B, Paretsky D (1955) Proteolytic enzymes in the house fly, *Musca domestica* (L.). Ann Entomol Soc Am 48:46–50

Hasanagic M, van Meel E, Luan S et al (2015) The lysosomal enzyme receptor protein (LERP) is not essential, but is implicated in lysosomal function in *Drosophila melanogaster*. Biol Open 4:1316–1325

Höllbacher B, Schmitt AO, Hofer H et al (2017) Identification of proteases and protease inhibitors in allergenic and non-allergenic pollen. Int J Mol Sci 18:1199

Houseman JG, Downe AER (1980) Endoproteinase activity in the posterior midgut of the *Rhodnius prolixus* Stal (Hemiptera: Reduviidae). Insect Biochem 10:363–366

Houseman JG, Downe AER (1982) Characterization of an acidic proteinase from the proteinase midgut of *Rhodnius prolixus* Stal (Hemiptera: Reduviidae). Insect Biochem 12:651–655

Illy C, Quraishi O, Wang J et al (1997) Role of occluding loop in cathepsin B activity. J Biol Chem 272:1197–1202

Kane SE, Troen BR, Gal S et al (1988) Use of cloned multidrug resistance gene for coamplification and overproduction of major excreted protein, a transformation–regulated secreted acid protease. Mol Cell Biol 8:3316–3321

Kitch LW, Murdock LL (1986) Partial characterization of a major thiol proteinase from larvae of *Callosobruchus maculatus* F. Arch Insect Biochem Physiol 3:561–575

Kodrick D, Vinokurov K, Tomcala A et al (2012) The effect of adipokinetic hormone on midgut characteristic in *Pyrrhocoris apterus* L. (Heteroptera). J Insect Physiol 58:194–204

Koo YD, Ahn D-J, Salzman RA et al (2008) Functional expression of an insect cathepsin B-like counter-defense protein. Insect Mol Biol 17:235–245

Kumar NS, Bhamidimarri PM (2015) Lysosomal enzymes and their receptors in invertebrates: an evolutionary perspective. Curr Protein Pept Sci 16:49–65

Kyte J, Doolittle RF (1982) A simple method for displaying the hydropathic character of a protein. J Mol Biol 157:105–132

Lemos FJA, Terra WR (1991) Properties and intracellular distribution of a cathepsin-D-like proteinase active at the acid region of *Musca domestica* midgut. Insect Biochem 21:457–465

Lemos FJA, Campos FAP, Silva CP et al (1990) Proteinases and amylases of larval midgut of *Zabrotes subfasciatus* reared on cowpea (*Vigna unguiculata*) seeds. Entomol Exp Appl 56:219–227

Lopes AR, Juliano MA, Marana SR et al (2006) Substrate specificity of insect trypsins and the role of their subsites in catalysis. Insect Biochem Mol Biol 36:130–140

Martynov AG, Elpidina EN, Perkin L et al (2015) Functional analysis of C1 family cysteine peptidases in the larval gut of *Tenebrio molitor* and *Tribolium castaneum*. BMC Genomics 16:75

McKenna DD, Shina S, Ahrens D et al (2019) The evolution and genomic basis of beetle diversity. Proc Natl Acad Sci U S A 116:24729–24737

Musil D, Zucic D, Turk D et al (1991) The refined 2.15 A X-ray crystal structure of human liver cathepsin B: the structural basis for its specificity. EMBO J 10:2321–2330

Nagler DK, Storer AC, Portaro FC et al (1997) Major increase in endopeptidase activity of human cathepsin B upon removal of occluding loop contacts. Biochemist 36:12608–12615

Nishimura Y, Kawabata T, Furuno K et al (1989) Evidence that aspartic proteinase is involved in the proteolytic processing event of procathepsin L in lysosomes. Arch Biochem Biophys 271:400–406

Padilha MH, Pimentel AC, Ribeiro AF et al (2009) Sequence and function of lysosomal and digestive cathepsin D-like proteinases of *Musca domestica* midgut. Insect Biochem Mol Biol 39:782–791

Pimenta DC, Oliveira A, Juliano MA et al (2001) Substrate specificity of human cathepsin D using internally quenched fluorescent peptides derived from reactive loop of kallistatin. Biochim Biophys Acta 1544:113–122

Pimentel AC, Fuzita FJ, Palmisano G et al (2017) Role of cathepsins D in the midgut of *Dysdercus peruvianus*. Comp Biochem Physiol B Biochem Mol Biol 204:45–52

Pimentel AC, Dias RO, Bifano TD et al (2020) Cathepsins L and B in *Dysdercus peruvianus*, *Rhodnius prolixus*, and *Mahanarva fimbriolata*. Looking for enzyme adaptations to digestion. Insect Biochem Mol Biol 127:103488

Prabhakar S, Chen MS, Elpidina EN et al (2007) Sequence analysis and molecular characterization of larval midgut cDNA transcripts encoding peptidases from the yellow mealworm, *Tenebrio molitor* L. Insect Mol Biol 16:455–468

Puzer L, Cotrin SS, Alves MFM et al (2004) Comparative substrate specificity of recombinanat cathepsin V and cathepsin L. Archs Biochem Biophys 430:274–283

Rispe C, Kutsukake M, Doublet V et al (2008) Large gene family expansion and variable selective pressures for cathepsin B in aphids. Mol Biol Evol 25:5–17

Schechter I, Berger A (1967) On the size of the active site in proteases. I Papain Biochem Biophys Res Commun 27:157–162

Shewry PR, Halford NG (2002) Cereal and seed storage proteins: structure,properties, utilization. J Exp Bot 53:947–958

Silva CP, Terra WR (1994) Digestive and absorptive sites along the midgut of the cotton seed sucker bug *Dysdercus peruvianus* (Hemiptera: Pyrrhocoridae). Insect Biochem Mol Biol 24:493–505

Silva CP, Xavier-Filho J (1991) Comparison between the levels of aspartic and cysteine proteinases of the larval midguts of *Callosobruchus maculatus* (F) and *Zabrotes subfasciatus* (Boh) (Coleoptera: Bruchidae). Comp Biochem Physiol B 99:529–533

Silva CP, Dias RO, Bernardes V et al (2022) Recruitment of lysosomal cathepsins B, L, and D as digestive enzymes in Coleoptera. Insect Mol Biol 31:225–240

Sparks ME, Bansal R, Benoit JB et al (2020) Brown marmorated stink bug, *Halyomorpha halys* (Stål), genome: putative underpinnings of polyphagy, insecticide resistance potential and biology of a top worldwide pest. BMC Genomics 21:227

Terra WR, Cristofoletti PT (1996) Midgut proteinases in three divergent species of Coleoptera. Comp Biochem Physiol B Biochem Mol Biol 113:725–730

Terra WR, Ferreira C, Garcia ES (1988) Origin, distribution, properties and functions of the major *Rhodnius-prolixus* midgut hydrolases. Insect Biochem 18:423–434

Terra WR, Dias RO, Oliveira PL et al (2018) Transcriptomic analyses uncover emerging roles of mucins, lysosome/secretory addressing and detoxification pathways in insect midguts. Curr Op Insect Sci 29:34–40

Terra WR, Dias RO, Ferreira C et al (2019) Recruited lysosomal enzymes as major digestive enzymes in insects. Biochem Soc Trans 47:615–623

Turk D, Guncar C, Podonik M et al (1998) Revised definition of substrate binding sites of papain-like cysteine proteases. Biol Chem 379:137–147

Wieman KF, Nielsen SS (1988) Isolation and partial characterization of a major gut proteinase from larval *Acanthoscelides obtectus* Say (Coleoptera: Bruchidae). Comp Biochem Physiol B 89:419–426

Withers SG, Rupitz K (1990) Measurement of active-site homology between potato and rabbit muscle alpha-glucan phosphorylases through use of a linear free energy relationship. Biochemist 29:6405–6409

Zemia A, Geisbrecht B, Smith J et al (2007) STRALCP structure alignment-based clustering of proteins. Nuc Acid Res 35:e150

Chapter 10
Plant, Bacterial, and Fungal Cell Wall-Degrading Enzymes

Abstract Most insects have plant material in their diets. Plant cell walls are broken by mastication or under the action of plant cell wall-degrading enzymes. Primary plant cell walls, as those of grasses, are composed of cellulose, pectin, and a network of polysaccharides named hemicelluloses. Secondary plant cell walls, as those in wood, are formed by lignocellulose, which are cellulose and hemicelluloses cross-linked by lignin. Insect cellulases are always β-1,4-endoglucanases of family GH9 or GH45. GH9 cellulases are widespread but were lost in dipterans and lepidopterans. Those of GH45 were acquired by beetles of the clade Phytophaga by horizontal transfer from fungi. Despite the existence of endogenous cellulases, in many insects microbiota cellulases also concur. Pectinases are particularly important in hemipterans to facilitate the insertion of their stylets in sap-conducting structures and to beetles that bore plant tissues. Hemicellulases hydrolyze hemicelluloses. They are licheninases, laminarinases, xylanases, and mannanases. Laminarinases are widespread among insects and hydrolyze β-1,3-glucans (laminarins) and some of them, like the one of *Tenebrio molitor*, also hydrolyze yeast β-1,3-1,6-glucans. Laminarinases are supposed to digest fungal cells in contaminated food (*T. molitor)* or callose (lepidopterans), which is deposited in response to wounding caused by the larvae and that impairs nutrient availability for the larvae. The degradation of lignin can only be efficiently performed by oxidative depolymerization catalyzed by laccases aided by redox mediators that usually are produced by microbes. Because of that, most insects only attack wood partly digested by microbes. Fungi are nutrients for detritivorous and stored product insects. Digestive chitinases lack chitin-binding domains, so they are efficient in digesting fungi cell walls but are harmless for the peritrophic membrane. Lysozyme catalyzes the hydrolysis of the peptidoglycan of the cell walls of many bacteria. In insects, midgut lysozyme active in low pH is characteristic of cyclorrhaphous dipterans, in agreement with the fact that most of their larvae feed largely on bacteria. Insect digestive cellulases were studied in detail, including crystallography and resolution of their 3D structures.

W. R. Terra et al., *Molecular Physiology and Evolution of Insect Digestive Systems*, Entomology in Focus 7, https://doi.org/10.1007/978-3-031-39233-7_10

10.1 Cellulases

Most insects have plant materials in their diets in the form of detritus or plant parts like leaves, trunks, and seeds (see Chap. 3). Plant cell walls are broken down by mastication, but frequently it is caused by plant cell wall-degrading enzymes. The major constituents of the plant's primary cell walls, such as those of grasses and plant leaves, are cellulose and pectin to which associate a network of polysaccharides named hemicelluloses. These include xylans, xyloglucans, and mannans. Secondary cell walls, like in wood, are strengthened by lignin cross-linked to cellulose and hemicelluloses, forming a composite named lignocellulose (Pettersen 1984; Cosgrove 2005; Terra et al. 2019; see also Chap. 3).

Cellulose is a non-ramified β-1,4-glucan chain forming compact aggregates arranged in crystalline form stabilized by hydrogen bonds. As a consequence, cellulose is not soluble and is difficult to disrupt. In microbial systems, cellulose is digested by the action of two enzymes. An endo-β-1,4-glucanase (EC 3.2.1.4) breaks internal bonds in the amorphous regions of cellulose. This step is followed by the action of an exo-β-1,4-glucanase (EC 3.2.1.91) (cellobiohydrolase), which releases cellobiose from the non-reducing end of the cellulose chain. The enzyme is processive, which means that it releases cellobiose in a series of catalytic rounds without dissociating from the cellulose chain. The cellobiohydrolase comprises a catalytic domain linked to a cellulose-binding module, which enhances the activity toward insoluble cellulose. Cellobiohydrolase has surface loops that prevent the displaced cellulose chains from re-associating to the crystal surface on progressing into crystalline cellulose (Rouvinen et al. 1990; Kleywegt et al. 1997; Linder and Teeri 1997).

The final products of cellulase are glucose and the disaccharide cellobiose, this one being hydrolyzed by β-glucosidases, also known as cellobiase. Although β-glucosidases are usually discussed as part of cellulase systems, actually they are also active on products of the digestion of hemicelluloses (see below) and also of plant glycosides. Because of that, β-glucosidases were discussed together with the ordinary enzymes in Chap. 4.

For a long time, it was believed that animals are not able to digest cellulose, and where cellulose digestion was found, it was supposed to be carried out by symbiotic organisms. After the finding of a cellulose-coding gene in termites (Watanabe et al. 1998), there were growing evidence that β-1,4-endoglucanases from GH9 family (see Chap. 4) occur, besides in termites (Watanabe et al. 1998), in crickets (Kim et al. 2008), in several families of beetles (Tenebrionidae: Willis et al. 2011; Cerambycidae: Busconi et al. 2014; McKenna et al. 2016), and in stick insects (Phasmatodea: Shelomi et al. 2016a, 2016b). Despite the existence of endogenous cellulases in many insects, in some of them, mainly termites, cellulases from symbiotic organisms also concur (Brune 2014; Tokuda 2019; see also Chap. 12).

The best-known insect β-1,4-endoglucanases are the ones from the termites *Reticulitermes speratus* and *Nasutitermes takasagoensis* and the one from the woodroach *Panesthia cribata* (Lo et al. 2000; Watanabe and Tokuda 2010). The β-1,4-endoglucanases from *Nasutitermes takasagoensis* had their 3D structure resolved (Khadeni et al. 2002). The enzyme has the fold $(\alpha/\alpha)_6$ barrel, common in GH9 enzymes, the conserved active site residues (Asp55 as a nucleophile and Glu424 as a proton donor), and the Ca^{2+}-binding site near its substrate binding cleft, but it lacks the cellulose-binding module. Stick insects have multiple copies of cellulase genes, and some of them code for bifunctional cellulases that are able to hydrolyze amorphous cellulose and xylan or xyloglucan (Shelomi et al. 2016a).

The β-1,4-endoglucanases of the GH9 family were lost independently in Diptera and Lepidoptera (Watanabe and Tokuda 2010), probably associated with the fact that these insects grow very fast, in which case their dietary requirements may be met by more easily digestible food constituents than crystalline cellulose. This is exemplified by the larva of the lepidopteran *Erinnyis ello* feeding on *Euphorbia pulcherrima* leaves. Those larvae do not digest cellulose nor pectin but increase their weight by 30% a day, consuming 2.6-fold their weight in leaves a day (Terra et al. 1987). Inspection of *E. ello* fecal boluses under the microscope reveals pieces of leaf tissue with empty cells, but maintaining apparently preserved cell walls. This means that mastication may be efficient enough to release the cell contents without the aid of cellulase. Thus, true cellulose digestion is necessary only in insects with low growth rates, mainly in nutritionally poor diets like humus and wood, exemplified by woodroaches, termites, and scarabaeid and cerambycid beetles.

Beetles of the Phytophaga clade (families Chrysomelidae, Cerambycidae, and Curculionidae) acquired genes coding for GH45 β-1,4-endoglucanase by horizontal transference from fungi (Busch et al. 2019). These genes went duplication and some of them evolved to code for bi-functional β-1,4-endoglucanase/glucomannase and others changed to code for xyloglucanases of GH 5 (Watanabe and Tokuda 2010).

Whichever GH family insect cellulases pertain, they are always β-1,4-endoglucanases, except for a single report of an exo-β-1,4-glucanase in beetles (Chang et al. 2012). Slaytor (1992) proposed that the low efficiency of endoglucanases from termites and woodroaches on crystalline cellulose is compensated by a large production of them. Anyhow, mechanistic details of how insect cellulases act on crystalline cellulose are lacking.

10.2 Pectinases

Pectin consists of several polysaccharides known as pectic polysaccharides. The most abundant is homogalacturonan formed by chains of galacturonic acids linked by α-1,4-glycosidic bonds, which may be highly methylated. Other pectic polysaccharides are rhamnogalacturonan-I, consisting of a chain of galacturonic and rhamnose residues with different monosaccharides as side chains and other more complex structures. Details on the structure of pectin may be found in Chap. 3. Pectin amounts to about 25–30% of the polysaccharides in primary cells, whereas in secondary cells, it makes up less than 10% of the total polysaccharides (Tokuda 2019).

Pectinases refer to the different enzymes that hydrolyze pectin, including polygalacturonidases (EC 3.2.1.15), pectin methylesterases, and pectin lyases. Pectinases are considered to be important for hemipterans, since they make easier the insertion of their stylets into sap-conducting structures and also for insects that bore plant tissues, like many beetles. Accordingly, pectinases have been found in hemipteran saliva (Vonk and Western 1984) and coleopteran guts. The pectinases from the coleopterans *S. oryzae* (Shen et al. 1996) and *Diaprepes abbreviatus* (Dootstar et al. 1997) were purified in several chromatographic steps and the one from *Phaedon cochleariae* was produced in recombinant form Girard and Jouanin (1999a). The pectinases have a pH optimum of 5.5 and the last one is from the GH28 family, conserves the signature centered in active site residue His223, and is similar to the fungal endogalacturonase (Girard and Jouanin 1999a; Markovic and Janecek 2001). The pectinase from *D. abbreviatus* is inhibited by a polygalacturonase-inhibitor protein thought to be associated with plant resistance to insects (Dootstar et al. 1997)

In the curculionid beetle *Sitophilus oryzae*, pectin is demethylated by a pectin methylesterase (with a 3D structure resolved, Teller et al. 2014) acting synergistically with a polygalacturonase on xylogalaturonan in a way similar with microbial enzymes. This finding together with transcriptomic and RNA-seq data supported the hypothesis that genes coding for pectinases were transferred from bacteria to insects before the evolution of stick insects (Kirsch et al. 2014; Shelomi et al. 2016b).

10.3 Hemicellulases

Hemicellulases hydrolyze hemicelluloses, which comprise licheninases, laminarinases, xylanases, and mannanases. Licheninases (EC 3.2.1.73) hydrolyze only β-1,3;1,4-glucans (lichenins), whereas laminarinases may hydrolyze β-1,3;1,4-glucans and also β-1,3-glucans (laminarins) or only the last glucan (EC 3.2.1.39). β-1,3;1,4-glucans are cereal β-glucans (Bacic et al. 1988) and β-1,3-glucans are present in callose, a polysaccharide deposited immediately after a tissue damage caused by the insect larvae themselves or by infection by pathogens (Radford et al. 1998) and in fungal cell walls (Bacic et al. 1988).

Laminarinases are widespread among insects (Terra and Ferreira 1994) and some of them were purified and characterized. Three salivary laminarinases were isolated from *Periplaneta americana:* two digest laminarin and lichenin and the other only laminarin (Genta et al. 2003). A major laminarinase was found in the midguts of *Abracris flavolineata* (Genta et al. 2007), *T. molitor* (Genta et al. 2009), and *Spodoptera frugiperda* (Bragatto et al. 2010). The enzymes from *A. flavolineata* and *S. frugiperda* digest only laminarin and that of *T. molitor* hydrolyses, besides laminarin, yeast β-1,3-β-1,6-glucan. The laminarinases of all those insects lyse *Saccharomyces cerevisiae* cells and have pH optima from 5.0 to 6.5, except the enzyme from *S. frugiperda*, which has a pH optimum of 9.0 and which does not lyse fungal cells. The exo-β-1,3-glucanase of *Abracris flavolineata* is processive, that is, they perform multiple rounds of catalysis while the enzyme remained attached to the substrate. Processivity is thought to result from consecutive transferences of substrate between a high-affinity accessory site and the active site (Genta et al. 2007).

Both insect laminarinases and Gram-negative bacteria-binding proteins (GNBP) and other β-glucan-binding proteins active in the insect innate immune system are derived from the laminarinase of the ancestor of mollusks and arthropods. β-glucan-binding proteins lost the catalytical residues and the insect laminarinases, an extended N-terminal region of the ancestral laminarinase (Bragatto et al. 2010).

Two roles are proposed for laminarinases in insect guts (Terra and Ferreira 2012). In insects feeding on contaminated food, like *T. molitor* and *P. americana*, β-1,3 glucanases are supposed to digest the fungal cell walls from their diet. Otherwise, in insects eating on plants, exemplified by lepidopterans, β-1,3-glucanase may be important to digest callose, which is deposited in response to wounding caused by the larvae themselves or by infection by pathogens. In the absence of a β-1,3-glucanase, callose deposition can impair nutrient digestibility. The increase in β-1,3-glucanase activity in *Helicoverpa armigera* after fungi or bacteria ingestion was supposed by Pauchet et al. (2009) to have a role in immunity. However, it is also probable that the enzyme activity increase is caused instead by an increase in substrate concentration. Further studies are necessary to clarify the question.

Xylanases are enzymes active on xylans, major hemicelluloses that account for one-third of all renewable organic carbon available on earth (Bacic et al. 1988). Xylans are major hemicelluloses in land plants, which main chains are usually made of β-1,4-xylopyranose units with various substitutions of arabinose, galactose, glucuronic acid, etc. (Pollet et al. 2010). Xylans in woods are called glucuronoxylans, because they contain large levels of 4-methyl glucuronic acid, and those of cereals are referred to as arabinoxylan, due to the large amount of arabinose they have attached. Xylan-degrading enzymes include xylanases (endo-β-1,4-xylanases, EC 3.2.1.8), β-D-xylosidases (EC 3.2.1.37), α-L-arabinofuranosidases (EC 3.2.1.55), α-D-glucuronidases (EC 3.2.1.139), and some esterases. Most xylanases are classified into the glycoside hydrolase families GH10 and GH11 and less frequently into GH5. One of these enzymes was recombinantly expressed from a beetle (*Phaedon cochlearae*) and shown to have a high sequence identity to fungal xylanases and to have the two conserved catalytic regions (Girard and Jouanin 1999a). In a cerambycid (*Apriona japonica*), xylanase (GH5) was heterologously expressed and enzymatically characterized (Pauchet et al. 2014). Whereas the previous enzymes were all endoxylanases, an exo-β-1,4-xylanase (EC 3.2.1.37) was partially purified from termites (Matoub and Rouland 1995), apparently acting synergistically with an endo-β-1,4-xylanase coming from a fungus ingested by the termites. Xylanase was also produced by symbiotic flagellates in the hindguts of termites (Arakawa et al. 2009, see also Chap. 13). Much more work is needed on this class of enzymes that may be important mainly for detritivorous insects.

Mannans are polymers of mannose units with β-1,4-glycoside bonds, whereas glucomannans include some glucan residues in the backbone and in galactomannans the main chain is substituted with α-1,6-bound galactose residues (Soni and Kango 2013). Few digestive insect mannanases have been described. The one found in the coffee berry borer beetle, *Hypothenemus hampei*, was recombinantly expressed and supposed to hydrolyze galactomannan, the major coffee storage polysaccharide (Acuña et al. 2012). The gene coding for this enzyme was thought to be acquired from bacteria by horizontal gene transfer. More recently, Busch et al. (2017) making use of recombinant expression and enzyme characterization, showed that several proteins from GH10 (subfamily 5) from the beetle *Gastrophysa viridula* are endo-β-1,4-mannanases, one of which has additional activity on carboxymethylcellulose that it is an endo-β-1,4-glucanase activity (Busch et al. 2017). In spite of the recent studies, insect mannanases and xylanases demand more kinetic and structural studies as done for xylanases from other sources (Pollet et al. 2010).

10.4 Laccases

Lignin is a biopolymer that is extensively cross-linked to both cellulose and hemicelluloses that strengthen the cell walls, mainly in wood (Pettersen 1984, Chap. 3). The degradation of lignin polymers can only be efficiently performed by oxidative depolymerization catalyzed by laccases (p-diphenol:dioxygen oxireductase, EC 1.10.3.2), which are member of the family of multicopper oxidases (Dittmer and Kanost 2010). Most laccases need extracellular redox mediators to disrupt lignocelluloses, that, as a rule, are not synthesized by insects, because of their aromatic rings (Coy et al. 2010; Eggert et al. 1996; Sethi et al. 2013). The lack of self-synthesizing redox mediators is circumvented by beetles attacking wood partially digested by wood-rotting microbes. However, the cerambycid beetle *Anoplophora glabripennis* is able to attack living trees and perform all necessary reaction steps within their guts (Scully et al. 2013). This apparent capacity of *A. glabripennis* demands further research.

10.5 Exinase

Pollen grains have a refractory exine (outer shell) that is resistant to digestion (Chap. 3). However, a Collembola was reported to have exinase active on exine (Scott and Stojanovich 1963). This finding demands further attention, as up till now no enzyme with an activity on exine was described. Actually, most authors suppose that pollen grains burst and open at the pores due to osmotic shock caused by the low osmotic pressure of the midgut, followed by the outgrowth of pollen tubes. The primary cell walls of the tubes are then attacked by endoglucanases and pectinases, liberating the pollen content (Kroon et al. 1974; Klungness and Peng 1984a, b).

10.6 Chitinases and Lysozymes

Fungi are nutrients especially for detritivorous and stored product insects (see Chap. 3). The fungi cell wall is made of chitin, a β-1,4-homopolymer of N-acetylglucosamine, which is hydrolyzed by chitinolytic enzymes. Chitinase (EC 3.2.1.14) catalyzes the random hydrolysis of internal bonds in chitin, forming smaller oligosaccharides, from which N-acetylglucosamine units are liberated from their nonreducing ends under the action of β-N-acetyl-D-glucosaminidases (EC 3.2.1.52) (Arakane and Muthukrishan 2010). The best-known insect chitinase (GH18) is the molting fluid chitinase from the lepidopteran *Manduca sexta.* The enzyme has an N-terminal catalytic domain and a C-terminal chitin-binding domain interconnected by a serine/threonine-rich O-glycosylated linker. The chitin-binding domain enhances activity toward insoluble chitin, whereas the linker region helps to stabilize the enzyme in the presence of proteolytic enzymes (Lu et al. 2002; Arakane et al. 2003). The gut chitinase from *Anopheles gambiae has* a structure similar to that of *M. sexta,* is secreted upon blood-feeding, and modulates the thickness and permeability of the chitin-containing peritrophic membrane (Shen and Jacobs-Lorena 1997; see also Chap. 7).

The insect chitinases that are really digestive lack a chitin-binding domain so that they are efficient in digesting fungi cell walls but are harmless for the peritrophic membrane. They have been characterized, for example, in *P. cochleariae* (Girard and Jouanin 1999b), *T. molitor* (Genta et al. 2006), and *Periplaneta americana* (Tamaki et al. 2014). All of them pertain to group IV of chitinases (Arakane and Muthukrishan 2010) that as a rule lack a chitin-binding domain.

Lysozyme (EC 3.2.1.17) catalyzes the hydrolysis of the 1,4-β-glycosidic linkage between N-acetyl-muramic acid and N-acetylglucosamine of the peptidoglycan present in the cell wall of many bacteria (muramidase activity), causing cell lysis. There are four types of lysozymes according to their sequences (Van Herreghe and Michiels 2012), from which only two are found in insects. Despite the fact that lysozyme is part of a widespread immune defense mechanism against bacteria, lysozyme has been implied in the digestion of bacteria cell walls in animals ingesting large amounts of bacteria like some insects (Espinoza-Fuentes and Terra 1987) and marine bivalves or in those that maintain a bacterial culture in their guts like ruminants (Callewaert and Michiels 2010. In insects, acidic midgut digestive lysozymes seem to be an ancestral trait of cyclorrhaphous dipterans (Lemos and Terra 1991; Regel et al. 1998), in agreement with the fact that most of their larvae are saprophagous, feeding largely on bacteria. These insects have midgut lysozymes (Lemos et al. 1993; Regel et al. 1998) similar to those of vertebrate fermenters (ruminants). Thus, these enzymes are active at very low pH and are resistant to the cathepsin D-like aspartic proteinase present in midguts and have a weak chitinase activity (Lemos et al. 1993; Regel et al. 1998; Fujita 2004; Cançado et al. 2007, 2008). The lysozyme from *M. domestica* was isolated, characterized (Lemos et al. 1993), produced in recombinant form, and had its 3D structure resolved (Cançado et al. 2007). Site-directed mutagenesis confirmed the proposal advanced from 3D studies regarding which residues around the catalytic groups lead to a decrease of the pH optimum, from the usual neutral pH, to about pH 3. This decrease depends also on a less positively charged surface (Cançado et al. 2010). It is remarkable, however, that the digestive lysozyme from ruminants is less well adapted to function at very acidic pH than the fly lysozyme (Nonaka et al. 2009)

A lysozyme found in the salivary glands of *Reticulitermes speratus* by Fujita et al. (2001) was produced in recombinant form and shown to be active at neutral pH (Fujita et al. 2002), thus differing from the digestive cyclorrhaphan lysozymes. The termite enzyme is thought to be secreted by the salivary glands and to digest hindgut bacteria ingested from the feces (proctodeal trophallaxis) (Fujita et al. 2001). Another lysozyme differing from the cyclorrhaphous lysozymes is that of *Rhodnius prolixus,* which is also proposed to be digestive (Ursic-Bedoya et al. 2008).

References

Acuña R, Padilla BE, Flórez-Ramos CP et al (2012) Adaptive horizontal transfer of a bacterial gene to an invasive insect pest of coffee. Proc Natl Acad Sci U S A 10:4197–4202

Arakane Y, Muthukrishan S (2010) Insect chitinase and chitinase-like proteins. Cell Mol Life Sci 67:201–216

Arakane Y, Zhu Q, Matsumiya M et al (2003) Properties of catalytic, linker and chitin-binding domains of insect chitinase. Insect Biochem Mol Biol 33:631–648

Arakawa G, Watanabe H, Yamasaki HY et al (2009) Purification and molecular cloning of xylanases from the wood-feeding termite, *Coptotermes formosanus* Shiraki. Biosci Biotechnol Biochem 73:710–718

Bacic A, Harris PJ, Stone BA (1988) Structure and function of plant cell walls. In: Stumpf PK, Conn EE (eds) Biochemistry of plants, vol 14. Academic, New York, pp 297–371

Bragatto I, Genta FA, Ribeiro AF et al (2010) Characterization of a β-1,3-glucanase active in the alkaline midgut of *Spodoptera frugiperda* larvae and its relation to β-glucan-binding proteins. Insect Biochem Mol Biol 40:861–872

Brune A (2014) Symbiotic digestion of lignocellulose in termite guts. Nat Rev Microbiol 12:168–180

Busch A, Kunert G, Heckel DG et al (2017) Evolution and functional characterization of CAZymes belonging to subfamily 10 of glycoside family 5 (GH5_10) in two species of phytophagous beetles. PLoS 12:e0184305

Busch A, Danchin EGJ, Pauchet Y (2019) Functional diversification of horizontally acquired glycoside hydrolase family 45 (GH45) proteins in Phytpphaga beetles. BMC Evol Biol 19:100

Busconi M, Berzolla A, Chiappini E (2014) Preliminary data cellulase encoding genes in the xylophagous beetle, Hylotrupes bajulus (Linnaeus). Int Biodeter Biodegr 86:92–95

Callewaert L, Michiels C (2010) Lysozymes in the animal kingdom. J Biosci 35:127–160

Cançado FC, Valério AA, Marana SR et al (2007) The crystal structure of a lysozyme c from housefly *Musca domestica*, the first structure of a digestive lysozyme. J Struct Biol 160:83–92

Cançado FC, Chimoy Effio P, Terra WR et al (2008) Cloning, purification and comparative characterization of two digestive lysozymes from *Musca domestica* larvae. Braz J Med Biol Res 41:969–977

Cançado FC, Barbosa ARG, Marana SR (2010) Role of the triad N46, S106 and T107 and the surface charges in the determination of the acidic pH optimum of digestive lysozyme from *Musca domestica*. Comp Biochem Physiol B 155:387–395

Chang CJ, Wu CP, Lu SC et al (2012) A novel exo-cellulase from the white spotted longhorn beetle (*Anoplophora malasiaca*). Insect Biochem Mol Biol 42:629–636

Cosgrove DJ (2005) Growth of the plant cell wall. Nat Rev Mol Cell Biol 6:850–861

Coy MR, Salem TZ, Denton JS et al (2010) Phenol-oxidizing laccases from the termite gut. Insect Biochem Mol Biol 40:723–732

Dittmer NT, Kanost MR (2010) Insect multicopper oxidases: diversity, properties, and physiological roles. Insect Biochem Mol Biol 40:179–188

Dootstar H, McCollum TG, Mayer RT (1997) Purification and characterization of an endo-polygalaturonase from the gut of West Indies sugarcane rootstalk borer weevil (*Diaprepes abbreviatus* L.) larvae. Comp Biochem Physiol B 118:861–867

Eggert C, Temp U, Dean JF et al (1996) A fungal metabolite mediates degradation of non-phenolic lignin structures and synthetic lignin by laccase. FEBS Lett 39:144–148

Espinoza-Fuentes FP, Terra WR (1987) Physiological adaptations for digesting bacteria. Water fluxes and distribution of digestive enzymes in *Musca domestica* larval midgut. Insect Biochem 17:809–817

Fujita A (2004) Lysozymes in insects: what role do they play in nitrogen metabolism? Physiol Entomol 29:305–310

Fujita A, Shimizu I, Abe T (2001) Distribution of lysozyme and protease, and amino acid concentration in guts of a wood-feeding termite, *Reticulitermes speratus* (Kolbe): possible digestion of symbiont bacteria transferred by trophallaxis. Physiol Entomol 26:116–123

Fujita A, Minamoto T, Shimizu I et al (2002) Molecular cloning of lysozyme-encoding cDNAs expressed in the salivary gland of a wood-feeding termite, *Reticulitermes speratus*. Insect Biochem Mol Biol 32:1615–1624

Genta FA, Terra WR, Ferreira C (2003) Action pattern, specificity, lytic activities, and physiological role of five digestive β-glucanases isolated from *Periplaneta americana*. Insect Biochem Mol Biol 33:1085–1097

Genta FA, Blanes L, Cristofoletti PT et al (2006) Purification, characterization and molecular cloning of the major chitinase from *Tenebrio molitor* larval midgut. Insect Biochem Mol Biol 36:789–800

Genta FA, Dumont AF, Marana SR et al (2007) The interplay of processivity, substrate inhibition and a secondary substrate binding site of an insect exo-β-1,3-glucanase. Biochim Biophys Acta 1774:1070–1091

Genta FA, Bragatto I, Terra WR et al (2009) Purification, characterization and sequencing of the major beta-1,3-glucanase from the midgut of *Tenebrio molitor* larvae. Insect Biochem Mol Biol 39:861–874

Girard C, Jouanin L (1999a) Molecular cloning of cDNAs encoding a range of digestive enzymes from a phytophagous beetle *Phaedon cochleariae*. Insect Biochem Mol Biol 29:1129–1142

Girard G, Jouanin L (1999b) Molecular cloning of a gut-specific chitinase cDNA from the beetle *Phaedon cochleariae*. Insect Biochem Mol Biol 29:549–556

Khadeni S, Guarino LA, Watanabe H et al (2002) Structure of an endoglucanase from termite, *Nasutitermes takasogoensis*. Acta Cryst D 58:653–658

Kim N, Choo YM, Lee KS et al (2008) Molecular cloning and characterization of a glycosyl family 9 cellulase distributed throughout the digestive tract of the cricket *Teleogryllus emma*. Comp Biochem Physiol B 150:368–376

Kirsch R, Gramzow L, Theissen G et al (2014) Horizontal gene transfer and functional diversification of plant cell wall degrading polygalaturonases: key events in the evolution of herbivory in beetles. Insect Biochem Mol Biol 52:33–50

Kleywegt GJ, Zou JY, Divne C et al (1997) The crystal structure of the catalytic core domain of endoglucanase I from *Trichoderma reesei* at 3.6 angstrom resolution, and a comparison with related enzymes. J Mol Biol 272:393–397

Klungness LM, Peng YS (1984a) Scanning electron microscope observations of pollen food in the alimentary canal of honeybees (*Apis mellifera* L.). Can J Zool 62:1316–1319

Klungness LM, Peng YS (1984b) A histochemical study of pollen digestion in the alimentary canal of honeybees (*Apis mellifera* L.). J Insect Physiol 30:511–522
Kroon GH, van Praag JP, Velthuis HHW (1974) Osmotic shock as a prerequisite to pollen digestion in the alimentary tract of the worker bee. J Agric Res 13:177–181
Lemos FJA, Terra WR (1991) Digestion of bacteria and the role midgut lysozyme in some insect larvae. Comp Biochem Physiol B 100:265–268
Lemos FJA, Ribeiro AF, Terra WR (1993) A bacteria-digesting midgut lysozyme from *Musca domestica* (Diptera) larvae: purification, properties and secretory mechanism. Insect Biochem Mol Biol 23:533–541
Linder M, Teeri TT (1997) The roles and function of cellulose-binding domains. J Biotechnol 57:15–28
Lo N, Tokuda G, Watanabe H et al (2000) Evidence from multiple genes sequences indicates that termites evolved from wood-feeding cockroaches. Curr Biol 10:801–804
Lu Y, Zen K-C, Muthukrishnan S et al (2002) Site-directed mutagenesis and functional analysis of active site acidic amino acid residues D142, D144 and E146 in *Manduca sexta* (tobacco hornworm) chitinase. Insect Biochem Mol Biol 32:1369–1382
Markovic O, Janecek S (2001) Pectin degrading glycoside hydrolases of family 28: sequence-structural features, specificities and evolution. Protein Eng 14:615–631
Matoub M, Rouland C (1995) Purification and properties of the xylanases from the termite *Macrotermes bellicosus* and its symbiotic fungus *Termitomyces sp.* Comp Biochem Physiol B 112:629–635
McKenna DD, Scully ED, Pauchet Y et al (2016) The genome of the Asian longhorned beetle (Anoplophora glabripennis), a globally significant invasive species, reveals key functional and evolutionary innovations at the beetle-plant interface. Genome Biol 17:227
Nonaka Y, Akieda D, Aizawa T, Watanabe N et al (2009) X-ray crystallography and structural stability of digestive lysozyme from cow stomach. FEBS J 276:2192–2200
Pauchet Y, Freitak D, Heidel-Fischer HM et al (2009) Immunity or digestion. Glucanase activity in a glucan-binding protein family from Lepidoptera. J Biol Chem 285:2214–2224
Pauchet Y, Kirsch R, Giraud S et al (2014) Identification and characterization of plant cell wall degrading enzymes from three glycoside hydrolase families in the cerambycid beetle *Apriona japonica*. Insect Biochem Mol Biol 49:1–13
Pettersen RC (1984) The chemical composition of wood. In: Rowell RM (ed) The chemistry of solid wood. Advances in chemistry series, vol 207. American Chemical Society, Washington, DC, pp 57–126
Pollet A, Delcourt JA, Coutin CM (2010) Structural determinants of the substrate specificities of xylanases from different glycoside hydrolase families. Critic Rev Biotech 30:176–191
Radford JE, Vesk M, Overall RL (1998) Callose deposition at plasmodemata. Protoplasma 201:30–37
Regel R, Matioli SR, Terra WR (1998) Molecular adaptation of *Drosophila melanogaster* lysozymes to a digestive function. Insect Biochem Mol Biol 28:309–319
Rouvinen J, Bergfors T, Teeri T et al (1990) Three-dimensional structure of cellobiohydrolase II from Trichoderma reesei. Science 249:380–386
Scott HG, Stojanovich CJ (1963) Digestion of Juniper pollen by Collembola. Flor Entomologist 16:189–191
Scully ED, Hoover K, Carlson JE et al (2013) Midgut transcriptome profiling of *Anoplophora glabripennis*, a lignocellulose degrading cerambycid beetle. BMC Genomics 14:1–26
Sethi A, Slack JM, Kovaleva ES et al (2013) Lignin-associated metagene expression in a lignocellulose-digesting termite. Insect Biochem Mol Biol 43:91–101
Shelomi M, Heckel DG, Pauchet Y (2016a) Ancestral gene duplication enabled the evolution of multifunctional cellulases in stick insects (Phasmatodea). Insect Biochem Mol Biol 71:1–11
Shelomi M, Danchin EGJ, Heckel D et al (2016b) Horizontal gene transfer of pectinases ffrom bacteria preceded the diversification of stick and leaf insects. Sci Rep 6:1–9

Shen Z, Jacobs-Lorena M (1997) Characterization of a novel gut-specific chitinase gene from the human malaria vector *Anopheles gambiae*. J Biol Chem 272:28895–28900

Shen Z, Reese JC, Reeck GR (1996) Purification and characterization of polygalacturonase from the rice weevil, *Sitophilus oryzae* (Coleoptera: Curculionidae). Insect Biochem Mol Biol 26:427–433

Slaytor M (1992) Cellulose digestion in termites and cockroaches: what role do symbionts play. Comp Biochem Physiol B 103:775–784

Soni H, Kango N (2013) Microbial mannanases: properties and applications. In: Shukla P, Pletschke BI (eds) Advances in enzyme biotechnology. Springer, New Delhi, pp 41–56

Tamaki FK, Pimentel AC, Dias AB et al (2014) Physiology of digestion and the molecular characterization of the major digestive enzymes from *Periplaneta americana*. J Insect Physiol 70:22–35

Teller DC, Behnke CA, Pappan K et al (2014) The structure of rice weevil pectin methylesterase. Acta Crystallogr 70:1480–1484

Terra WR, Ferreira C (1994) Insect digestive enzymes: properties, compartmentalization and function. Comp Biochem Physiol B 109:1–62

Terra WR, Ferreira C (2012) Biochemistry and molecular biology of digestion. In: Gilbert LI (ed) Insect molecular biology and biochemistry. Academic/Elsevier, London, pp 365–418

Terra WR, Valentin A, Santos CD (1987) Utilization of sugars, hemicellulose, starch, protein, fat and minerals by *Erinnyis ello* larvae and digestive role of their midgut hydrolases. Insect Biochem 17:1143–1147

Terra WR, Barroso IG, Dias RO et al (2019) Molecular physiology of insect midgut. Adv Insect Physiol 56:117–163

Tokuda G (2019) Plant cell wall degradation in insects: recent progress on endogenous enzymes revealed by multi-omics technologies. Adv Insect Physiol 57:97–136

Ursic-Bedoya RJ, Nazzari H, Cooper D et al (2008) Identification and characteriazation of two novel lysozymes from *Rhodnius prolixus*, a vector of Chagas disease. J Insect Physiol 54:503–603

Van Herreghe JM, Michiels CW (2012) Invertebrate lysozymes: diversity and distribution, molecular mechanism and in vivo function. J Biosci 37:327–348

Vonk HJ, Western JRH (1984) Comparative biochemistry and physiology of enzymatic digestion. Academic, New York

Watanabe H, Tokuda G (2010) Cellulolytic systems in insects. Annu Rev Entomol 55:609–632

Watanabe H, Noda H, Tokuda G et al (1998) A cellulase gene of termite origin. Nature 394:330–331

Willis JD, Oppert B, Oppert C et al (2011) Identification, cloning, and expression of a GHF9 cellulose from *Tribolium castaneum* (Coleopter: Tenebrionidae). J Insect Physiol 57:300–306

Chapter 11
Mechanisms of Avoiding the Action of Plant Inhibitors on Digestion

Abstract In this chapter, we further explore the biochemical adaptations of insects to avoid the inhibitors of endopeptidase, α-amylase, polygalacturonase, and lipase inhibitors. How herbivorous insects deal with the proteinaceous inhibitors of digestive enzymes found in host plants is a key question in the evolution and in the ecology of insect-plant interactions. Those types of inhibitors are not the only ones found in plants, but they are by far the most abundant and for which there is more data in the literature. We will begin by discussing adaptations to endopeptidase inhibitors, which are the best-known and most diverse. We will then turn to carbohydrate digestion inhibitors, particularly how insects adapt to the ingestion of α-amylase, polygalacturonase, and lipase inhibitors.

11.1 Introduction

Herbivorous insects are the most diverse animal group in terrestrial environments and represent a considerable part of herbivory pressure in nature and in human plantations. Because of this, the physical and chemical defenses of plants to tolerate or avoid insect attack are varied. In this chapter, our emphasis will be on the various mechanisms by which insects adapt to the expression of potentially toxic proteins and whose main targets are the digestive enzymes responsible for initiating the digestion of proteins, starch, and pectin by herbivorous insects. Insect mechanisms to avoid inhibition of β-glucosidases by plant components are discussed in Chap. 4.

Plant defense proteins that interact specifically with enzymes involved in initial digestion leading to total or partial inhibition of their activities will be called here endopeptidases inhibitors (PIs), α-amylases inhibitors (AIs), or polygalacturonase-inhibiting proteins (PGIPs) (see Chap. 15). The distribution, diversity, and expression control of these insect digestive enzyme inhibitors attest that they are the result of an "arms race" between herbivorous insects and plants. As a result, in addition to the diverse forms of plant defenses, we also witness a great diversity of insect adaptations to these defensive arsenals.

W. R. Terra et al., *Molecular Physiology and Evolution of Insect Digestive Systems*, Entomology in Focus 7, https://doi.org/10.1007/978-3-031-39233-7_11

Adaptation strategies of herbivorous insects to inhibitors present in their diets involve behavioral, genetic, and biochemical aspects. Behavioral adaptations allow insects to avoid overconsumption of inhibitors or nutritional compensation due to inhibition of plant nutrient digestion. Biochemical and genetic adaptations, on the other hand, involve, for example, the expression of digestive enzymes that are insensitive to inhibitors, of enzymes capable of degrading or inactivating inhibitory proteins, or the overexpression of native digestive enzymes leading to overcoming the inhibitory activity. These biochemical adaptations are the result of millions of years of coevolution, and, on some occasions, it is possible to witness the counteradaptation of insects through the observation of structural modifications in the active sites of enzymes, which result in a lower interaction with the inhibitors found in plants, as is the case with inhibitor-insensitive peptidases.

11.2 Adaptations to the Ingestion of Plant Inhibitors of Digestive Endopeptidases

The landmark discovery in the early 1970s that plants can increase expression of trypsin inhibitor genes in response to mechanical injury caused by an insect bite revolutionized our understanding of how plants react chemically to herbivory (Green and Ryan 1972). These observations also showed the importance of inhibiting the digestion of dietary proteins as a plant defense strategy against herbivores in general and against insects, in particular (Ryan 1990). Endopeptidase inhibitors (PIs) are found in virtually all living beings, as they are involved in diverse functions in the control of proteolytic activity. However, here we are referring to PIs that are generally found in relatively high concentrations in certain plant tissues, such as those found in seeds, leaves, or roots, and that are part of the defensive arsenal of plants. In this case, these inhibitors are considered constitutive. As mentioned above, plants are also capable of overexpressing PIs in response to insect attack. In this case, these PIs are considered inducible, and their accumulation only occurs under specific conditions.

Although proteinaceous inhibitors of aspartic and metallic endopeptidases have already been found in plants, serine and cysteine endopeptidase inhibitors are by far the most abundant and about which there is much more data in the literature. The greater occurrence of these inhibitors as plant defense macromolecules also reflects the greater occurrence and importance of these peptidases in the main groups of insects that attack plants, which are the beetles of the Cucujiformia clade, which rely on digestive cysteine and/or serine endopeptidases, or lepidopteran species, that make use of digestive serine endopeptidases like trypsin or chymotrypsin (Silva et al. 2022; Terra and Ferreira 2020). The adaptations of insects to the ingestion of plant inhibitors of digestive enzymes are multifaceted, as are the strategies of plants to defend themselves through the expression of genes that encode these defensive proteins. Insect adaptations to these inhibitors range from behavioral responses, which are mainly aimed at reducing the intake of inhibitors, to adaptations at the molecular level, that is, biochemical adaptations, which reflect the results of the arms race expressed in differences in the regions of interaction between inhibitors and digestive enzymes.

11.2.1 Behavior Adaptation to Avoid Ingestion of PIs

Insects can avoid eating PI-containing plant tissues as soon as they are able to detect the PI after ingestion or their antinutritional effects. Herbivorous insects respond in part by seeking more poorly defended tissues within individual plants, such as leaves with different ages, or by exploiting the plant at a stage when PIs are absent or present at lower levels, or even move to other plants that are expressing less PIs. Examples of behavior\al adaptations have been described in different species that feed on older leaves, which have lower concentrations of PIs. Savala and co-authors showed in elegant experiments that the Japanese beetle (*Popillia japonica*) systematically moves around the plants of soybean looking for older leaves with lower levels of cysteine endopeptidase inhibitors (Savala et al. 2009). Behavioral adaptation to minimize PIs intake has also been recorded in Lepidoptera. Larval *Manduca sexta* actively seek leaves with lower PI content when fed on *Nicotiana attenuata* plants, which have been modified to express different levels of a trypsin inhibitor (Paschold et al. 2007; Zavala et al. 2008). Another striking example of behavioral adaptation to the ingestion of PIs was described in larvae of *Spodoptera frugiperda*, which start to have a cannibalistic behavior, feeding on other caterpillars when submitted to an artificial diet containing different concentrations of trypsin inhibitor or nicotine (Roy 2020). The author attributed this behavior as a way of compensating for the lower digestion of dietary proteins with the ingestion of animal tissues richer in proteins and free of PIs.

11.2.2 Metabolic Responses to the Ingestion of PIs

Regardless of behavioral adaptations, metabolic responses to the presence of PIs in the diet of adapted insects involve: (a) overexpression of multiple digestive endopeptidases (Bown et al. 1997; Broadway 1997; Gatehouse et al. 1997; Oppert et al. 2005); (b) proteolytical inactivation of PIs mediated by the insect digestive endopeptidases (Giri et al. 1998; Ahn and Zhu-Salzman 2009); (c) by expressing new endopeptidases that are resistant to the inhibitor (Jongsma et al. 1995; Jongsma and Bolter 1997; Mazumdar-Leighton and Broadway 2001; Brito et al. 2001; de Oliveira et al. 2013); (d) recruitment of lysosomal proteins as digestive cathepsins (Pimentel et al. 2020; Silva et al. 2022; see also Chap. 9); (e) by expressing pseudogenes that codifies inactive peptidases capable of binding to the PIs (Shibao et al. 2021). Over the past decades, the number of published studies describing these different adaptive strategies has grown year after year.

11.2.3 Structural Differences Between Sensitive and Insensitive Digestive Peptidases Toward PIs

The molecular basis of the difference between sensitive and inhibitor-insensitive trypsins, as well as the regulation of their expression, has been extensively investigated since the discovery of this phenomenon (Jongsma et al. 1995). Structural molecular studies provide an important tool for the detailed characterization of interactions between PIs and their target enzymes at the atomic level. PIs have a region, named the reactive site or reactive loop, that interacts with the active site of their target enzymes resulting in lower or abolishing enzymatic activity. The amino acid composition of the reactive site contributes to the functional diversity of PIs. The amino acid residue in the P_1 position, which corresponds to the primary cleaving site of the substrate (see numbering of endopeptidase subsites in Chap. 9), is of particular importance to determine the specificity and efficiency of the inhibitor. The reactive sites of many trypsin inhibitors are hydrophilic loops with a lysine residue at P_1 (Lopes et al. 2004; Zhou et al. 2020). As lepidopteran trypsins have hydrophobic subsites and have preference for Lys, instead of the more usual Arg at $P1_1$, they are usually more resistant to PIs than the coleopteran insect trypsins, for example (Lopes et al. 2006). In this respect, it is interesting to note that PI-insensitive trypsins from *Heliothis virescens* bind more tightly to a hydrophobic chromatographic column than sensitive trypsins do (Brito et al. 2001). Tamaki and Terra (2015) by using biochemical experiments and computer 3D modeling demonstrated that lepidopteran trypsins have more hydrophobic surface clusters than coleopteran ones. According to these authors, clusters of hydrophobic amino acid residues on the surface of endopeptidases insensitive to PIs cause oligomerization of these enzymes, which leads to less interaction with inhibitors and, therefore, less inhibition or complete resistance to inhibitors. Such observations lead to the hypothesis that the molecular differences between sensitive and insensitive trypsins must rely on the interactions of PIs with residues in and around the enzyme active site.

An interesting approach to study insect trypsin-PI interactions was introduced by Volpicella et al. (2003). They compared the sequence of a sensitive trypsin from *Helicoverpa armigera* with the insensitive trypsin from the closely related species *Helicoverpa zea*. The 57 different amino acids observed between the two enzymes were superimposed on the porcine trypsin crystal structure, where the residues known to be in contact to a Kunitz-type inhibitor (Song and Suh 1998) were identified. The residues at positions (chymotrypsin numbering) 41, 57, 60, 95, 99, 151, 175, 213, 217, and 220 were considered by Volpicella et al. (2003) to be important in *H. zea* trypsin-PI interaction. However, some of the interacting residues may have been misidentified because trypsins from different species were compared. In a similar approach, Lopes et al. (2004) aligned all available trypsin sequences characterized as sensitive or insensitive to Kunitz-type inhibitor (Bown et al. 1997; Mazumdar-Leighton and Broadway 2001) with porcine trypsin. After discounting conserved positions and positions not typical of sensitive or insensitive trypsin, the remaining positions that agree with those involved in porcine trypsin-PI (Bowman-Birk type, Lin et al. 1993; Kunitz type, Song and Suh 1998) or substrate (Koepke et al. 2000) interactions were: 60, 94, 97, 98, 99, 188, 190, 213, 215, 217, 219, 228. These positions support the tree branches in a neighbor-joining analysis of sensitive (I, III) and insensitive (II) trypsin sequences (Lopes et al. 2004). Site-directed mutagenesis of trypsin, followed by the determination of the binding constants of mutated trypsins with PIs, may help to resolve the discrepancy.

The molecular mechanism of chymotrypsin-PI inhibition was also investigated. Two chymotrypsins were purified from the midgut of *Helicoverpa punctigera*, one PI-sensitive and the other PI-insensitive. After their corresponding cDNAs were cloned and sequenced, molecular modeling revealed that a Phe → Leu substitution at position 37 in the chymotrypsin results in the loss of important contacts with the PI. This was confirmed by site-directed mutagenesis of chymotrypsin molecules, followed by inhibition tests (Dunse et al. 2010). Chymotrypsins from insects that routinely ingest ketone-releasing compounds (like several plant glycosides) (see Chap. 3) are not affected much by these compounds and others that react with His 57. Thus, in comparison with bovine chymotrypsin, the chymotrypsin from polyphagous lepidopteran insects reacts slowly with chloromethyl ketones, whereas those of oligophagous pyralid insects react rapidly (Lopes et al. 2009). Modeling *S. frugiperda* (Noctuidae) chymotrypsin, based on its sequence and on crystallographic data of bovine chymotrypsin, showed that the neighborhood of His 57 differs from bovine chymotrypsin, thus affecting His reactivity (Lopes et al. 2009). These adaptations are new examples of the interplay between insects and plants during their evolutionary arms race and deserve more attention through site-directed mutagenesis of recombinant chymotrypsin.

11.2.4 Expression of Insensitive or PI-Metabolizing Endopeptidases

Herbivorous insects respond to dietary PIs by both qualitative and quantitative changes in expression of their digestive endopeptidases. Higher expression of genes encoding insensitive or PI-metabolizing peptidases is a very common cause of PI resistance in insects. Some insects can combine multiple strategies in response to dietary PIs by increasing total proteolytic activity or shifting toward insensitive digestive endopeptidases. The underlying regulatory mechanisms of these compensatory responses remain largely elusive, although it was found that the first step in the process is the expression of the whole set of midgut trypsins and chymotrypsins in larval *S. frugiperda* (Brioschi et al. 2007). Giri et al. (1998) demonstrated that some of the overexpressed trypsins from *H. armigera* can hydrolyze the ingested PIs from chickpea (*Cicer arietinum*), while other trypsins are insensitive to the inhibitor. Overexpression of these "detoxifying" serine endopeptidases, capable of degrading PIs, can result in higher levels of metabolic resistance to plant inhibitors in Lepidoptera (Hafeez et al. 2021). Consistent with these observations on the adaptability of the expression of different endopeptidases in response to PI ingestion in Lepidoptera, the occurrence of multigenic families of trypsins and chymotrypsins has been described, mainly in polyphagous species. These multigenic endopeptidase families contain multiple forms with overlapping substrate specificities and reactivity to inhibitors (Jongsma et al. 1995; Giri et al. 1998; Lokya et al. 2020; Hafeez et al. 2021). Knowledge about the degradation mechanisms of the PI molecule or about the mechanisms of evasion of the inhibitory activity helps us to understand the biochemical details underlying the resistance of insects to the ingestion of PIs. This type of information is important to guide integrated pest management strategies.

Although the induction of PI-insensitive serine endopeptidases in response to the ingestion of inhibitors was first described in lepidopterans (Jongsma et al. 1995; Jongsma and Bolter 1997; Brito et al. 2001; Zhu-Salzman and Zeng 2015), this phenomenon has also been described in insects of other orders and induction of cysteine endopeptidases as well. Digestive flexibility and expression of genes related to counterdefense toward PIs were previously documented as adaptive mechanisms in numerous phytophagous insect species. The goal of protecting crops through the manipulation of insect digestion via PI-expressing transgenic plants has generated considerable interest in the characterization of digestive peptidases from agricultural insect pests from the orders Coleoptera, Orthoptera, and Hemiptera. The records of adaptation of insects belonging to other orders and which do not have pest status are also found in the literature. For example, in *Drosophila melanogaster* (Diptera) larvae, several alterations are observed in the midgut epithelium when these animals are fed PIs in an artificial diet (Li et al. 2010).

11.2.5 Recruitment of Lysosomal Proteins as Digestive Cathepsins

In terms of agronomic importance and in ecological terms, the order Coleoptera represents a great pressure of herbivory, having several species with important pest status. Consequently, there are many articles dedicated to understanding how herbivorous beetles deal with the PIs present in their diets. An important finding is that most beetles that feed primarily on living plant tissues depend on cysteine and aspartic digestive peptidases, in addition to the serine endopeptidases trypsin and chymotrypsin. In beetles of the infraorder Cucujiformia, which includes the superfamilies Tenebrionoidea, Curculionoidea, and Chrysomeloidea (which make up 90% of the species that feed mainly on plants), the recruitment of lysosomal proteins as digestive cathepsins is interpreted as an evolutionary adaptation that helped these insects to explore serine endopeptidase inhibitor-containing plants, mainly

Angiosperms, as food (Silva et al. 2022). Among the digestive cathepsins of beetles, the cathepsins L and B and the aspartic cathepsin D stand out (see details in Chap. 9).

The flexibility of expression of serine endopeptidases in response to the ingestion of PIs described in Lepidoptera is also observed in the expression of digestive cathepsins in several species of beetles. It is important to note that plants are also capable of accumulating cysteine peptidase inhibitors. The adaptation of the beetle *Diabrotica virgifera* (Chrysomelidae) to soybean plants, which produce serine and cysteine endopeptidases inhibitors, depends on the increase in cathepsin L expression (Curzi et al. 2012). In this case, the overexpressed cathepsin L is insensitive to both serine and cysteine endopeptidases inhibitors. In larvae of the seed-beetle *Callosobruchus maculatus* (Chrysomelidae: Bruchinae), the ingestion of the soybean cysteine peptidase inhibitor soyacystatin N causes an increase in the expression of cathepsin B. The expression of this cathepsin compensates for the observed inhibition of cathepsins L, which are predominant in these larvae (Koo et al. 2008; Zhu-Salzman et al. 2003). Still on this seed-beetle species, Nogueira et al. (2012) showed that feeding larvae with cystatin from egg whites (an inhibitor of cysteine endopeptidases) led to increased expression of many genes related to stress response, in addition to induction of different cathepsins L.

Recruitment of lysosomal proteins as digestive cathepsins followed by gene expansion has not only occurred in Coleoptera but it is also observed in many species of the order Hemiptera. However, in this case, the selective pressure for that recruitment was the return to a protein diet of insects which ancestors feed low-protein plant saps (see details in Chap. 9).

Curiously, the expansion in cathepsin L genes in respect to cathepsin B genes is a common observation both in Coleoptera and in most Hemiptera (Silva et al. 2022; Sardoy et al. 2021; Pimentel et al. 2020). It is worth noting that endopeptidase activities from the cysteine cathepsin L and B are frequently complemented in Coleoptera and Hemiptera by activity from the aspartic cathepsin D, which can digest cathepsin L or cathepsin B inhibitors and amplify the overall efficiency of dietary protein hydrolytic processes of the insect (Sardoy et al. 2021; Pimentel et al. 2020; Nogueira et al. 2012; Ahn et al. 2007; Amirhusin et al. 2007; Brunelle et al. 1999) (see also Chap. 9).

11.2.6 Expression of Pseudoendopeptidases

Pseudoenzymes are proteins that share similar amino acid sequences to wild-type or native enzymes, but that are proven or predicted to lack enzyme activity due to substitutions in one or more conserved catalytic amino acids.

A mechanism of adaptation of insects to the ingestion of PIs, still little studied and understood, is based on the expression of pseudoendopeptidases that, hypothetically, bind to the inhibitors, promoting a kind of sequestration of these inhibitors during the formation of the pseudoendopeptidase/PI complex. This kind of sequestration or scavenger of PIs results in the release of enzymatically active endopeptidases to further digest dietary proteins. Expression in the midgut of pseudogenes encoding trypsin-like serine endopeptidases or digestive cysteine cathepsins has been described in Lepidoptera and Coleoptera. The proposition of this possible mechanism of adaptation to the ingestion of PIs is relatively recent since it emerged after the annotations and validations carried out from transcriptomes and proteomes of tissues originating from the midgut of different insects. Data from transcriptomes and proteomes of midgut tissues have shown that the expression of inactive peptidases in herbivorous insects is relatively common. Although there is little direct experimental data demonstrating the formation of such pseudoenzyme/inhibitor complexes in the intestinal lumen, there is a kind of common sense that the sequestration of PIs must be one of the main adaptive functions due to the relatively high expression of these proteins in the midgut of herbivorous insects.

Serine pseudoendopeptidases, which are expressed at high levels in midgut, have already been identified in several lepidopteran species such as *H. armigera* (Noctuidae) (Bown et al. 1997; Velasquez-Vasconez et al. 2022), *Epiphyas postvittana* (Tortricidae) (Simpson et al. 2007), *Mamestra configurata* (Noctuidae) (Erlandson et al. 2010), *Bombyx mori* (Bombycidae) (Liu et al. 2017), and *Manduca sexta* (Sphingidae) (Miao et al. 2020). Among coleopteran species where inactive endopeptidases expressed in the midgut have been described, we have *Tenebrio molitor* (Tenebrionidae) (Prabhakar et al. 2007), *Tribolium castaneum* (Tenebrionidae) (Morris et al. 2009), and *Sphenophorus levis* (Curculionidae) (Shibao et al. 2021). While effective binding with inhibitors by pseudoenzymes is not yet confirmed, a certain level of inhibitor sequestration likely protects active enzymes from inhibition and the insect from nutritional deficiency.

All these types of adaptations to cope with the ingestion of plant peptidase inhibitors described here are summarized in Table 11.1. We have grouped part of the most interesting records available in the literature to date, but we cannot rule out the discovery of new mechanisms with advances in research in the post-genomic era. It is also important to note that many of these adaptations occur simultaneously in many species of herbivorous insects and that they are not mutually exclusive; on the contrary, the insect counterattack is often multifactorial.

11.3 Mechanisms of Avoiding the Action of Plant α-Amylases Inhibitors (AIs)

11.3.1 Overview

To protect themselves from insect attack and other herbivorous animals, plants produce proteinaceous inhibitors of α-amylases (AIs), which are considered to be part of constitutive or inducible defense mechanisms (Svensson et al. 2004; Franco et al. 2000; Gatehouse et al. 1986). Once ingested, AIs can impair starch digestion in the insect midgut, which results in limited availability of energy from dietary

Table 11.1 Adaptive mechanisms observed in herbivorous insect toward plant peptidase inhibitor (PI)

Type of inhibitor	Adaptive mechanism	References
Serine PI; potato inhibitor II; Merops family I20	Behavior adaptation to avoid ingestion of PIs	Zavala et al. (2008)
Cysteine PI; phytocystatin; I25	Behavior adaptation to avoid ingestion of PIs	Zavala et al. (2009)
Serine PI; potato inhibitor II; Merops family I20	Overexpression of insensitive peptidases	Jongsma et al. (1995), Jongsma and Bolter (1997), Mazumdar-Leighton and Broadway (2001), and de Oliveira et al. (2013)
Serine PI; chickpea trypsin-chymotrypsin inhibitor; Merops family I12	Proteolytical inactivation of PIs	Giri et al. (1998) and Ahn and Zhu-Salzman (2009)
Cysteine PI; Soyacystatin N; Merops family I25	Overexpression of multiple digestive endopeptidases	Ahn et al. (2007)
Serine PIs	Recruitment of digestive cathepsins	Silva et al. (2022)
Serine PIs	Selection of hydrophobic surface clusters in digestive trypsin	Tamaki and Terra (2015)
Serine and cysteine PIs	Sequestration of PI by pseudoendopeptidases	Shibao et al. (2021)
Serine PIs	Structural changes in resistant enzymes	Lopes et al. (2019)

carbohydrates leading to high mortality rates, delayed development, and reduced fecundity. To illustrate the defensive potential of AIs, the successful use of AIs for controlling insect infestation in transgenic plants has already been demonstrated in laboratory-scale and field studies (Barbosa et al. 2010; Morton et al. 2000; Schroeder et al. 1995). α-Amylase inhibitors have been classified into seven different protein structural families that show remarkable structural diversity (Li et al. 2021; Franco et al. 2002). They are quite specific to their target enzyme, leading to different modes of inhibition and different specificity profiles against various insect α-amylases. Therefore, one AI that inhibits the activity of one amylase may not have the same effect on other amylases from the same insect or from α-amylases of a closely or distant insect species (Li et al. 2021; Da Lage 2018; Svensson et al. 2004).

In the phenomenon of coevolution along with plants, insects adapted in response to ingestion of AIs by changing their midgut physiology and expression pattern of α-amylases. The molecular mechanisms of avoiding the action of AIs involve (a) the occurrence of multigene families of digestive α-amylases; (b) overexpression of AI-insensitive α-amylases; (c) degradation of AIs by peptidases; (d) the prevalent in vivo physical conditions in the intestinal lumen; and d) a combination of all these factors.

11.3.2 Adaptation Based on the Occurrence of Multigene Families of Digestive α-Amylases

One of the adaptive responses to overcome AIs is to express several α-amylase paralogs with different sensitivities to plant inhibitors (González-Ruiz et al. 2021; Da Lage et al. 2018; Pytelkova et al. 2009; Silva et al. 2001a). Thus, it has been demonstrated that the expression of different α-amylase isoforms may be correlated with the ingestion of AIs in the Mediterranean flour moth *Ephestia kuehniella* (Pyralidae) when challenged with wheat extract containing a cereal-type inhibitor family (Pytelkova et al. 2009). In *E. kuehniella*, the three different expressed α-amylase isoenzymes differed in sensitivity to the AIs, minimizing the inhibitory effects as part of the strategy to hydrolyze starch and obtain the needed monosaccharides for energy extraction. In the coleopteran *Rhyzopertha dominica* (Bostrichidae), three α-amylase isoforms are also expressed when the larvae of this animal are fed wheat kernel (González-Ruiz et al. 2021).

11.3.3 Overexpression of AI-Sensitive and AI-Insensitive α-Amylases

Markwick and co-authors are among the first researchers to report overexpression of digestive α-amylases in response to ingestion of AIs (Markwick et al. 1998). They have demonstrated that wheat amylase inhibitors increased α-amylase activity up to fivefold when administered singly or even tenfold when combined in diet with the potato peptidase inhibitor 2 (a serine endopeptidase inhibitor) in the leafroller species *Ctenopseustis obliquana*, *Epiphyas postvittana*, and *Planotortrix octo* (Lepidoptera: Tortricidae). Because larval growth rate was not significantly affected when inhibitors were added to the diet, whereas α-amylase activity was significantly enhanced, the authors suggested that the overexpression of digestive α-amylases was elicited to compensate for inhibition of their activities. In larval Colorado potato beetle *L. decemlineata*, the overexpression of digestive α-amylase genes in third instar larvae is elicited to compensate for inhibition of enzyme activity caused by AIs present in white bean and rapeseed protein extracts (Ashouri and Pourabad 2021). According to these authors, the increased amylase gene expression at the L3 stage was a counter-response to the AIs ingested from the two plant extracts.

In the lepidopteran species *E. kuehniella* and in the beetle *R. dominica* cited above, in which multiple isoforms of amylases occurred, overexpression of α-amylases insensitive to the tested inhibitor was also observed (González-Ruiz et al. 2021; Pytelkova et al. 2009). In *E. kuehniella*, in vivo administration of the wheat α-amylase inhibitor resulted in a significantly 4.5-fold up-regulation of α-amylase genes (Pytelkova et al. 2009). According to the authors, the selective blocking of two sensitive digestive α-amylase isoenzymes (EkAmy2 and EkAmy3) resulted in overexpression of the α-amylase genes, including the expression of an insensitive isoform. In larval *R. dominica*, the activities from three digestive α-amylases differ in their affinity to the substrate and in the interaction with wheat kernel inhibitors (González-Ruiz et al. 2021). One isoform in particular (RdA70) has a high affinity for the AIs leading to its neutralization and allowing the other two more sensitive isoforms (RdA79 and RdA90) to hydrolyze dietary starch.

Another case where increased expression of α-amylase isoforms was observed when the insect fed on a diet containing AI is the bruchid *Zabrotes subfasciatus* (Chrysomelidae). In this bean-feeding species, two isoforms are overexpressed when larvae develop in common bean *Phaseolus vulgaris* seeds, which contain amylase inhibitor 1 (αAI-1) (Silva et al. 1999, 2001a, b). When the larvae develop in cowpea (*Vigna unguiculata*) seeds, which do not express this inhibitor, the larvae have a major and constitutive α-amylase isoform, but when they are transferred to *P. vulgaris* seeds, they increase the expression of two minor isoforms. Interestingly, there is evidence that the two induced isoforms form a heterodimer that is insensitive to the αAI-1 (Silva et al. 2001b). The induction of the two inhibitor-insensitive isoforms in *Z. subfasciatus* is irreversible (Bifano et al. 2010).

Therefore, digestive α-amylase multiplicity in herbivorous insects ensures that only one amylase inhibitor cannot collapse the entire amylase digestive system. Such a counter-response might help to overcome the inhibition effect. Moreover, as different isoenzymes can be overexpressed when amylases are inhibited in vivo, the adaptability of these insects to AI ingestion is also increased. Experimental evidence has shown that the induction of insensitive α-amylases causes insects to develop at rates like those fed on control diets without the presence of AIs. As for PIs, further studies are needed to identify how the inhibitor-containing diets stimulate the mechanism of gene expression by which insect digestive α-amylases are synthesized and secreted in response to the ingestion of AIs. The complex regulation of α-amylase genes in different developmental stages suggests the existence of mechanisms involved to detect nutrient imbalance caused by the inhibition of starch digestion (Ashouri and Pourabad 2021; González-Ruiz et al. 2021; Da Lage et al. 2018).

11.3.4 Cleavage of AIs by Digestive Endopeptidases and Stability of AIs in the Physical-Chemical Conditions Prevalent in the Intestinal Lumen

The first record that AIs can be inactivated by proteolytic degradation was made by Ishimoto and Chrispeels (1996). They have suggested that the adaptation of larval *Z. subfasciatus* to the *P. vulgaris* α-amylase 1 inhibitor (αAI-1) might be multifactorial, involving both the occurrence of inhibitor-insensitive α-amylases and the proteolytic degradation of the inhibitor. The authors demonstrated that the cleavage of the inhibitor is not caused by the action of digestive cathepsins, but by a minor serine endopeptidase. A single proteolytic cleavage of αAI-1 occurs at the reactive site of the inhibitor resulting in its inactivation and allowing the animal to fully develop in seeds rich in this potent α-amylase inhibitor. This serine endopeptidase involved in αAI-1 detoxification was later isolated and partially characterized by Silva and co-authors (Silva et al. 2001c). Giri and Kachole (1998) attributed the partial in vivo inhibition and partial effects of pigeon pea (*Cajanus cajan*) amylase inhibitors on the growth of *H. armigera* to the proteolytic degradation caused by the larvae's own digestive endopeptidases.

It is important to note that the physicochemical conditions prevailing in the midgut lumen have a profound influence on the stability as well as on the interaction of AIs with insect digestive amylases. The inhibitory activity of AIs is pH dependent, which means that there is an optimal range of inhibitory activity (Li et al. 2021; Da Lage et al. 2018; Franco et al. 2000, 2002). As the intestinal pH values of the insects that exert the most herbivory pressure (Coleoptera, Leipdoptera and Hemiptera) differ greatly from each other, it is unlikely that an inhibitor has a range of efficiency that covers all these targets. For example, the acidic optimum pH activity of the inhibitor found in wild varieties of *P. vulgaris* called αAI-2 is effective against Coleoptera, such as the bean-feeding beetles *C. maculatus* and *Z. subsfasciatus*, which have slightly acidic pH values in the midgut lumen (Barbosa et al. 2010; Ishimoto and Chrispeels 1996). However, such amylase inhibitor may not be so effective against lepidopteran species, whose midgut contents are very alkaline. There is also the possibility that certain pH values have an influence on the susceptibility of AIs to be hydrolyzed by digestive peptidases.

11.4 Mechanisms of Avoiding the Action of Plant Polygalacturonase Inhibitors (PGIPs)

11.4.1 Overview

It has been reported in the last two decades that certain herbivorous beetles, hemipterans, and stick insects (Phamatodea) can use plant cell wall-degrading enzymes (PCWDEs), which genes were horizontally transferred from bacteria and fungi to the insect genome, to aid digestion (Kirsch et al. Kirsch et al. 2022; McKenna et al. 2019; Shelomi et al. 2016; Wybouw et al. 2016; see also Chap. 10). All insect-acquired polygalacturonases (PGs) belong to family GH28 of glycoside hydrolases, which are enzymes capable of hydrolyzing the polygalacturonic acid backbone in pectin either terminally (exo-PGs) or internally (endo-PGs) (McKenna et al. 2019; Pauchet et al. 2010; Pedra et al. 2003). Insects use PCDWEs to compromise the structure of the plant cell wall, which gives access to intracellular nutrients. In larval mustard leaf beetle *Phaedon cochleariae* (Chrysomelidae), gene silencing of these enzymes has a strong negative effect on larval performance. Interestingly, the supplementation of partially digested cell wall fragments does not reverse this negative effect, while the addition of PG to the diet mitigates the negative effects of gene silencing (Kirsch et al. 2022). These data reinforce the hypothesis that PGs are essential for the disruption of the cell wall and subsequent access to nutrients contained within plant cells.

In response to this herbivory pressure, plants also have in their defense arsenal proteins that inhibit the activity of PGs, the so-called PG-inhibiting proteins (PGIPs). These inhibitors belong to the leucine-rich repeat (LRR) protein family, and they can selectively inhibit the activity of PGs from bacteria, fungi, and insects (Haeger et al. 2020). PGIPs are constitutively expressed in the cell wall of many dicotyledonous plants, or they can be induced in response to various environmental stimuli such as mechanical wounding, infection by bacteria or fungi, and insect herbivory, reinforcing their role in plant defense (Kirsch et al. 2020; Maulik et al. 2012; D'Ovidio et al. 2004). There is an increasing number of articles attributing plant resistance to insects based on the expression of these inhibitors and their potential to plant transformation to get resistance against insect pests (Gamage et al. 2022; Zhang et al. 2021; Kirsch et al. 2020; Haeger et al. 2018; Kalunke et al. 2015; D'Ovidio et al. 2004).

In a very similar way to what we saw in relation to the adaptations of herbivorous insects to PIs and AIs, insects adapt to the ingestion of PGIPs according to the following strategies: (a) first expressing multigene families of PGs; (b) expressing PGIP-insensitive PGs; (c) expressing pseudoenzymes capable of binding to PGIPs.

11.4.2 Adaptation Based on the Expression of Multigene Families of PGs

Both PGs and PGIPs belong to multi-gene families believed to have been shaped by an evolutionary arms race between plants against pathogens and herbivores. In insects, pectin-degrading enzymes belonging to GH28 family have conserved sequences following the initial gene transfer event, and after millions of years of coevolution, it is possible to recognize the occurrence of gene duplication and subfunctionalization (McKenna et al. 2019; Haeger et al. 2018; Shelomi et al. 2016; Kirsch et al. 2014; Calderón-Cortés et al. 2012; Pauchet et al. 2010). This functional redundancy of PGs observed in herbivorous beetles has been attributed to an adaptation to the ingestion of PGIPs and reflects the result of the evolutionary arms race between beetles and Angiosperm (Haeger et al. 2021; McKenna et al. 2019; Tokuda 2019). As discussed earlier for digestive endopeptidases and α-amylases, many PG isoenzymes, each showing differences in specificity both with respect to their substrates and interaction with inhibitors, may confer an adaptive advantage to exploit a broader range of hosts. The gene expansion of PGs observed in the Phytophaga clade comprising weevils (Curculionidae), leaf beetles (Chrysomelidae), and long-horned beetles (Cerambycidae) seems to be directly involved in the great diversification of these insects (McKenna et al. 2019). The changes in specificity in relation to different substrates (neofunctionalization) and the greater ability to evade the different PGI isoforms conferred a great adaptive advantage to these herbivorous beetles and boosted the coevolution between beetles and angiosperms.

In mirid bugs, there is a positive correlation between the number of expressed PGs and the number of plant hosts (Xu et al. 2019). In these hemipterans, there is a clear discrepancy in the number of expressed PGs between bugs that preferentially feed on plants (phytozoophagous) and those that are preferentially predators (zoophytophagous). In the case of phytozoophagous mirids, the number of PG isoenzymes is greater than those that feed on other animals. These data suggest that a broad expression of PG may be related to the higher probability of ingesting PGIPs from host plants (Xu et al. 2019).

11.4.3 Adaptation Based on the Expression of Pseudo-PGs

It has been demonstrated that some beetles from the Phytophaga clade express both active and catalytically inactive (pseudoenzymes) pectin-degrading PGs. These pseudo-PGs lost their hydrolytic activities due to a substitution of one or more of the three catalytic amino acid residues directly involved in catalysis or substrate binding (Kirsch et al. Kirsch et al. 2012, 2014, 2019). Despite the lack of pectinolytic activity, these pseudoenzymes were shown to be equally expressed and secreted into the midgut lumen together with the active counterparts and may have a different function (Kirsch et al. Kirsch et al. 2012, 2019, 2022; Pauchet et al. 2014). The gene silencing of GH28 pseudoenzymes results in a poorer performance of the insects, lower food conversion efficiency, lower PG total activity, despite overexpression of active PGs being observed (Kirsch et al. 2014, 2019). All these results led to the suggestion that the pseudo-PGs can bind to PGIPs, releasing the active enzymes to act in the degradation of the plant cell wall. As in the case of pseudopeptidases, there is yet little direct experimental evidence that this phenomenon occurs in physiological conditions. However, complex formation between pseudoenzymes and PGIPs in the midgut lumen was reported by Kirsch et al. (2012) with the mustard leaf beetle, *Phaedon cochleariae* (Chrysomelidae).

In fact, the relationships between insect digestive PGs, pseudo-PGs, and PGIPs appear to be more complex than initially imagined. The occurrence of pseudo-PGs has been correlated with the pectinolytic pathway, even if indirectly, by decreasing the inhibitory effect of PGIPs. To further enrich this issue, it was discovered that certain proteins from Chinese cabbage (*Brassica rapa*), called PGI-like proteins, have a great similarity in amino acid sequence to "classical" PGIPs, clustering together in phylogenetic trees as members of the leucine-rich repeat family. The PGI-like proteins are considered new players in this arm race. These proteins also interact with PG pseudoenzymes in *P. cochleariae* (Haeger et al. 2021). It is still not known exactly how these PGI-like proteins act, but they are among the proteins that are overexpressed in response to plant chewing by insects and must somehow have a defensive function (Haeger et al. 2021).

11.5 Adaptations to the Ingestion of Plant Inhibitors of Digestive Lipases

Adaptation to plant inhibitors involving pseudoenzymes is proposed for the interaction between the silkworm *B. mori* pseudo lipases and mulberry lipase inhibitors (Wen et al. 2023). In the case of the silkworm, the gene encoding a non-classical form of lipase harboring an S144N substitution in the Ser-Asp-His catalytic triad underwent gene amplification during the domestication of this moth species (Wen et al. 2023). Pseudo lipase has much lower lipase activity than the wild-type forms, but it has reduced sensitivity to the mulberry lipase inhibitor morachalcone A. The cost-benefit between having lower catalytic activity from a pseudo lipase and reducing the effects of mulberry inhibitors is positive. Combining the expression of active but inhibitor-sensitive lipases with the high expression of the inhibitor-insensitive pseudo lipase is a result of silkworm artificial selection. This mechanism to deal with the lipase inhibitor may be of high adaptive value and an interesting trait for an improvement in the breeding of these insects.

References

Ahn JE, Zhu-Salzman K (2009) CmCatD, a cathepsin D-like protease has a potential role in insect defense against a phytocystatin. J Insect Physiol 55:678–685. https://doi.org/10.1016/j.jinsphys.2009.04.016

Ahn JE, Lovingshimer MR, Salzman RA et al (2007) Cowpea bruchid *Callosobruchus maculatus* counteracts dietary protease inhibitors through modulating propeptides of major digestive enzymes. Insect Mol Biol 16:295–304. https://doi.org/10.1111/j.1365-2583.2007.00726.x

Amirhusin B, Shade RE, Koiwa H et al (2007) Protease inhibitors from several classes work synergistically against *Callosobruchus maculatus*. J Insect Physiol 53:734–740. https://doi.org/10.1016/j.jinsphys.2007.03.008

Ashouri S, Pourabad RF (2021) Regulation of gene expression encoding the digestive α-amylase in the larvae of Colorado potato beetle, *Leptinotarsa decemlineata* (Say) in response to plant protein extracts. Gene 766:145159. https://doi.org/10.1016/j.gene.2020.145159

Barbosa AE, Albuquerque ÉV, Silva M et al (2010) α-Amylase inhibitor-1 gene from *Phaseolus vulgaris* expressed in *Coffea arabica* plants inhibits α-amylases from the coffee berry borer pest. BMC Biotechnol 10:1–8. https://doi.org/10.1186/1472-6750-10-44

Bifano TD, Samuels RI, Alexandre D, Silva CP (2010) Host-mediated induction of α-amylases by larvae of the Mexican bean weevil *Zabrotes subfasciatus* (Coleoptera: Chrysomelidae: Bruchinae) is irreversible and observed from the initiation of the feeding period. Arch Insect Biochem Physiol 74:247–260. https://doi.org/10.1002/arch.20375

Bown DP, Wilkinson HS, Gatehouse JA (1997) Differentially regulated inhibitor-sensitive and insensitive protease genes from the phytophagous insect pest, *Helicoverpa armigera*, are members of complex multigene families. Insect Biochem Mol Biol 27:625–638

Brioschi D, Nadalini LD, Bengtson MH, Sogayar MC, Moura DS, Silva-Filho MC (2007) General up regulation of *Spodoptera frugiperda* trypsins and chymotrypsins allows its adaptation to soybean proteinase inhibitor. Insect Biochem Mol Biol 37:1283–1290. https://doi.org/10.1016/j.ibmb.2007.07.016

Brito LO, Lopes AR, Parra JRP et al (2001) Adaptation of tobacco budworm *Heliothis virescens* to proteinase inhibitors may be mediated by the synthesis of new proteinases. Comp

Biochem Physiol Part B Biochem Mol Biol 128B:365–375. https://doi.org/10.1016/S1096-4959(00)00325-0

Brunelle F, Nguyen-Quoc B, Cloutier C, Michaud D (1999) Proteinhydrolysis by Colorado potato beetle, *Leptinotarsa decemlineata*, digestive proteases: the catalytic role of cathepsin D. Arch Insect Biochem Physiol 42:88–98

Calderón-Cortés N, Quesada M, Watanabe H, Cano-Camacho H, Oyama K (2012) Endogenous plant cell wall digestion: a key mechanism in insect evolution. Annu Rev Ecol Evol Syst 43:45–71. https://doi.org/10.1146/annurev-ecolsys-110411-160312

Curzi MJ, Zavala JA, Spencer JL, Seufferheld MJ (2012) Abnormally high digestive enzyme activity and gene expression explain the contemporary evolution of a *Diabrotica* biotype able to feed on soybeans. Ecol Evol 2:2005–2017. https://doi.org/10.1002/ece3.331

D'Ovidio R, Raiola A, Capodicasa C, Devoto A et al (2004) Characterization of the complex locus of bean encoding PG-inhibiting proteins reveals subfunctionalization for defense against fungi and insects. Plant Physiol 135:2424–2435. https://doi.org/10.1104/pp.104.044644

Da Lage JL (2018) The amylases of insects. Int J Insect Sci:10. https://doi.org/10.1177/1179543318804783

de Oliveira CF, de Paula ST, Parra JR et al (2013) Insensitive trypsins are differentially transcribed during *Spodoptera frugiperda* adaptation against plant protease inhibitors. Comp Biochem Physiol Part B Biochem Mol Biol 165B:19–25. https://doi.org/10.1016/j.cbpb.2013.02.008

Dunse KM, Kaa Q, Guarino RF, Barton PA et al (2010) Molecular basis for the resistance of an insect chymotrypsin to a potato type II proteinase inhibitor. Proc Natl Acad Sci U S A 107:15016–15021. https://doi.org/10.1073/pnas.1009327107

Erlandson MA, Hegedus DD, Baldwin D et al (2010) Characterization of the *Mamestra configurata* (Lepidoptera: Noctuidae) larval midgut protease complement and adaptation to feeding on artificial diet, *Brassica* species, and protease inhibitor. Arch Insect Biochem Physiol 75:70–91. https://doi.org/10.1002/arch.20381

Franco OL, Rigden DJ, Melo FR et al (2000) Activity of wheat α-amylase inhibitors towards bruchid α-amylases and structural explanation of observed specificities. Eur J Biochem 267:2166–2173. https://doi.org/10.1046/j.1432-1327.2000.01199.x

Gamage SI, Kaewwongwal A, Laosatit K et al (2022) Tandemly duplicated genes encoding PG inhibitors are associated with bruchid (*Callosobruchus chinensis*) resistance in moth bean (*Vigna aconitifolia*). Plant Sci 323:111402. https://doi.org/10.1016/j.plantsci.2022.111402

Gatehouse AM, Fenton KA, Jepson I, Pavey DJ (1986) The effects of α-amylase inhibitors on insect storage pests: inhibition of α-amylase in vitro and effects on development in vivo. J Sci Food Agric 37:727–734

Giri AP, Kachole MS (1998) Amylase inhibitors of pigeonpea (*Cajanus cajan*) seeds. Phytochemistry 47:197–202

Giri AP, Harsulkar AM, Deshpande VV et al (1998) Chickpea defensive proteinase inhibitors can be inactivated by podborer gut proteinases. Plant Physiol 116:393–401

González-Ruiz CR, Del Toro-Sánchez CL, Cornejo-Ramírez YI et al (2021) Differential biochemical and kinetic properties of α-amylases from *Rhyzopertha Dominica* (F.) progenies reared on wheat varieties differing in α-amylase inhibitory activity. J Stored Prod Res 90:101748. https://doi.org/10.1016/j.jspr.2020.101748

Green TR, Ryan CA (1972) Wound-induced proteinase inhibitors in plant leaves: a possible defense mechanism against insects. Science 175:776–777

Haeger W, Pauchet Y, Kirsch R (2018) Evolution of PGs and PG-inhibiting proteins: a view beyond the classical perspective. Ann Plant Rev 15:81–122. https://doi.org/10.1002/9781119312994.apr0776

Haeger W, Henning J, Heckel DG et al (2020) Direct evidence for a new mode of plant defense against insects via a novel PG-inhibiting protein expression strategy. J Biol Chem 295:11833–11844. https://doi.org/10.1074/jbc.RA120.014027

Hafeez M, Li XW, Zhang JM et al (2021) Role of digestive protease enzymes and related genes in host plant adaptation of a polyphagous pest, *Spodoptera frugiperda*. Insect Sci 28:611–626. https://doi.org/10.1111/1744-7917.12906

Ishimoto M, Chrispeels MJ (1996) Protective mechanism of the Mexican bean weevil against high levels of α-amylase inhibitor in the common bean. Plant Physiol 111:393–401

Jongsma MA, Bolter C (1997) The adaptation of insects to plant protease inhibitors. J Insect Physiol 43:885–895

Jongsma MA, Bakker PL, Peters J, Bosch D, Stiekema WJ (1995) Adaptation of *Spodoptera exigua* larvae to plant proteinase inhibitors by induction of gut proteinase activity insensitive to inhibition. Proc Natl Acad Sci U S A 92:8041–8045

Kalunke RM, Tundo S, Benedetti M et al (2015) An update on PG-inhibiting protein (PGIP), a leucine-rich repeat protein that protects crop plants against pathogens. Front Plant Sci 6:146. https://doi.org/10.3389/fpls.2015.00146

Kirsch R, Wielsch N, Vogel H et al (2012) Combining proteomics and transcriptome sequencing to identify active plant-cell-wall-degrading enzymes in a leaf beetle. BMC Genomics 13:1–5. https://doi.org/10.1186/1471-2164-13-587

Kirsch R, Gramzow L, Theißen G et al (2014) Horizontal gene transfer and functional diversification of plant cell wall degrading PGs: key events in the evolution of herbivory in beetles. Insect Biochem Mol Biol 52:33–50. https://doi.org/10.1016/j.ibmb.2014.06.008

Kirsch R, Kunert G, Vogel H, Pauchet Y (2019) Pectin digestion in herbivorous beetles: impact of pseudoenzymes exceeds that of their active counterparts. Front Physiol 10:685. https://doi.org/10.3389/fphys.2019.00685

Kirsch R, Vurmaz E, Schaefer C et al (2020) Plants use identical inhibitors to protect their cell wall pectin against microbes and insects. Ecol Evol 10:3814–3824. https://doi.org/10.1002/ece3.6180

Kirsch R, Okamura Y, Haeger W et al (2022) Metabolic novelty originating from horizontal gene transfer is essential for leaf beetle survival. Proc Natl Acad Sci U S A 119:e2205857119. https://doi.org/10.1073/pnas.2205857119

Koepke J, Ermler U, Warkentin E et al (2000) Crystal structure of cancer chemopreventive Bowman-Birk inhibitor in ternary complex with bovine trypsin at 2.3Ä resolution. Structural basis of Janus-faced serine protease inhibitor specificity. J Mol Biol 298:477–491. https://doi.org/10.1006/jmbi.2000.3677

Koo YD, Ahn JE, Salzman RA et al (2008) Functional expression of an insect cathepsin B-like counter-defence protein. Insect Mol Biol 17:235–245. https://doi.org/10.1111/j.1365-2583.2008.00799.x

Li HM, Sun L, Mittapalli O et al (2010) Bowman-Birk inhibitor affects pathways associated with energy metabolism in *Drosophila melanogaster*. Insect Mol Biol 19:303–313. https://doi.org/10.1111/j.1365-2583.2009.00984.x

Li H, Zhou H, Zhang J et al (2021) Proteinaceous α-amylase inhibitors: purification, detection methods, types and mechanisms. Int J Food Prop 24:277–290. https://doi.org/10.1080/10942912.2021.1876087

Lin G, Bode W, Huber R et al (1993) The 0.25-nm X-ray structure of Bowman-Birk-type inhibitor from mung bean in ternary complex with porcine trypsin. Eur J Biochem 212:549–555

Liu HW, Li YS, Tang X et al (2017) A midgut-specific serine protease, BmSP36, is involved in dietary protein digestion in the silkworm, *Bombyx mori*. Insect Sci 24:753–767. https://doi.org/10.1111/1744-7917.12369

Lokya V, Swathi M, Mallikarjuna N, Padmasree K (2020) Response of midgut trypsin-and chymotrypsin-like proteases of *Helicoverpa armigera* larvae upon feeding with peanut BBI: biochemical and biophysical characterization of PnBBI. Front Plant Sci 11:266. https://doi.org/10.3389/fpls.2020.00266

Lopes AR, Juliano MA, Juliano L, Terra WR (2004) Coevolution of insect trypsins and inhibitors. Arch Insect Biochem Physiol 55:140–152. https://doi.org/10.1002/arch.10134

Lopes AR, Juliano MA, Marana SR et al (2006) Substrate specificity of insect trypsins and the role of their subsites in catalysis. Insect Biochem Mol Biol 36:130–140. https://doi.org/10.1016/j.ibmb.2005.11.006

Lopes AR, Sato PM, Terra WR (2009) Insect chymotrypsins: Chloromethyl ketone inactivation and substrate specificity relative to possible coevolutional adaptation of insects and plants. Arch Insect Biochem Physiol 70:188–203. https://doi.org/10.1002/arch.20289

Markwick NP, Laing WA, Christeller JT et al (1998) Overproduction of digestive enzymes compensates for inhibitory effects of protease and α-amylase inhibitors fed to three species of leafrollers (Lepidoptera: Tortricidae). J Econ Entomol 91:1265–1276

Maulik A, Sarkar AI, Devi S, Basu S (2012) PG-inhibiting proteins–leucine-rich repeat proteins in plant defence. Plant Biol 14:22–30. https://doi.org/10.1111/j.1438-8677.2011.00501.x

Mazumdar-Leighton S, Broadway RM (2001) Transcriptional induction of diverse midgut trypsins in larval *Agrotis ipsilon* and *Helicoverpa zea* feeding on the soybean trypsin inhibitor. Insect Biochem Mol Biol 31:645–657. https://doi.org/10.1016/S0965-1748(00)00169-7

McKenna DD, Shin S, Ahrens D et al (2019) The evolution and genomic basis of beetle diversity. Proc Natl Acad Sci U S A 116:24729–24737. https://doi.org/10.1073/pnas.1909655116

Miao Z, Cao X, Jiang H (2020) Digestion-related proteins in the tobacco hornworm, *Manduca sexta*. Insect Biochem Mol Biol 126:103457. https://doi.org/10.1016/j.ibmb.2020.103457

Morris K, Lorenzen MD, Hiromasa Y et al (2009) *Tribolium castaneum* larval gut transcriptome and proteome: a resource for the study of the coleopteran gut. J Proteome Res 8:3889–3898. https://doi.org/10.1021/pr900168z

Morton RL, Schroeder HE, Bateman KS et al (2000) Bean α-amylase inhibitor 1 in transgenic peas (*Pisum sativum*) provides complete protection from pea weevil (*Bruchus pisorum*) under field conditions. Proc Natl Acad Sci U S A 97:3820–3825. https://doi.org/10.1073/pnas.070054597

Nogueira FC, Silva CP, Alexandre D et al (2012) Global proteome changes in larvae of *Callosobruchus maculatus* (Coleoptera: Chrysomelidae: Bruchinae) following ingestion of a cysteine proteinase inhibitor. Proteomics 12:2704–2715. https://doi.org/10.1002/pmic.201200039

Oppert B, Morgan TD, Hartzer K, Kramer KJ (2005) Compensatory proteolytic responses to dietary proteinase inhibitors in the red flour beetle, *Tribolium castaneum* (Coleoptera: Tenebrionidae). Comp Biochem Physiol Part C Toxicol Pharmacol 140C:53–58. https://doi.org/10.1016/j.cca.2005.01.006

Paschold A, Halitschke R, Baldwin IT (2007) NaCOI1 mediates herbivore-induced resistance in *Nicotiana attenuata* and reveals the role of herbivore movement in avoiding defenses. Plant J 51:79–91. https://doi.org/10.1111/j.1365-313X.2007.03119.x

Pauchet Y, Wilkinson P, Chauhan R, Ffrench-Constant RH (2010) Diversity of beetle genes encoding novel plant cell wall degrading enzymes. PLoS One 5:e15635. https://doi.org/10.1371/journal.pone.0015635

Pauchet Y, Kirsch R, Giraud S et al (2014) Identification and characterization of plant cell wall degrading enzymes from three glycoside hydrolase families in the cerambycid beetle *Apriona japonica*. Insect Biochem Mol Biol 49:1–13. https://doi.org/10.1016/j.ibmb.2014.03.004

Pedra JH, Brandt A, Westerman R et al (2003) Transcriptome analysis of the cowpea weevil bruchid: identification of putative proteinases and α-amylases associated with food breakdown. Insect Mol Biol 12:405–412. https://doi.org/10.1046/j.1365-2583.2003.00425.x

Pimentel AC, Dias RO, Bifano TD et al (2020) Cathepsins L and B in *Dysdercus peruvianus*, *Rhodnius prolixus*, and *Mahanarva fimbriolata*. Looking for enzyme adaptations to digestion. Insect Biochem Mol Biol 127:103488. https://doi.org/10.1016/j.ibmb.2020.103488

Prabhakar SH, Chen MS, Elpidina EN et al (2007) Sequence analysis and molecular characterization of larval midgut cDNA transcripts encoding peptidases from the yellow mealworm, *Tenebrio molitor* L. Insect Mol Biol 16:455–468. https://doi.org/10.1111/j.1365-2583.2007.00740.x

Pytelkova J, Hubert J, Lepšík M et al (2009) Digestive α-amylases of the flour moth *Ephestia kuehniella* – adaptation to alkaline environment and plant inhibitors. FEBS J 276:3531–3546. https://doi.org/10.1111/j.1742-4658.2009.07074.x

Roy R (2020) Influence of plant induced defenses on cannibalistic behavior of *Spodoptera frugiperda* caterpillars. Western Illinois University PhD thesis

Sardoy P, Ilina N, Borniego L et al (2021) Proteases inhibitors-insensitive cysteine proteases allow *Nezara viridula* to feed on growing seeds of field-grown soybean. J Insect Physiol 132:104250. https://doi.org/10.1016/j.jinsphys.2021.104250

Schroeder HE, Gollasch S, Moore A et al (1995) Bean α-amylase inhibitor confers resistance to the pea weevil (*Bruchus pisorum*) in transgenic peas (*Pisum sativum* L.). Plant Physiol 107:1233–1239

Shelomi M, Danchin EG, Heckel D et al (2016) Horizontal gene transfer of pectinases from bacteria preceded the diversification of stick and leaf insects. Sci Rep 23:1–9. https://doi.org/10.1038/srep26388

Shibao PY, Ferro M, de Paula FF et al (2021) Identification and functional analysis of a pseudo-cysteine protease from the midgut transcriptome of *Sphenophorus levis*. Int J Mol Sci 22:11476. https://doi.org/10.3390/ijms222111476

Silva CP, Terra WR, Xavier-Filho J et al (1999) Digestion in larvae of *Callosobruchus maculatus* and *Zabrotes subfasciatus* (Coleoptera: Bruchidae) with emphasis on α-amylases and oligosaccharidases. Insect Biochem Mol Biol 29:355–366. https://doi.org/10.1016/S0965-1748(99)00010-7

Silva CP, Terra WR, De Sá MG et al (2001a) Induction of digestive α-amylases in larvae of *Zabrotes subfasciatus* (Coleoptera: Bruchidae) in response to ingestion of common bean α-amylase inhibitor 1. J Insect Physiol 47:1283–1290. https://doi.org/10.1016/S0022-1910(01)00115-9

Silva CP, Terra WR, Lima RM (2001b) Differences in midgut serine proteinases from larvae of the bruchid beetles *Callosobruchus maculatus* and *Zabrotes subfasciatus*. Arch Insect Biochem Physiol 47:18–28. https://doi.org/10.1002/arch.1031

Silva CP, Terra WR, Xavier-Filho J et al (2001c) Digestion of legume starch granules by larvae of *Zabrotes subfasciatus* (Coleoptera: Bruchidae) and the induction of α–amylases in response to different diets. Insect Biochem Mol Biol 31:41–50. https://doi.org/10.1016/S0965-1748(00)00103-X

Silva CP, Dias RO, Bernardes V et al (2022) Recruitment of lysosomal cathepsins B, L and D as digestive enzymes in Coleoptera. Insect Mol Biol 31:225–240. https://doi.org/10.1111/imb.12753

Simpson RM, Newcomb RD, Gatehouse HS et al (2007) Expressed sequence tags from the midgut of *Epiphyas postvittana* (Walker) (Lepidoptera: Tortricidae). Insect Mol Biol 16:675–690. https://doi.org/10.1111/j.1365-2583.2007.00763.x

Song HK, Suh SW (1998) Kunitz-type soybean trypsin inhibitor revisited: refined structure of its complex with porcine trypsin reveals an insight into the interaction between a homologous inhibitor from *Erythrina caffra* and tissue-type plasminogen activator. J Mol Biol 275:347–363

Svensson B, Fukuda K, Nielsen PK, Bønsager BC (2004) Proteinaceous α-amylase inhibitors. Biochim Biophys Acta 1696:145–156. https://doi.org/10.1016/j.bbapap.2003.07.004

Tamaki FK, Terra WR (2015) Molecular insights into mechanisms of lepidopteran serine proteinase resistance to natural plant defenses. Biochem Biophys Res Commun 467:885–891. https://doi.org/10.1016/j.bbrc.2015.10.049

Terra WR, Ferreira C (2020) Evolutionary trends of digestion and absorption in the major insect orders. Arthropod Struct Dev 56:100931. https://doi.org/10.1016/j.asd.2020.100931

Tokuda G (2019) Plant cell wall degradation in insects: recent progress on endogenous enzymes revealed by multi-omics technologies. Adv Insect Physiol 57:97–136. https://doi.org/10.1016/bs.aiip.2019.08.001

Velasquez-Vasconez PA, Hunt BJ, Dias RO et al (2022) Adaptation of *Helicoverpa armigera* to soybean peptidase inhibitors is associated with the transgenerational upregulation of serine peptidases. Int J Mol Sci 23:14301. https://doi.org/10.3390/ijms232214301

Volpicella M, Ceci LR, Cordewener J et al (2003) Properties of purified gut trypsin from *Helicoverpa zea*, adapted to proteinase inhibitors. Eur J Biochem 270:10–19. https://doi.org/10.1046/j.1432-1033.2003.03368.x

Wen F, Wang J, Shang D et al (2023) Non-classical digestive lipase BmTGL selected by gene amplification reduces the effects of mulberry inhibitor during silkworm domestication. Int J Biol Macromol 229:589–599. https://doi.org/10.1016/j.ijbiomac.2022.12.294

Wybouw N, Pauchet Y, Heckel DG, Van Leeuwen T (2016) Horizontal gene transfer contributes to the evolution of arthropod herbivory. Genome Biol Evol 8:1785–1801. https://doi.org/10.1093/gbe/evw119

Xu P, Lu B, Liu J et al (2019) Duplication and expression of horizontally transferred PG genes is associated with host range expansion of mirid bugs. BMC Evol Biol 19:1–9. https://doi.org/10.1186/s12862-019-1351-1

Zavala JA, Giri AP, Jongsma MA, Baldwin IT (2008) Digestive duet: midgut digestive proteinases of *Manduca sexta* ingesting *Nicotiana attenuata* with manipulated trypsin proteinase inhibitor expression. PLoS One 3:e2008. https://doi.org/10.1371/journal.pone.0002008

Zavala JA, Casteel CL, Nabity PD et al (2009) Role of cysteine proteinase inhibitors in preference of Japanese beetles (*Popillia japonica*) for soybean (*Glycine max*) leaves of different ages and grown under elevated CO_2. Oecologia 161:35–41. https://doi.org/10.1007/s00442-009-1360-7

Zhang Q, Yan Q, Yuan X et al (2021) Two PG-inhibiting proteins (VrPGIP) of *Vigna radiata* confer resistance to bruchids (*Callosobruchus* spp.). J Plant Physiol 258:153376. https://doi.org/10.1016/j.jplph.2021.153376

Zhou J, Li C, Chen A et al (2020) Structural and functional relationship of *Cassia obtusifolia* trypsin inhibitor to understand its digestive resistance against *Pieris rapae*. Int J Biol Macromol 148:908–920. https://doi.org/10.1016/j.ijbiomac.2020.01.193

Zhu-Salzman K, Zeng R (2015) Insect response to plant defensive protease inhibitors. Annu Rev Entomol 60:233–252. https://doi.org/10.1146/annurev-ento-010814-020816

Zhu-Salzman K, Koiwa H, Salzman RA et al (2003) Cowpea bruchid *Callosobruchus maculatus* uses a three-component strategy to overcome a plant defensive cysteine protease inhibitor. Insect Mol Biol 12:135–145. https://doi.org/10.1046/j.1365-2583.2003.00395.x

Chapter 12
Role of Microorganisms in Digestion and Nutrition

Abstract Microorganisms play several roles in insects. However, in this chapter, we will consider only the roles of microbiota in insect digestion and nutrition. Bacteria are food for insects, exemplified by larval houseflies and blowflies, living in feces, decaying fruits, and corpses that are rich in bacteria. Microbiota, particularly bacteria, also play a role in the digestion of recalcitrant materials such as tannins, lignins, and humic substances. The last ones are complex molecules derived from residues occurring in soils. Bacterial action on those recalcitrant molecules is observed in termites and some beetles. Fungi are important players in lignocellulose digestion in termites of the subfamily Macrotermitinae. Sap-feeding insects usually lack essential nutrients, especially amino acids, which are supplied by microbial symbionts, such as those found in pea aphids. In diets poor in vitamins, symbiont microbiota may provide them for the insect, exemplified by the blood-feeder *Rhodnius prolixus*. Finally, gut microbiota help their hosts in detoxifying molecules that are characteristic of plant defenses or insecticides used in insect control.

12.1 Introduction

The association of microorganisms, mainly bacteria, with an immense range of functions in mammals led them to be considered a bacterial organ integrated into the host system (Bäcked et al. 2005). It is likely that the microbiota in insects are as important as in mammals. Insect microbiota have been implicated in immune system interactions, nutritional symbioses, effects on insect development, protection against parasite invasion, and insect communication by affecting the release of volatiles (Engel and Moran 2013; Gurung et al. 2019; Zhang et al. 2022). The microbiota also affect insect resistance to pesticides (Siddiqui et al. (2022) and in manipulating both plant and host to favor their dissemination (Franco et al. 2021). In this chapter, however, we consider only the role of microbiota in insect digestion and nutrition.

W. R. Terra et al., *Molecular Physiology and Evolution of Insect Digestive Systems*, Entomology in Focus 7, https://doi.org/10.1007/978-3-031-39233-7_12

The microbiota of insects include bacteria, fungi, archaea, and protozoa (protists), from which bacteria, as a rule, is the most abundant. Fungi are usually found in the guts of insects that digest wood or detritus. The microbiota are found in the gut lumen, attached to the peritrophic membrane, adhering to the midgut cells or inside special cells, the mycetocytes, frequently organized in groups named mycetomes. Intracellular symbionts inhabiting mycetocytes are common in sap-feeding insects.

The largest microbiota are usually associated with more compartmentalized guts or expanded hindguts, which are characteristic of some detritivores and wood-feeders, including termites, crickets, cockroaches, and beetles (Cazemier et al. 1997; Dillon and Dillon 2004). Diagrammatic representation of the guts of those insets is shown in Fig. 2.2 (Chap. 2). Microbiota may be acquired from the environment; in this case, it depends largely on diet, may be transferred from one insect to another by coprophagy or proctodeal trophallaxis, or transmitted by inheritance (endosymbionts) (Engel and Moran 2013).

There are many roles proposed for the insect gut microbiota. Bacteria may be used as food, may help digesting recalcitrant polymers, may supplement essential nutrients, and may help in the detoxification of harmful ingested compounds.

12.2 Bacteria as Food

Bacteria that grow in feces and on decaying fruits and corpses may be used as food by several insects. The natural food of the larvae of *Musca domestica* is dung (Brues 1946) and bacteria may be their sole food (Levinson 1960). Actually, bacteria were shown to be the sole or at least a major food for all cyclorrhaphous Diptera (see Chap. 10).

Bacteria are very abundant in humus, which is a detritus-rich material that mainly combines decaying plant debris, fungi, and soil minerals. Bacteria can make up to a considerable part of this necromass, which is part of the soil organic matter (Liang et al. 2019). Some insects, such as some dermestid beetles, have adapted to digest lignocellulosic material, pectin, and bacterial and fungal mass. Thus, these beetles have access to nitrogen-enriched nutrients such as amino acids, proteins, peptidoglycans, and chitin, in addition to vitamins, lipids, carbohydrates, and several micronutrients (Li and Brune 2005). Another example of bacterial biomass that serves as food for insects is those that grow on decaying plant and animal material. In fact, the blowfly larvae of *Lucilia sericata* (Diptera: Calliphoridae) have been used for centuries to aid wound healing, which they do by ingesting and digesting bacteria in necrotic tissue (Greenberg 1968; Mumcuoglu et al. 2001; Harvey et al. 2021). Bacteria that reside internally in plant tissues, called endophytes, can also serve as a source of nutrients for herbivorous insects (Martınez-Romero et al. 2021).

12.3 Digestion of Recalcitrant Compounds

Digestion of cellulose may be carried out by several insects (see Chap. 10), although in many of them there are a contribution of microbial enzymes in a variable extension (Engel and Moran 2013). It is important to stress that the finding of microbiota-producing digestive enzymes in the gut of an insect does not necessarily mean that those enzymes are really contributing to the digestion of the host. For example, despite the finding of cellulolytic enzymes in the microbiota of lepidopterans (Anand et al. 2010; Dantur et al. 2015), no cellulolytic enzymes have been assayed in midgut samples of lepidopterans, and in fact, nutritional data showed that lepidopterans do not digest cellulose and pectin (Terra et al. 1987).

In contrast to the digestion of cellulose, digestion of lignocellulose from wood and humus is restricted to a few insects that rely on endogenous enzymes and with the exception of *Anoplophora* and *Zootermophis* that also have the concur of symbionts (Geib et al. 2008; see also Chap. 11).

The best-studied lignocellulose-digesting insects are the termites (Watanabe and Tokuda 2010; Brune 2014; Tokuda 2019; Arora et al. 2022) which comprise a taxon of eusocial cockroaches, sister of the wood roach family Cryptocercidae (Lo et al. 2007; Chouvenc et al. 2021; see also Chap. 2). Termites are usually separated into lower and higher termites. The lower termites include several families, exemplified by Kalotermitidae (dry-wood termites) and Rhinotermitidae (like the subterranean termites *Coptotermes, Heterotermes*, and *Reticulitermes*) characterized by cellulolytic protozoa in the paunch, an enlarged region in the anterior midgut. The products of microbial processing of cellulose are short-chain fatty acids, mainly acetate (Hungate 1943; Yamin 1980; Odelson and Breznak 1983; Breznak and Switzer 1986; Warnecke et al. 2007), which are absorbed by the paunch (Hogan et al. 1985).

The higher termites (family Termitidae) lack protozoa but, except for the fungus-growing termites (subfamily Macroterminitinae), have bacteria playing several roles in their highly differentiated hindgut (see Fig. 2.2d in Chap. 2). Macrotermitinae rely in wood and have an obligate mutualism with fungi that are involved in the digestion of lignocellulose and provision of essential amino acids for the host. As the digested products are supplied by fungi before ingestion, Macrotermitinae digestion is external (Chouvenc et al. 2021). There are some non-Macrotermitinae Termitidae that rely on wood, but most of them feed on soil that is a layer on the surface of the ground impoverished of easily digestible organic compounds and enriched in recalcitrant molecules such as tannins, lignins, and humic substances associated with inorganic particles. Humic substances are complex molecules derived from aromatic substances, carbohydrates, and proteins which enter the soil system as plant residues or remains of animals (Schulten and Schnitzer 1998). Soil feeders have a stable community of gut bacteria and are able to digest lignocellulose, but most of their diet is the microbial biomass of the soil and its nitrogen-rich organic residues (Brune 2001, 2005, 2014; Ngugi et al. 2011; Ngugi and Brune 2012).

The dung-feeding beetle *Oryctes borbonicus* (Scarabaeidae) has no genes coding for enzymes involved in the degradation of cellulose, hemicellulose, and pectin, but is able to digest these carbohydrates. This suggests the concur of microbiota in digestion. Gut bacteria also seem to play an important role in lignocellulose digestion in Passalidae (Scarabaoidea) (Tokuda 2019).

12.4 Nutrient Provisioning

Microbiota nutrient provisioning refers to the production of essential nutrients absent or difficult to obtain from the original diet. This was the case in the synthesis de novo or recycling of nitrogen-containing wastes from the host and the production of absorbable fatty acids from lignocellulose by termite microbiota discussed in the previous item.

The diet of hemipterans usually lacks essential compounds that are supplied by microbial symbionts. For example, the pea aphid *Acyrthosiphon pisum*, which feeds on phloem sap, has the bacteria *Buchnera* in the mycetocytes of the mycetomes occurring in the aphid hemocoel. This bacterium uses non-essential amino acids absorbed by the aphid in the synthesis of essential amino acids (Shigenobu et al. 2000; Douglas 2006). Essential amino acids are also supplied by symbionts in the case of the stink-bug *Ishikawaella capsulatus* (Nikoh et al. 2011), whereas the xylem sap-feeder *Homalodisca vitripennis* has amino acids provided by one bacterium and vitamins and cofactors and by another one (Wu et al. 2006). In diets poor in vitamins, the symbiotic microbiota may provide them, as in the case of the blood-feeder *Rhodnius prolixus,* which obtains B-complex vitamins from *Rhodococcus rhodnii* (Eichler and Schaub 2002).

Microbiota also provides nutrients for insects other than hemipterans. Thus, the olive oil fruit fly *Bactrocera oleae* has an obligate symbiont supplying essential amino acids that lack in unripe olives (Ben-Yousef et al. 2014) and cockroaches use nitrogenous wastes stored as uric acid during dietary lack of nitrogen for the synthesis of essential amino acids with the aid of their obligate symbiont *Blattabacterium* (Sabree et al. 2009).

12.5 Detoxification of Harmful Ingested Compounds

Insect gut microbiota helps their host in detoxifying both molecules that are characteristic of plant defenses as those introduced by men, exemplified by insecticides (Van den Bosch and Welte 2016). Insecticides are outside the scope of this book. So we will bring here only some examples of the study of the role of microbiota in the detoxification of plant toxic compounds in insect hosts. Plant β-glucosides are toxic for many insects, mainly by their aglycones, which are their non-glucoside moieties. The addition of saligenin, the aglycone of the plant β-glucoside salicin in the diet of *Tenebrio molitor* larvae, has more deleterious effects on microbiota-free larvae than on the conventionally reared larvae (Genta et al. 2006). The metabolism of phenolic aglycones is well known in microorganisms (Pillai et al. 2002), and some insect gut bacteria use phenolic aglycones as a carbon source (Dillon and Charnley 2002). Thus, aglycone detoxification is probably carried out by the midgut microbiota, and insects with a reduced or absent gut microbiota may be subject to more deleterious effects of aglycones such as saligenin, as observed in the mentioned study of *Tenebrio molitor* larvae. In the same direction, Jurjitza (1979) found that cultures of yeast isolated from the cigarette beetle *Lasioderma serricorne* may use salicin as a sole carbon source. Thus, the data suggest a detoxifying role of midgut microbiota relative to toxic β-glucosides.

Other examples of detoxifying symbiosis (detoxification provided by symbionts) (Van den Bosch and Welte 2016) are known. The detoxification of oxalate present in some legumes is the result of an oxalate decarboxylase expressed by a bacterial symbiont of the legume pest *Megacopta* spp. (kudzu bug) (Nikoh et al. 2011). Bacterial symbionts are also responsible for the detoxification of terpenes in the pine weevil (*Dendroctonus ponderosae*) (Adams et al. 2013), isotiocyanates in the cabbage root fly (*Della radicum*) (Welte et al. 2016), and caffeine in coffee borer beetle (*Hypothenemus hampei*) (Ceja-Navarro et al. 2015).

References

Adams AS, Aylward FO, Adams SM et al (2013) Mountain pine beetles colonizing historical and naïve host trees are associated with a bacterial community highly enriched in genes contributing to terpene metabolism. Appl Environ Microbiol 79:3468–3475

Anand AAP, Vennison SJ, Sankar SG (2010) Isolation and characterization of bacteria from the gut of *Bombyx mori* that degrade cellulose, xylan, pectin and starch and their impact in digestion. J Insect Sci 10:107

Arora J, Kinjo Y, Sobotnik J et al (2022) The functional evolution of termite gut microbiota. Microbiome 10:78

Bäcked F, Ley RE, Sonnenberg IL et al (2005) Host-bacterial mutualism in the human intestine. Science 307:1915–1920

Ben-Yousef M, Pasternak Z, Jurkevitch E et al (2014) Symbiotic bacteria enable olive flies (*Bactrocera oleae*) to exploit intractable sources of nitrogen. J Evol Biol 27:2695–2705

Breznak JA, Switzer JM (1986) Acetate synthesis from H (2) plus CO (2) by termite gut microbes. Appl Environ Microbiol 52:623–630

Brues CT (1946) Insect dietary. Harvard University Press, Cambridge

Brune A (2001) Transformation and mineralization of 14C-labeled cellulose, peptidoglycan, and protein by the soil-feeding termite *Cutermes orthognatus*. Biol Fertil Soils 33:166–174

Brune A (2005) Digestion of peptidic residues in humic substances by an alkali-stable and humic-tolerant proteolytic activity in the gut of soil-feeding termites. Soil Biol Biochem 37:1648–1655

Brune A (2014) Symbiotic digestion of lignocellulose in termite guts. Nature Rev Microbiol 12:168–180

Cazemier AE, Hackstein JHP, Op den Camp HJM et al (1997) Bacteria in the intestinal tract of different species of arthropods. Microb Ecol 33:189–197

Ceja-Navarro JA, Vega FE, Karaoz U et al (2015) Gut microbiota mediate caffeine detoxification in the primary insect pest of coffee. Nat Commun 6:7618. https://doi.org/10.1038/ncomms8618

Chouvenc T, Sobotnik J, Engel MS et al (2021) Termite evolution: mutualistic associations, key innovations, and the rise of Termitidae. Cell Mol Life Sci 78:2749–2769

Dantur KI, Enrique R, Welin B et al (2015) Isolation of cellulolytic bacteria from the intestine of *Diatraea saccharalis* larvae and evaluation of their capacity to degrade sugarcane biomass. AMB Express 5:15

Dillon RJ, Charnley AK (2002) Mutualism between the desert locust *Schistocerca gregaria* and its gut microbiota. Res Microbiol 153:503–509

Dillon RJ, Dillon VM (2004) The gut bacteria of insects: nonpathogenic interactions. Annu Rev Entomol 49:71–92

Douglas AE (2006) Phloem-sap feeding by animals: problems and solutions. J Exp Bot 57:747–754

Eichler S, Schaub GA (2002) Development of symbionts in Triatomine bugs and the effects of infections with trypanosomatids. Exp Parasitol 100:17–27

Engel P, Moran NA (2013) The gut microbiota of insects-diversity in structure and function. FEMS Microbiol Rev 37:699–735

Franco FP, Túler AC, Gallan DZ et al (2021) Fungal phytopathogen modulates plant and insect responses to promote its dissemination. Isem J 15:3522–3533

Geib SM, Filley TR, Hatcher PG et al (2008) Lignin degradation in wood-feeding insects. Proc Natl Acad Sci U S A 105:12932–12937

Genta FA, Dillon RJ, Terra WR et al (2006) Potential role for gut microbiota cell wall digestion and glucoside detoxification in *Tenebrio molitor* larvae. J Insect Physiol 52:593–601

Greenberg B (1968) Model for destruction of bacteria in the midgut of blow fly maggots. J Med Entomol 5:31–83

Gurung K, Wertheim B, Salles JF (2019) The microbiome of pest insects: it is not just bacteria. Ent Exp Appl 167:1–15

Harvey ML, Dadour IR, Gasz NE (2021) Maggot therapy in chronic wounds: new approaches to historical practices. Ann Entomol Soc Am 114:415–424

Hogan ME, Slaytor M, O'Brien RW (1985) Transport of volatile fatty acids across the hindgut of the cockroach Panesthia cribrata, Sausurre and the temite, *Mastotermes darwinensis* Frogatt. J Insect Physiol 31:587–591

Hungate RE (1943) Quantitative analyses on the cellulose fermentation by termite protozoa. Ann Entomol Soc Am 36:730–739

Jurjitza G (1979) The fungi symbiotic with anobiid beetles. In: Batra LR (ed) Insect fungus symbiosis. Allenhald and Osmun, New Jersey

Levinson ZH (1960) Food of housefly larvae. Nature 188:427–428

Li X, Brune A (2005) Digestion of microbial mass, structural polysaccharides, and protein by the humivorous larva of *Pachnoda ephippiata* (Coleoptera: Scarabaeidae). Soil Biol Biochem 37:107–116

Liang C, Amelung W, Lehmann J et al (2019) Quantitative assessment of microbial necromass contribution to soil organic matter. Glob Change Biol 25:3578–3590

Lo N, Beninati T, Stone F et al (2007) Cockroaches that lack *Blattabacterium* endosymbionts: the phylogenetic divergent genus *Nocticola*. Biol Lett 3:327–330

Martınez-Romero E, Aguirre-Noyola JL, Bustamante-Brito R et al (2021) We and herbivores eat endophytes. Microb Biotechnol 14:1282–1299

Mumcuoglu KY, Miller J, Mumcuoglu M et al (2001) Destruction of bacteria in the digestive tract of the maggot of *Lucilia sericata* (Diptera: Calliphoridae). J Med Entomol 38:161–166

Ngugi DK, Brune A (2012) Nitrate reduction, nitrous oxide formation, and anaerobic ammonia oxidation to nitrite in the gut of soil-feeding termites (*Cubitermes* and *Ophiotermes spp.*). Environ Microbiol 14:860–871

Ngugi DK, Ji R, Brune A (2011) Nitrogen mineralization, denitrification, and nitrate ammonification by soli-feeding termites: a ^{15}N-based approach. Biogeochemistry 103:355–369

Nikoh N, Hosokawa T, Oshima K et al (2011) Reductive evolution of bacterial genome in insect gut environment. Genome Biol Evol 3:702–714

Odelson DA, Breznak JA (1983) Volatile fatty acid production by the hindgut microbiota of xylophagous termites. Appl Environ Microbiol 45:1602–1613

Pillai BVS, Bhinu VS, Swarup S (2002) Elucidation of the flavonoid catabolism pathway in *pseudomonas putida* PML2 by comparative metabolic profiling. Appl Environ Microbiol 68:143–151

Sabree ZL, Kambhampati S, Moran NA (2009) Nitrogen recycling and nutritional provisioning by *Blattabacterium*, the cockroach endosymbiont. Proc Natl Acad Sci U S A 106:19521–19526

Schulten HR, Schnitzer M (1998) The chemistry of soil organic nitrogen: a review. Biol Fertil Soils 26:1–15

Shigenobu S, Watanabe H, Hattori M et al (2000) Genome sequence of the endocellular bacterial symbiont of aphis *Buchenera* sp. APS Nature 407:81–86

Siddiqui JA, Khan MM, Bamisile S et al (2022) Role of insect gut microbiota in pesticide degradation: a review. Front Microbiol 13:870462

Terra WR, Valentin A, Santos CD (1987) Utilization of sugars, hemicellulose, starch, protein, fat and minerals by *Erinnyis ello* larvae and digestive role of their midgut hydrolases. Insect Biochem 17:1143–1147

Tokuda G (2019) Plant cell wall degradation in insects: recent progress on endogenous enzymes revealed by multi-omics technologies. Adv Insect Physiol 57:97–136

Van den Bosch TJM, Welte CU (2016) Detoxifying symbionts in agriculturally important pest insects. Microbiol Biotechnol 10:531–540

Warnecke F, Luginbuhl P, Ivanova N et al (2007) Metagenomic and functional analysis of hindgut microbiota of wood-feeding higher termite. Nature 450:560–565

Watanabe H, Tokuda G (2010) Cellulotic systems in insects. Annu Rev Entomol 55:609–632

Welte CU, de Graaf RM, van den Bosch TJM et al (2016) Plasmids from the gut microbiome of cabbage root fly larvae encode SaxA that catalyzes the conversion of the plant toxin 2-phenylethyl isothiocyanate. Environ Microbiol 18:1379–1390

Wu D, Daugherty SC, Van Aken SE et al (2006) Metabolic complementarity and genomics of the dual bacterial symbiosis of sharpshooters. PLoS Biol 4:e188

Yamin MA (1980) Cellulose metabolism by the termite flagellate *Trichomitopsis termopsidis*. Applied Env Microbiol 39:859–863

Zhang X, Zhang F, Lu X (2022) Diversity and functional roles of the gut microbiota in lepidopteran insects. Microorganisms 10:1234

Chapter 13
Molecular View of Digestion and Absorption in the Major Insect Orders

Abstract The digestive physiology of representative insects is described with emphasis on the molecular aspects. The groups studied included Polyneoptera (grasshoppers, crickets, cockroaches, and termites), Condynognatha + Psocodea (aphids, bugs, and lice), and Holometabola (beetles, bees and ants, mosquitoes and flies, butterflies, and moths). For this, a combination of techniques was employed, such as digestive enzyme assays, isolation of cell fractions, determination of the absorption of amino acids, glucose, water fluxes, enzyme excretory rates, and luminal pH values. In addition to that, the major nutrient and ion transporters were identified by transcriptomics and proteomics, and their role in vivo was tested with inhibitors. Finally, gut ultrastructure was described to support the findings obtained by cell fractionation and hypothesis on the role of specific cells in buffering, digestive enzyme secretion, and excretion of ions. Few insects were subjected up till now to all research techniques described above, but the results were clear to show that the features of insect digestion have a strong correlation with insect phylogeny.

13.1 Introduction

The spatial organization of digestion and absorption among insects vary depending mainly on their phylogenetic position and secondarily on their diets. In this chapter, we discuss the digestive and absorptive processes in the insect groups for which there are more data. The insects are separated into Polyneoptera (A), Condylognatha + Phthiraptera (B), and Holometabola (C). See Chap. 1 for phylogenetic and habit details.

A. Polyneoptera

W. R. Terra et al., *Molecular Physiology and Evolution of Insect Digestive Systems*, Entomology in Focus 7, https://doi.org/10.1007/978-3-031-39233-7_13

13.2 Orthoptera

13.2.1 Introduction

Orthoptera has more than 25700 species. It is the most diverse order among Polyneoptera (Grimaldi and Engel 2005), and based on molecular phylogeny with fossil calibration, it is evolving for at least 250 million years (Song et al. 2015). Orthoptera comprises two suborders: Ensifera, which includes the family Grillidae (the crickets) and Gryllotalpidae (mole crickets), and Caelifera, which includes the family Acrididae (grasshoppers and locusts). Locusts are grasshoppers which are able to periodically form large populations that denuge an area before swarm another. Locust species vary depending on the continent.

13.2.2 Caelifera

There are numerous papers dealing with the morphology of grasshopper gut (Hodge 1936, 1939; Heinrich and Zebe 1973; Marana et al. 1997) and some that described the ultrastructural features of the gut cells involved in water fluxes (see more details on those features in Chap. 5) and reported the occurrence of digestive enzyme exocytosis in midgut ceca and anterior ventriculus, giving origin to dark granules among microvilli, and cell extrusions corresponding to apocrine secretion. The gut of grasshoppers (Chap. 2, Fig. 2.2a), exemplified by *Abracris flavolineata*, is formed by a large crop, a short ventriculus with six midgut ceca having an anterior and a posterior lobe (named anterior and posterior ceca, respectively) and a hindgut. The pH values of gut contents change from acid (crop, 6.1; ceca, 6.3; anterior ventriculus, 6.5) to alkaline (posterior ventriculus, 7.4; hindgut, 7.0) (Marana et al. 1997).

A peritrophic membrane is found in the ventriculus of all grasshoppers, but its presence in midgut ceca is controversial (see references in Baines 1979). Probably the controversy resulted from the fact that the midgut ceca peritrophic membrane may be very thin in some cases and embedded in mucus, which may be misinterpreted as a fluid, or the peritrophic membrane may be incomplete or absent (Dias et al. 2018).

According to earlier work done in grasshoppers (Wigglesworth 1972), digestion occurs mainly in their large crops by saliva ingested with food and enzymes moved forward from the midgut. This description was based on microscopic observation of gut contents and semiquantitative determination of digestive enzymes determinations. More recent work only partly agreed with that description. With the exception of a minor activity of amylase, salivary enzymes have activities too low to play a role in the digestion of at least *Locusta migratoria* (Droste and Zebe 1974) and *Abracris flavolineata* (Ferreira et al. 1990). Most digestion in those insects occur in the crop under the action of midgut enzymes (Droste and Zebe 1974; Anstee and Charnley 1977; Ferreira et al. 1990).

The activity of digestive enzymes in crop, midgut caeca, ventriculus, and at the midgut cell fraction corresponding to the secretory vesicles of *A. flavolineata* was determined at different times after a meal. From the data, a model of the digestive events occurring in *A. flavolineata* was proposed (Marana et al. 1997). According to this model, soon after the meal, the anterior midgut ceca cells secrete digestive enzymes that are moved into the crop by contractions of the ceca. Digestion starts in the crop, and after 3 h, the contents of the crop moved to the ceca. Digestion in the midgut ceca occurs between 3 and 5 h, after when the midgut ceca empty their contents into the ventriculus, where digestion is finished. Based on the specificities of the enzymes occurring at the different gut compartments, the initial digestion of carbohydrates is carried out chiefly in the crop; of protein, mainly in ceca lumina, and the final digestion of proteins taking place at the surface of cecal cells by membrane-bound enzymes (Ferreira et al. 1990). Anterior ceca are richer in enzymes than the posterior ceca. As the relative size of the posterior ceca differs among grasshoppers with diets with different allelochemical contents, these ceca were supposed to have roles other than digestion, such as detoxification (Chapman 1988). For example, the removal of phenolics from the gut may be caused by the epithelial pockets seen in the posterior midgut ceca of several grasshoppers (Bernays 1981; Chapman 1988), including *A. flavolineata* (Marana et al. 1997).

The absorption of glucose and amino acids was studied in locusts. For this, the locusts were fed a mixture of radiolabeled glucose (or amino acid) and the non-absorbable dye amaranth. The changes in the ratio of the concentrations of radiolabeled compounds and dye along the gut showed that most absorption of glucose and amino acids occur in the midgut, especially from the ceca lumen (Treherne 1958a, 1959, 1967).

The concentration of digestive enzymes in the posterior ventriculus is similar to that found in the hindgut of feeding *A. flavolineata*, meaning that they have no midgut counter fluxes of fluid that would recycle digestive enzymes, thus avoiding their excretion (Ferreira et al. 1990). This is similar to what was found in *Schistocerca gregaria*, where abundant saliva saturates the absorbing midgut ceca sites, thus hindering the countercurrent fluxes of fluid propelled by fluid excretion by the Malpighian tubules. On the contrary, midgut counter fluxes of fluids were observed in fasting grasshoppers (Dow 1981a, b).

Ongoing transcriptomic research in the authors' lab provided details of the gut molecular physiology of *A. flavolineata*. Thus, most amylase, cellulase, and phospholipase A2 are expressed at *A. flavolineata* salivary glands; α-glucosidases at the ventriculus, whereas the other digestive enzymes have the highest expression in the midgut ceca. Regarding nutrient transporters, amino acid and fatty acid transporters are similarly expressed at the midgut ceca and ventriculus, whereas sugar transporter expression predominates at the ventriculus, as expected from the in vivo absorption data (see above). The transporters involved in water transport (see details in Chap. 6), aquaporins, KCC, and NKCC are more expressed at the midgut ceca and the Malpighian tubules, in agreement with the proposed water-absorbing function of the midgut ceca and water secretory activity of the Malpighian tubules.

13.2.3 Ensifera

Crickets are omnivorous or predatory and have a spacious crop, a proventriculus, a pair of anteriorly midgut caeca, a short ventriculus, a long hindgut, and a ureter collecting the Malpighian tubules inserted between the ventriculus and the hindgut (Chap. 2, Fig. 2.2b) (Thomas and Nation 1984; Teo and Woodring 1985; Woodring and Lorenz 2007; Biagio et al. 2009). A peritrophic membrane encloses the food in the ventriculus, but not in the midgut ceca. The pH of gut contents varies from acid in crop to alkaline in ceca and ventriculus (Cooper and Vulcano 1997; Biagio et al. 2009). The distribution of enzyme activities led to the conclusion that the initial digestion of starch occurs in crop; of protein, in midgut caeca lumina; and final digestion of starch and protein in midgut ceca lumina, extending also to ventriculus, in the case of protein. The enzymes involved in the final digestion (aminopeptidase and maltase) are mostly membrane bound, probably microvillar. Crop enzymes come from midgut caeca, which emptying is driven by peristalsis (*G. sigillatus*) or by the relative pressure produced by proventriculus and ceca (*G. bimaculatus*) (Woodring and Lorenz 2007; Biagio et al. 2009).

The absorption of radioactive glucose, glycine, and palmitic acid was determined by the injection in vitro of those compounds and a non-absorbable dye in isolated midgut ceca plus anterior midgut and posterior midgut, which the authors defined as specialized segment of the anterior hindgut. The ratio of the concentrations of those compounds and dye after a while showed that they were absorbed mainly from the anterior rather than the posterior midgut (Thomas and Nation 1984). The cuticle described by Thomas and Nation (1984) covering the posterior ventriculus is probably the peritrophic membrane to account for the fact that injected glycine at the lumen of the anterior and posterior ventriculus appears in the incubation medium at the same rate.

After ingestion of a dye, the luminal side of the anterior midgut of *G. sigillatus* is heavily stained, and following dye injection in the hemocoel, the hemal side of the posterior midgut is stained. Hence, the anterior midgut is water-absorbing and the posterior midgut is water-secreting in both feeding and starving crickets. In agreement with the dye results, the columnar cells of the anterior and posterior midgut have morphological features corresponding to a function in water absorption and secretion, respectively. As expected, digestive enzymes are excreted at a low rate in both feeding and starving insects. In addition, the inspection of midgut cell morphology along the midgut showed that merocrine secretory processes occur mainly in the ventriculus and apocrine processes occur chiefly in the midgut ceca (Biagio et al. 2009).

13.3 Dictyoptera

13.3.1 Introduction

Dictyoptera includes Blattodea (cockroaches and termites) and Mantodea (mantids) (Chap. 1). Blattodea comprises all cockroaches, one group, the Cryptocercidae (wood roach)-termites branch, of which is characterized by the acquisition of gut protists and proctodeal trophallaxis (Chouvenc et al. 2021). Thus, termites may be considered as insects with the same ancestor as wood roaches, but more adapted that the latter to feed on wood and humus. Mantids are carnivorous feeding on grasshoppers, caterpillars, flies, etc.

13.3.2 Blattodea: Cockroaches

The cockroach digestive tube comprises a large crop, a relatively small midgut with associated midgut ceca and a well-developed hindgut (Chap. 2, Fig. 2.2e). The salivary glands are large with noticeable reservoirs (Bignell 1981; Ma et al. 2009; Tamaki et al. 2014). Ultrastructural research on the gut of *P. americana* showed the gastric ceca and anterior midgut have apical extrusions (apocrine secretion), whereas dense secretory granules (exocytic secretion) are visible along the whole midgut. Immunocytochemical research revealed that trypsin is released by apocrine secretion and amylase by exocytosis. Gastric ceca cells have basal infoldings characteristic of absorptive cells (Tamaki et al. 2014).

The salivary glands of *P. americana* secrete two cellulases and three laminarinases able to lyse fungal cells and open plant cells (Genta et al. 2003) and a liquefying (non-processive) amylase (Tamaki et al. 2014). It is interesting to add that *P. americana* has another amylase, a saccharifying (processive) one in the midgut (Tamaki et al. 2014). The saliva goes to the crop, which also receives enzymes propelled forward by antiperistalsis from the gastric ceca and the anterior midgut. The crop, and secondarily the ceca and anterior ventriculus, are the site where most digestion takes place. However, the terminal digestion of proteins, under the action of partly membrane-bound enzymes, occurs in the posterior midgut. Digestion takes place under mildly acidic pH along the whole midgut with digestive enzymes which have a low excretory rate. This low excretory rate is supposed to be due to enzyme recycling propelled by a countercurrent flux of fluid powered by fluid secretion by the Malpighian tubules and absorption by the ceca cells with developed basal membrane infoldings (Tamaki et al. 2014).

Although there is less information on cockroaches other than *P. americana*, the available data reveal some variability among them. Thus, the blaberid *N. cinerea* has the anterior midgut acid and the posterior midgut alkaline, and the wood roaches, like *Panesthia* and *Cryptocercus*, possess an enlarged hindgut harboring cellulolytic microbiota (Cleveland et al. 1934). In accordance with that, the products of microbial cellulolytic activity, mainly acetate and butyrate (see Chap. 12), are more transported by the hindgut of *Panesthia cribrata* than of *P. americana* (Hogan et al. 1985).

A large number of *P. americana* digestive enzymes have been kinetically studied and many of them have had their coding genes cloned and sequenced (see Tamaki et al. 2014 and references therein). It is interesting to comment here on the peptidase inhibitors found in the lumen of midgut ceca and the anterior midgut of *Leucophea maderae* (Engelmann and Geraerts 1980) and other cockroaches (Vinokurov et al. 2007). Those inhibitors seem to be a device to decrease the proteolytic inactivation of glycosidases in the anterior midgut, where they predominate. More evolved insects lack those inhibitors because their digestive glycosidases are stable in the presence of their own peptidases (Terra 1988).

A quantitative study of the absorption of fatty acids and glucose was conducted in *P. americana* with the use of radioactively labeled nutrients and a non-absorbable dye as detailed above. Although tripalmitin hydrolysis partially occurs in the crop, fatty acid absorption was found only in the ceca and ventriculus (Treherne 1958b), the same being true for glucose (Treherne 1957). Phenylalanine is absorbed with Na^+ in the midgut of *Blabera gigantea*, and this process is inhibited by ouabain, the inhibitor of the NK-ATPase. This suggests that the transport of Na^+ from the cell to the hemolymph by the NK-ATPase powers the symport phenylalanine-Na^+. The existence of a symport phenylalanine-Na^+ was confirmed in vitro with purified midgut microvillar membranes (Parenti et al. 1986).

The capacity of *P. americana* to live in a very toxic environment depends on special proteins. Thus, it was not unexpected that Zhang et al. (2016), with a transcriptomic approach, described a remarkable number of proteins involved in detoxification coded by *P. americana* endogenous genes or from those of their microbiota. These proteins include cytochrome P450 monooxigenases, able to detoxify many kinds of xenobiotics by redox reactions; glutathione S-transferases that catalyze the addition of reduced glutathione to oxidized lipids or other toxins, reducing their toxicity or making them more soluble and easily to be excreted; and carboxylesterases that catalyze the hydrolysis of insecticides like pyrethroids and organophosphates (Zhang et al. 2016 and references therein).

13.3.3 Blattodea: Termites

Termites are major cellulose-feeders that together with *Cryptocercus* have as ancestor a gregarious wood-feeding roach, which progressively evolved from a detritivorous diet to a xylophagous one (Chouvenc et al. 2021). As mentioned before (Chap. 12), termites are separated into lower and higher termites, which differ in their gut morphology and contribution of gut microbiota in digestion.

The lower termites have a gut characterized by a short foregut (a crop is absent), a midgut without gastric ceca, and a hindgut having an enlarged region named paunch. The paunch harbors cellulolytic protozoa that act on wood after mechanical grinding by mandibles and the gizzard. The whole luminal gut has a pH value around neutrality (Noirot and Noirot-Timothe 1969; Fujita et al. 2010; Brune 2014; Costa-Leonardo and Silva 2022). The cellulolytic protozoa have an endosymbiont, which genome annotation unveiled its capacity to fix nitrogen and recycle host nitrogen wastes (Hongoh et al. 2008). This confirmed the early studies of Benemann (1973) that showed that nitrogen fixation by microbial symbionts occurs in termites and clarified how termites may survive ingesting food poor in nitrogen. However, the products of microbial action are formed in the hindgut, and hence, they cannot pass through the midgut for further digestion and absorption. This problem is circumvented by passing the hindgut mass from one individual to another by proctodeal (anal) trophallaxis, which is a phenomenon characteristic of termites (Chouvenc et al. 2021).

The evolution of higher termites that have a wide diversity of diets occurred with the loss of protozoa symbionts and the acquisition of alternative bacterial symbionts. This transition apparently occurred as a result of the externalization of digestion in a fungal or bacterial comb among the fungus-growing termites (subfamily Macrotermitinae) or on adaptation to feed soil (most higher termites) (Chouvenc et al. 2021; see also Chap. 12).

The gut of higher termites includes a short foregut (a crop is absent, like in lower termites), a midgut without ceca, a mixed segment between the midgut and hindgut P1, and a hindgut differentiated in five regions (P1-5) (see Chap. 2, Fig. 2.2d) (Köhler et al. 2012; Brune 2014; Souza et al. 2017). The enzymes necessary to digest lignocelluloses are endogenous (present in the salivary glands and midgut) and associated with the hindgut microbiota (Gelb et al. 2008; Watanabe and Tokuda 2010; see also Chaps. 10 and 12). Except for Macrotermitinae, all higher termites have the mixed segment that secretes an alkaline solution that increases the pH of segment P1 to about 10-11 (Brune 2014). As there is a high expression of H^+-V-ATPase, K^+/H^+ antiporter, and carbonic anhydrase in the mixed segment (Kumara et al. 2016), it is likely that the alkaline secretion is a carbonate solution produced by a mechanism reminiscent of that of lepidopterans (see Chap. 6). The probable function of the high pH of P1 is to extract lignocelluloses from plant cell walls (Terra 1988).

Digestion of cellulose, with accompanying accumulation of fermentation products, is thought to occur in reducing conditions and at neutral pH in P3 (also known as paunch) (Tokuda et al. 1997; Tokuda and Watanabe 2007; Brune 2014). The deficiency of nitrogen in wood is circumvented by the higher termites by anal trophallaxis, like the lower termites. The digestion of the microbial mass rich in nitrogen-containing compounds, once ingested with the feces, is accomplished by midgut lysozyme, glycosidases, and peptidases (Hogan et al. 1988; Fujita and Abe 2002). Midgut enzymes are supposed to be retained in major amounts in the midgut of at least in higher termites by enzyme recycling (Terra and Ferreira 1994; Fujita et al. 2010), instead of enzyme resorption, as proposed by Hogan et al. (1988). There are no data regarding midgut fluxes in termites, but due to their phylogenetic proximity with cockroaches, it is probable that their Malpighian tubules secrete a fluid that is partially directed to the midgut, which is eventually absorbed in the anterior midgut.

13.3.4 Mantodea

Despite most Mantids are carnivorous, some of them use pollen as a supplement to prey ingestion (Beckman and Hurd 2003). This means that mantids maintain part of the cell wall-degrading capacity of other Dictyoptera, further supported by the finding of cellulase in these insects (Shemoli et al. 2020). Mantids have a large crop (Chap. 2, Fig 2.2e), where most digestion is supposed to occur under the action of enzymes passed forward from the midgut, as suggested by ion and water movement data (Hatle et al. 2002).

13.4 Phasmatodea

Phasmatodea comprises the stick and leaf insects that mimic the stems and leaves where they live and feed (Grimaldi and Engel 2005). The alimentary canal of Phasmatodea consists of a tubular foregut, a midgut with appendices in its posterior region formed by small protuberances linked to thin tubules (midgut tubules), resembling Malpighian tubules, followed by the hindgut (see Chap. 2, Fig. 2.2f) (Azevedo et al. 2013; Monteiro et al. 2014; Shelomi et al. 2015). Midgut tubules seem to be anatomical structures characteristic of phasmids thought to alkalize the posterior midgut (see below).

Both starved and fed stick bugs (*C. phyllinus*), after being orally fed with amaranth, showed the luminal side of the anterior midgut stained. Dye injected into the hemocoel of stick bugs does not stain the hemal side of the midgut, but it was cleared by the Malpighian tubules, which secretion is directed both backward to the hindgut and forward to the midgut (Monteiro et al. 2014). This means that the anterior midgut is absorbing and the Malpighian tubules, secretory, giving origin to a countercurrent flux of water powering enzyme recycling. Enzyme recycling is confirmed by the low rate of enzyme excretion Monteiro et al. 2014). The foregut and anterior midgut are acidic and the posterior midgut is highly alkaline, putatively a result of bicarbonate secretion by the carbonic anhydrase-rich midgut tubule cells, which in contrast to the Malpighian cells are not involved in hemolymph dye clearance (Monteiro et al. 2014). Midgut tubules may have excretory functions other than bicarbonate secretion (Shelomi and Kimsey 2014).

The initial digestion of both carbohydrates (except cellulose) and proteins occur at the foregut and anterior midgut by the action of enzymes secreted by anterior midgut cells through an exocytic mechanism. There are no salivary digestive enzymes. Initial digestion of carbohydrates occurs in the foregut and anterior midgut and intermediate and final digestion, along the midgut by the action of glycocalyx-associated enzymes. Digestion of proteins starts in the middle midgut and finishes in the middle and posterior midgut by membrane-bound enzymes (Monteiro et al. 2014; Shelomi et al. 2014a, b). The highly alkaline medium of the posterior midgut is expected to make available hemicelluloses of the cell wall of vegetables (Monteiro et al. 2014).

B. Condylognatha and Psocodea

13.5 Phthiraptera

The digestive tube of the louse has a short foregut, followed by a midgut where three regions may be recognized: a distensible anterior midgut (also known as stomach), an elongated posterior midgut, and a short region between the two, the middle midgut. At the anterior midgut, mycetocytes harbor symbiotic bacteria supposed to aid in lice nutrition, furnishing B-vitamins, as an example. Finally, after the insertion of the Malpighian tubules, the gut ends in the hindgut (Gonçalves 2002; Waniek 2009). Like other Psocodea (Peters 1992; Silva et al. 2004), no peritrophic membrane was observed in micrographs of the midgut of the louse *Haematopinus suis*, even with the aid of a fluorescent wheat germ agglutinin, which is able to bind to chitin structures in the presence of excess N-acetyl glucosamine (Gonçalves 2002). Therefore, the presence of genes coding for a midgut chitin synthase (CSM) and some PM peritrophins is intriguing (Dias et al. 2019).

The first studies of initial protein digestion in lice revealed the role of trypsin (Borovsky and Schlein 1988; Waniek et al. 2002). By assaying different fractions of the *H. suis* midgut, it was possible to show that trypsin, chymotrypsin, and α-glucosidase are soluble enzymes in the midgut, whereas aminopeptidase is mostly membrane bound (Table 13.1). Cysteine endopeptidase has a low activity and may correspond to a contaminating lysosomal enzyme and not to a recruited lysosomal enzyme to act as digestive enzyme. This is confirmed by the occurrence of a single gene coding for cysteine endopeptidase in the louse genome (Pimentel et al. 2020). The predominance of serine as major endopeptidases in the midgut of *H. suis* was assured by ion-exchange chromatography. The four peaks of activity of general

Table 13.1 Enzyme activities at different fractions of the midgut homogenates of *H. suis*

Enzyme	Substrate	Soluble (S1)	Glycocalyx (S2)	Membrane-bound
Trypsin	Z-Arg-MCA	70.1 (243)	8.8 (30.5)	21.1 (75.6)
Chymotrypsin	SAAP-MCA	89.1 (48.7)	4.9 (2.7)	6.1 (3.4)
Aminopeptidase	LpNA	5.1 (0.31)	5.8 (0.4)	98.1 (5.4)
α-glucosidase	MU-α-Glu	84.8 (4.4)	6.8 (0.4)	8.4 (0.4)

Figures refer to percentage of total activity and activities expressed in mUnits per animal (parentheses). Soluble (S1) supernatant of the centrifugation of the midgut. It corresponds mostly to the midgut contents. Glycocalyx fraction is the supernatant of the centrifugation of the resuspended pellet of the first centrifugation after three cycles of freezing and thawing. It corresponds to the enzymes trapped in the glycocalyx (see Chap. 5). Data calculated from the results published by Gonçalves (2002). LpNA, Leucine p-nitroanilide; MU-α-Glu, 4-methylumbelliferyl α-D-glucoside; SAAP-MCA-succinyl-Ala-Ala-Phe-4 methyl coumarin; Z-Arg-MCA, carbobenzoxy-Arg-7 amide-4 methyl coumarin endopeptidase activity (assayed with fluorescent casein) resolved by ion chromatography coincided with the activity peaks on carbobenzoxy-Arg-7-amido-4 methyl coumarin (substrate for trypsin) or on N-succinyl-Ala-Ala-Phe 4-nitroanilide (substrate for chymotrypsin) (Gonçalves 2002). The aminopeptidase activity apparently is due to at least two enzymes, one of which was characterized, but has a low expression (Ochanda et al. 2000), and another that is mainly expressed at the posterior midgut (Kollien et al. 2007). The genes coding for trypsin and chymotrypsin are more expressed in the anterior midgut and one of the trypsins have an unusual chymotrypsin-activation-peptide cleavage site, thus differing from the usual auto-activated process (Kollien et al. 2004; Waniek et al. 2005).

Electron micrographs showed that in the anterior and middle midgut of *H. suis*, there are vesicles budding from the cell microvilli (Gonçalves 2002). These vesicles are thought to correspond to microapocrine secretion (see Chap. 5). It is probable that those secretory vesicles are carrying serine endopeptidases that are more expressed at the same region (see above).

The lack of PM in lice deserves attention. We discussed previously that Condylognatha lost PM because they no longer have polymeric food in their food demanding digestion. However, lice digesting blood, which contains polymeric food molecules, are expected to have a PM, as mosquitoes have. The production of PM has a cost of both chemical compounds and energy that is usually advantageously compensated by the gain in digestive efficiency provided by PM (Terra 2001; Bolognesi et al. 2008). Perhaps in tiny insects like lice (and fleas), the balance of digestive enhancement/cost of PM production is negative, becoming an evolutionary pressure leading to the elimination of the PM inherited from Psocodea ancestors. The mechanical and chemical protection of the surface of midgut cells in Phthiraptera was thought to depend on the peritrophic gel (Silva et al. 2004). However, with the widespread finding in the insect midguts of a vertebrate-like mucus formed by mucus-forming mucins (Dias et al. 2018; see also Chap. 7), it is more probable that the so-called peritrophic gel is actually a mucus layer. The genes coding for CSM and some PM peritrophins in Psocodea are supposed to be vestigial, non-functional genes.

13.6 Thysanoptera

Thysanoptera are tiny insects which comprise two suborders: Terebrantia, with many families, and Tubulifera, with a single family. The mouthparts of Thysanoptera are usually too short to directly tap into the vascular system of plants. Feeding thrips perforate the surface of plant tissues (their most common food) and insert the mouthparts in them. Then, in back-and-forth movements, which may eventually attain the vascular system (phloem), fluids, particles of plants or prey (in predatory thrips), are ingested (Grimaldi and Engel 2005). This type of feeding is named *scratch-and-suck* and thought to be the ancestor of all hemipteran types of feeding.

The midgut of Terebrantia, exemplified by *Frankliniella occidentalis,* is a long tube without ceca varying somewhat among species. The midgut cells of the anterior midgut have their microvilli associated with membranes described as usual perimicrovillar membranes (Del Bene et al. 1991; Silva et al. 2004). However, close inspection shows that actually the microvilli-associated membranes often enclose several microvilli in a bundle. Thus, they are more appropriately named bundle-forming perimicrovillar membranes thought to aid in absorption (see Chap. 5). At the posterior midgut, microvilli lack associated membranes, but have rod-like projections (Del Bene et al. 1991; Leal 2018) with unknown function. These microvillar structures have been described before by Kitajima (1975) but, probably by mistake, the organization he described was the reverse of that presented above. Results similar to those found by Del Bene et al. (Del Bene et al. 1991) were found for *Franklinothrips vespiformis* (Leal 2018). The apical structures of Tubulifera, exemplified by *Liothrips* sp., are like those described for Terebrantia, except that the bundle-forming perimicrovillar membranes of anterior midgut cells seem to be collapsed at the cell surface, resembling an almost continuous membrane (Leal 2018). If the collapsed membranes are not the result of fixation artifacts, their surfaces exposed to the midgut lumen are smaller than those not collapsed. As a consequence, they are thought to be less efficient in absorption and be one of the reasons why Tubulifera are much less numerous than Terebrantia.

Thysanoptera lacks PMs and associated with this have no midgut chitin synthase and a reduced number of peritrophins, including of PM peritrophins, in relation to their Polyneopteran-like ancestors (Dias et al. 2019).

Except for the determination of α-glucosidase bound to the microvilli-associated membranes (Silva et al. 2004), no other enzyme was assayed in Thysanoptera. However, like hemipterans (see below), Thysanoptera has an amplification of cathepsin B and cathepsin L coding genes. Thus, Thysanoptera probably had recruited some of these lysosomal proteins as digestive enzymes (unpublished results of the authors).

A transcriptomic analysis of the salivary gland of *F. occidentalis* revealed several genes coding for digestive enzymes (Stafford-Banks et al. 2014), but without further study, including of the midgut, it is not possible to evaluate their role in digestion. It should be noticed that many cases of extra-oral digestion carried out by salivary digestive enzymes are actually only dispersion of tissues, caused by digestion of the cement linking together the cells. Actual digestion in these conditions occurs in the midgut under the action of midgut enzymes (Fialho et al. 2012; see also Hemiptera below).

13.7 Hemiptera

Hemiptera includes the major suborder Sternorrhyncha (aphids and whiteflies), Auchenorrhyncha (cicadas, spittlebugs, leafhoppers, and planthoppers), and Heteroptera (true bugs). Hemipterans have several strategies to acquire food. In *salivary (or stylet)-and-sheath* feeding, a discharged gelling saliva solidifies, forming a tube that encapsulates the mouthparts (stylets) from the surface of the plant to the target vascular tubes (phloem or xylem). *Lacerate-and-flush* feeding consists in movements of the stylets so that they tear apart the vegetal tissue aided by a flush of watery saliva, followed by sucking the dispersed material. At last, in *macerate-and-flush* feeding, the movements of stylets are not intense, and tissue breakdown is caused by enzymes that attack the intercellular cement, leading to the disaggregation of vegetal (pectinase) or animal (collagenase or elastin) tissue. Sternorrhynchans and auchenorrhynchans usually use salivary-and-sheath feeding, whereas heteropterans, the other feeding strategies described above (Panizzi et al. 2021).

13.7.1 Sternorrhyncha

The gut of Sternorrhyncha, exemplified by that of aphids, comprises a short and slender esophagus, a midgut and a dilated rectum and lacks Malpighian tubules. The midgut starts with a dilated anterior midgut (V1), frequently known as stomach, which is followed by a long posterior midgut, also named intestine, which may be divided by convenience into three regions (V1, V2, V3) (Fig. 2.2g) (O'Loughlin and Chambres 1972; Ponsen 1987; Cristofoletti et al. 2003). The columnar cells of the anterior midgut have in their apex lamellae (apical lamellae), which consist of microvilli-like structures, without core filaments, associated with one another and with lamellae of adjacent cells through trabeculae (Ponsen 1991; Cristofoletti et al. 2003). Associated with the microvilli-like structures, there are amorphous membrane masses projecting into the lumen. These membranes, named modified perimicrovillar membranes, are originated from the Golgi and rough endoplasmic reticulum and move through the apical lamellar system before being linked at the tips of the lamellae (Cristofoletti et al. 2003). Sternorrhynchans do not have a peritrophic membrane and, like Thysanoptera, lack a midgut chitin synthase and have a reduced number of peritrophins, including PM peritrophins, in relation to their Polyneopteran-like ancestors (Dias et al. 2019 see Chap. 7 for other details).

Aphids may suck continuously phloem sap with up to three times the osmolarity of their hemolymph (Ashford et al. 2000), causing a high hydrostatic pressure in the anterior midgut, which stretching resistance is thought to be helped by the links between apical lamellae (Cristofoletti et al. 2003). As the ingested sap moves along the aphid midgut, its osmolarity decreases and a honeydew isosmotic with the hemolymph is delivered (Fisher et al. 1984). As expected from the resulting smaller hydrostatic pressure, the number of trabeculae decreases along the midgut (Ponsen 1991; Cristofoletti et al. 2003). The observed decrease in osmolarity of the ingested sap is caused by the action of an α-glucosidase that releases fructose from sucrose by transglucosylation (transference of the glucosyl group from sucrose to another sucrose molecule), followed by fast fructose absorption. The α-glucosidase is bound to the modified perimicrovillar membranes, what avoids its lost in the honeydew (Cristofoletti et al. 2003). Other devices to avoid osmotic stress caused by phloem feeding are the isomerization of sucrose into trehalulose, which is not a substrate for the α-glucosidase, thus avoiding the increase in osmolarity (Salvucci 2003) and occasional xylem feeding to dilute phloem sap (Pompon et al. 2011).

Phloem sap may contain considerable amounts of proteins, and it is known that some phloem-feeding hemipterans (white flies) are able to use dietary proteins (Salvucci et al. 1998). Hence, active endopeptidases are expected to be found in aphids, which lost their serine endopeptidases as all hemipterans. A cathepsin L was the only endopeptidase activity detected in *A. pisum* midguts and was found associated with the modified perimicrovillar membranes, from where it is set free in the presence of sodium dodecyl sulfate, but not of Triton X-100 (Cristofoletti et al. 2003). A similar cathepsin L was described in *Aphis gossypii* (Deraison et al. 2004), and preliminary results also showed the existence of membrane-bound cysteine endopeptidases in the midgut of *Brevicoryne brassicae* (Silva et al. 2000). These results suggest that cathepsin L bound to modified perimicrovillar membranes is characteristic of the columnar cells of aphid midguts.

It is interesting to note that the genomes of *A. pisum* and *A. gossypii* code for 11 and 12 cathepsins B, respectively, and one cathepsin L (Pimentel et al. 2020), but only one cathepsin L activity is detected in their midguts. It is not clear the meaning of the expansion of CAB genes, which is usually associated with the recruitment of lysosomal proteins as digestive enzymes (Terra et al. 2019), if the only cysteine endopeptidase activity used as the digestive enzyme in aphids is encoded by a single CAL gene.

An aminopeptidase N, supposed to finish the processing of phloem proteins started by CAL, is associated with the modified perimicrovillar membranes and is a major binding site for toxic mannose lectins (Cristofoletti et al. 2006). Lectin binding does not inhibit the aminopeptidase, but it decreases aphid performance, arguably because it affects amino acid absorption by nearby amino acid transporters (Cristofoletti et al. 2006). It is hypothesized that non-essential amino acids absorbed by aphids are used by the bacteria *Buchnera*, occurring in mycetomes in the aphid hemocoel, to synthesize essential amino acids (Shigenobu et al. 2000; Douglas 2006).

13.7.2 Auchenorrhyncha

The midgut of members of the infraorder Cicadomorpha may be exemplified by that of *Bucephalogonia xanthopis* (Utiyama et al. 2016), which is similar to other Cicadomorpha (Goodchild 1966; Herbert et al. 1989; Tsai and Perrier 1966; Zhang et al. 2012) (see Fig. 2.2h). The midgut starts with the filter chamber, which is a structure formed by the anterior and posterior ends of the midgut associated with the Malpighian tubules. The filter chamber permits excess water pass directly from the anterior midgut to the hindgut, leaving the concentrate sap in the middle midgut. The high capacity of the filter chamber membranes results from the water channels (aquaporins, see Chap. 6) they have, mainly in xylem feeders (Le Caherec et al. 1997). The filter chamber can concentrate ingested xylem sap more than 10-fold

(Cheung and Marshall 1973a, b) and phloem sap about 2.5-fold (Lindsay and Marshall 1981). Following the filter chamber, there are the dilated anterior midgut (also called conical segment) and the middle and posterior midgut (also known as tubular segment).

Auchenorrhynchans do not have a peritrophic membrane and associated proteins, as described for Sternorrhynchans above, and lost their digestive serine endopeptidases and recruited lysosomal cathepsins B as digestive enzymes (Pimentel et al. 2020). *Mahanarva fimbriolata* (Cicadomorpha: Cercopidae), for example, has 5 CALs and 9 CABS. Based on the expression of their encoding genes in different tissues and on their phylogenetic relationships with known lysosomal and digestive enzymes (see details in Chap. 9), it was concluded that all CALs are lysosomal and three CABs are digestive. Protein digestion in *M. fimbriolata* is thought to be negligible and be restricted to inactivating noxious sap proteins (Pimentel et al. 2020).

At the midgut cell apexes, membranes enclosing several microvilli (bundle-forming perimicrovillar membranes) are easier to see, if they are short as in *Lepyronia coleopterata* (Zhong et al. 2013), *Bucephalogonia xanthopis* (Utiyama et al. 2016), and *Mahanarva fimbriolata* (Dias et al. 2019), than in the longer bundle-forming perimicrovillar membranes seen in *Fulgora candelaria* (Marshal and Cheung 1970).

The bundle-forming perimicrovillar membranes isolate a space between them and the microvillar membranes, as shown with the aid of lanthanun nitrate used as a tracer in electron micrographs of midguts from *B. xanthopis* (Utiyama et al. 2016). Thus, those membranes probably are involved in the absorption of dilute nutrients, as proposed for the perimicrovillar membranes (see Chap. 5 and also Heteroptera below). Xylem feeders like *Homalodisca coagulata* (Cicadellidae) absorb as much as 99% of dietary amino acids and carbohydrates (Andersen et al. 1989). The columnar cells of *B. xanthopis* show secretory vesicles at their apexes, with apparent involvement in merocrine secretory processes, and basal membrane invaginations with associated mitochondria, usually thought to aid in water and ions transport (Utiyama et al. 2016).

13.7.3 Heteroptera

The suborder Heteroptera comprises two infraorders: Cimicomorpha and Pentatomomorpha (see Chap. 1). The midgut of Cimicomorpha consists of a dilated anterior midgut named stomach and a long posterior midgut named intestine. The molecular mechanisms underlying the digestive and absorptive processes are best known among Cimicomorpha in the blood-feeder *Rhodnius prolixus* (Cimicomorpha: Reduviidae).

The salivary secretion of *Rhodnius prolixus* (Reduviidae) is not involved in digestion. The saliva contains substances that make easier the location of vessels in the host and the displacement of blood up to the midgut (Ribeiro 1987). The ingested blood induces the production of endopeptidases (Garcia and Garcia 1977) and glycosidases (Ribeiro and Pereira 1984). The blood is stored in the anterior midgut from where water absorption occurs (Barret 1982) and erythrocyte hemolysis is carried out by a hemolytic factor (Azambuja et al. 1983). The necessity to excrete most blood water and salts, to decrease its weight (several times that of the insect), explains why the anterior midgut is the storage site of blood and not the non-permeable foregut crop, as in grasshoppers (see Chap. 2). The anterior midgut harbors a population of *Rhodococcus rhodnii* which may attain 2.5% of the insect dry weight (Hill et al. 1976). In the middle midgut, protein digestion starts under the action of cathepsin B and L (Pimentel et al. 2020), releasing heme from the hemolymph and digesting *R. rhodnii* with lysozyme and glycosidases (Ribeiro and Pereira 1984; Ursic-Bedoya et al. 2008). Protein digestion continues in the posterior midgut mainly by cathepsin L (Pimentel et al. 2020) and finishes under the action of luminal serine carboxypeptidases (S10) (Ferreira et al. 2015) and an aminopeptidase restricted to the perimicrovillar space (Terra et al. 1988a, b). This suggests that the small oligopeptides resulting from the action of cathepsins and serine carboxypeptidases were transported into the perimicrovillar space to be substrates of aminopeptidase, which products are absorbed.

Despite *R. prolixus* has genes coding for amylases predicted to be activated by chloride (Ribeiro et al. 2014), the faint amylase activity not responsive to chloride found in *R. prolixus* is mostly, if not completely, derived from the *R. rhodnii* (Terra et al. 1988a, b). In addition to cathepsins B and L, a cathepsin D activity amounting to about 15% of the total proteolytic activity was found in the posterior midgut (Terra et al. 1988a, b). A microvillar membrane-bound α-mannosidase and a perimicrovillar membrane-bound α-glucosidase release monosaccharides probably from carbohydrates associated with the peptides resulting from the digestion of blood glycoproteins (Ferreira et al. 1988). The heme molecule derived from hemoglobin is toxic, causing oxidative stress in midgut cells, which is avoided by several mechanisms (Graça-Souza et al. 2006).

Another taxon of Cimicomorpha is that of the bedbugs, exemplified by Cimicidae, which are obligatory blood-feeding insects. The anatomy and ultrastructure of the midgut of *Cimex* are similar to that of *R. prolixus* described above. The remarkable difference is the apparent absence of perimicrovillar membranes in fasting insects (Azevedo et al. 2009). Except for the finding of an expansion of the genes coding for cathepsins B and L (Pimentel et al. 2020), which are expected to include digestive enzymes, there are no other reports on *Cimex* digestive enzymes.

Pentatomorphan midguts comprise three chambers (V1, V2, V3), from which V1 is the largest and a short segment (V4) between V3 and the hindgut. In sap-feeding pentatomomorphans, like Lygaeidae, there are posterior midgut ceca that contact the anterior midgut (Goodchild 1966), functioning as the filter chamber described above (see Fig. 2.2i). The molecular organization of digestion is best known among the pentatomomorphans in *D. peruvianus*.

D. peruvianus (Pyrrhocoridae) lack true salivary digestive enzymes (Silva and Terra 1994), as earlier proposed by Saxena (1963), and acquire food with the lacerate-and-flush strategy. The ingested food is stored in the anterior midgut (V1), where sucrose is digested by an α-glucosidase, water is absorbed, and glucose is transported into cells with the aid of sugar porters (Silva and Terra 1994; Bifano et al. 2010). Proteins are initially digested by luminal endopeptidases, in the anterior midgut (V1) by cathepsins B and in the middle (V2) and posterior midgut (V3) by cathepsins L, in a reducing and acidic medium (pH around 5.8) (Pimentel et al. 2020). A cathepsin D, which is more active at the anterior midgut, removes cathepsin L inhibitors from seeds ingested by the bugs and, because of its substrate specificity, complements the action of cathepsin L. Protein digestion is finished by the action of an aminopeptidase restricted to the perimicrovillar space. The resulting amino acids are absorbed along the middle and posterior midgut with the aid of transporters activated by potassium and inhibited by sodium ions. Both α-glucosidase and cathepsins are increased on feeding (Silva and Terra 1994; Silva et al. 1996; Pimentel et al. 2017, 2020). Digestive enzyme activity increase on feeding was previously described in *Dysdercus cingulatus* by Muraleedharan and Prabhu (1979). According to them, the stimulation of enzyme secretion occurred through a secretogue mechanism. *D. peruvianus* have bacterial symbionts but, as they are situated at the posterior end of the posterior midgut, they are probably not involved in digestion, although they may be of physiological importance (Silva et al. 1995), like in *R. prolixus* described above.

Other well-known pentatomomorphan seed-sucker is *Nezara viridula* (Pentatomidae). On this insect, there are studies on digestive enzymes (Cantón and Bonning 2020) and proteomic and transcriptomica data associated with the compartmentalization of the digestive process (Denecke et al. 2020). The results obtained are similar to those described above for *D peruvianus*.

An example of a predatory pentatomomorpha is *Podisus nigrispinus* (Pentatomidae). It is usually considered to rely on salivary enzymes for extra-oral digestion. In this case, salivary enzymes are injected into the prey, causing the digestion of its tissues, before the ingestion of the digested material (Cohen 1993, 1995). However, unless a quantitative study is done comparing salivary and midgut enzyme activities, it is not possible to rule out the possibility that prey tissues are pre-orally disrupted, but truly digested only in the midgut. This was clarified by a study on *P. nigrispinus*. Collagenase is the only endopeptidase which activity is high in salivary glands in comparison with midgut activities (Fialho et al. 2012). This suggests that the prey tissues are being pre-orally dispersed (not digested), confirmed by the finding of prey muscle fibers inside the midgut of *P. nigrispinus*. These fibers are seen in the anterior midgut, but no longer found in the posterior midgut, indicating that they were digested. Enzyme assays along the midgut of *P. nigrispinus* showed that amylase and aminopeptidase are more active in the anterior and middle midgut, the major endopeptidases (cathepsin L1 and cathepsin L2) at the middle and posterior midgut, and, finally, an α-glucosidase similarly active along the whole midgut (Fialho et al. 2012). The sites of synthesis of amylase, cathepsin L, α-glucosidase, and aminopeptidase were identified by ultrastructural immunocytolocalization and are in agreement with previous distribution of digestive enzymes described before (Fialho et al. 2013).

The proposal that lepidopteran digestive enzymes play a role in the lepidopteran digestion by its predator, *P. maculiventris*, is based on assays performed at pH 10

(Pascual-Ruiz et al. 2009). As pH 10 favors lepidopteran digestive enzymes (Terra and Ferreira 1994, 2012) and maintain inactive *Podisus* enzymes (Fialho et al. 2012), their proposal must be re-evaluated.

C. Holometabola

13.8 Megaloptera

Megaloptera includes the alderflies and dobson flies, from which the best known is the Dobson fly *Corydalus.* Megaloptera is among the most primitive Holometabola, and the major part of it is predators of small invertebrates (Grimaldi and Engel 2005). The gut of *Corydalus* comprises a large crop, a small midgut with anteriorly placed midgut ceca and a large hindgut. Based on enzyme assays of different sections of the gut of *Corydalus* sp., it was proposed that most of the digestion occur in the crop, under the action of trypsin, the major endopeptidase, aminopeptidase, amylase, and maltase. Less than 3.3% of midgut amylases, maltase, and aminopeptidase is excreted at each midgut emptying, implying in the existence of digestive enzyme recycling. The high excretory rate of trypsin (27%) is probably a consequence of excess dietary protein that binds to trypsin (A.B Dias, W R Terra, unpublished data as previously cited in Terra and Ferreira 2012).

13.9 Coleoptera

13.9.1 Introduction

On adapting to subsurface habitats, the beetle ancestors gained a mechanically resistant cuticle and maintained their primitive biting mouthparts. Thanks to these characteristics, beetles occupied niches available to insects in surface and subsurface regions, becoming the largest group of animals. This led the British biologist J.B.S. Haldane to answer to a group of theologians that one of the characteristics of the Creator is "an inordinate fondness for beetles" (quoted by Evans 1977). Coleoptera comprises the major suborders: Adephaga and Polyphaga.

13.9.2 Adephaga

The most important family of Adephaga is Carabidae, the predaceous ground beetles. The gut of *Pheropsophus aequinoctialis* (Coleoptera: Carabidae) has a capacious crop, a long midgut without ceca and a relatively short hindgut (Fig. 2.2k). The pH of crop contents is 5.9 and that of the midgut is 6.4-6.6. The surface of the anterior midgut has a fluid material instead of a peritrophic membrane found in the posterior midgut, named at that time peritrophic gel (Ferreira and Terra 1989). This fluid material is now thought to be a vertebrate-like mucus (Dias et al. 2018). Aminopeptidase, carboxypeptidase A, trypsin (the major endopeptidase found), amylase, and cellobiase were assayed in crop and midgut contents and also in midgut cell subcellular fractions. The data led to the proposal that most protein and carbohydrate digestion occur in the crop carried out by enzymes passed forward from the midgut, with part of the intermediate and final digestion taking place under the action of enzymes immobilized at the midgut cell surface (Ferreira and Terra 1989). These findings confirm and extend previous results of Cheeseman and Pritchard (1984) based only on endopeptidase assays.

13.9.3 Polyphaga

The major infraorders of Polyphaga are Scarabaeiformia, Elateriphormia, Bostrichiformia, and Cucujiformia (see details in Chap. 1).

The largest family of Scarabaeiformia is Scarabaeidae, which includes the well-known dung beetles (see Chap. 1). The gut of these insects comprises a short foregut, a large midgut having three rows of ceca, and a hindgut with a large fermentation chamber (Fig. 2.21). The midgut pH increases up to 12 along a midgut groove, found between the middle and posterior row of the ceca, decreasing forward and backward (Bayon 1981). This high pH is thought to release cellulose from the biomass which, on passing to the hindgut fermentation chamber, is acted by endogenous and bacterial endoglucanases (cellulases). The products of those enzymes are short-chain fatty acids, chiefly acetic acid, eventually absorbed by hindgut cells (Lemke et al. 2003; Huang et al. 2010). The microbial biomass of the fermentation chamber is converted into larval biomass, but is uncertain if this occurs after feces ingested or by passing the bacterial biomass to the midgut for further digestion (Li and Brune 2005). Keratin-ingesting beetles (e.g., *Trox*, Scarabaeiodea: Trogidae) digest keratin-like Dermestidae (see below), with serine endopeptidases (Hughes and Vogler 2006).

Elateriformia comprises several superfamilies from which Elateroidea includes bioluminescent beetles like *Pyrearinus termitilluminans* (Elateridae). This species, as is common in the family, feed using a reflux extra-oral mechanism (see details in Chap. 2). For this, *P. termitilluminans* empties onto the prey its midgut contents, which enzymes accomplish initial digestion. The dispersed material, once ingested, finishes initial digestion before intermediary digestion, both inside the peritrophic membrane. Final digestion is carried out by enzymes immobilized at the midgut cell surface (Colepicolo-Neto et al. 1986).

Bostrichiformia includes the family Dermestidae like *Dermestes maculatus* that feed on very dry proteinaceous remains of carcasses. Baker (1981a, b) described the properties of several endopeptidases and showed that they are mostly found in midgut contents of *D. maculatus*. *D. maculatus* has a vertebrate-like mucus (earlier described as peritrophic gel) in the anterior midgut, whereas in the middle and posterior midgut, a peritrophic membrane is found. Serine endopeptidases and glycosidases assayed along the midgut led to the proposal that protein digestion starts in the anterior midgut and finishes at the surface of posterior midgut cells. Glycogen is completely digested at the anterior midgut. Electron micrographs showed that both merocrine secretion and apocrine secretion occur in the midgut, although apocrine secretion is more frequent in ceca than in the anterior midgut. The electron micrographs also showed that columnar cell basal infoldings suggest that the ceca are water-absorbing, which is confirmed by dye data (see details in Chap. 5). The combined results indicate *D. maculatus* has midgut countercurrent fluxes and, hence, digestive enzyme recycling, in agreement with the decreasing gradient of enzymes along the midgut and their low excretory rate (Caldeira et al. 2007).

The infraorder Cucujiformia contains most beetles distributed in the superfamilies Cucujoidea, Tenebrionoidea, Chrysomeloidea, and Curculionoidea (the last two correspond to Phytophaga) (see details in Chap. 1). All Cucujiformia use cathepsin B (CAB) and cathepsin L (CAL) recruited from lysosomes as major midgut endopeptidases. The selective pressures that resulted in the recruitment of lysosomal CALs and CABs are now supposed to be an adaptation to ingesting pollen in the beginning of Cucujiformia diversification as pollinators (Silva et al. 2022; Chap. 9).

The largest family of Tenebrionoidea is Tenebrionidae, which includes the flour beetles *Tribolium castaneum* and *Tenebrio molitor*. *Tenebrio molitor* has a cylindrical midgut without midgut ceca, but with a peritrophic membrane along the length of the midgut. The pH of the contents is 5.6 in the anterior and middle midgut and 7.9 in the posterior midgut (Terra et al. 1985). Most digestive enzymes are found in the midgut contents, except aminopeptidase, which is bound to the microvillar membranes of the midgut. Amylase, maltase, cellobiase (β-glucosidase), and cysteine endopeptidases (cathepsin L and B, see Chap. 9) predominate in the anterior and middle midgut; serine endopeptidase (trypsin and chymotrypsin) predominate in the middle and posterior midgut. It is interesting to note that digestion in *T. molitor* occurs with little hydration of the food – from 15.8 % in bran to 19 % in the midgut contents (Terra et al. 1985)

Several digestive enzymes from *T. molitor* larvae were purified and characterized, and for some of them, there are descriptions of their 3D structures. They are: amylase (Buonocore et al. 1976; Strobl et al. 1998), aminopeptidase (Cristofoletti and Terra 1999), cathepsin L (Cristofoletti et al. 2005; Beton et al. 2012; Damasceno et al. 2017), chitinase (Genta et al. 2006), chymoptrypsin (Elpidina et al. 2005; Sato et al. 2008), dipeptidyl peptidase (Tereshchenkova et al. 2016), β-galactosidase (Ferreira et al. 2003), β-glucanase (Genta et al. 2009), β-glucosidase (Ferreira et al. 2001), α-mannosidases (Moreira et al. 2015), prolyl carboxypeptidase (Goptar et al. 2013), prolidase (Tereshchenkova et al. 2017), trehalase (Gomez et al. 2013), and trypsin (Levinsky et al. 1997; Lopes et al. 2006).

Based on the enzyme distribution along the midgut, carbohydrate digestion occurs mainly in the anterior midgut and protein digestion, part in the contents (initial and intermediary digestion) and part in the surface of cells of the posterior midgut. Only about 5% of the midgut enzymes are excreted after midgut emptying, suggesting the existence of enzyme recycling, powered by a countercurrent flux of fluid caused by the secretion of water at the posterior midgut and its absorption back at the anterior midgut. The existence of these fluxes is supported by the use of dyes (Terra et al. 1985, see also technical details in Chap. 2), theoretical models based on midgut distribution and trypsin site of secretion identified by immunocytochemistry (Bolognesi et al.,2008), and the finding by RNA-seq that the genes coding for the appropriate transporters of water are expressed at the anterior and posterior midgut (Moreira et al. 2017). RNA-seq data also led to the conclusion that sugar and amino acids are absorbed along the whole midgut and that the anterior midgut is buffered at the acid zone by the transport of NH_4^+ and chloride and alkalinized at the posterior midgut by bicarbonate (Moreira et al. 2017).

The superfamily Chrysomeloidea includes Chrysomelidae and Cerambycidae, exemplified by *Zabrotes subfasciatus* and *Migdolus fryanus*, respectively. The midgut of *Z. subfasciatus* has a diameter that enlarges from the anterior to the middle midgut, then decreasing to the posterior midgut. The most important digestive endopeptidases are the cysteine and aspartic endopeptidases (Lemos et al. 1990; Silva and Xavier-Filho 1991), later on shown to be cathepsin L (CAL) and cathepsin D (CAD) derived from lysosomal enzyme ancestors and adapted to a digestive function, as is usual in Cucujiformia insects (Silva et al. 2022). CALs are expressed mainly at the anterior midgut, like in *T. molitor*, and CADs are expressed similarly along the whole midgut. Assays of different enzymes along the midgut led to the proposal that carbohydrates are digested at the midgut lumen together with the initial digestion of proteins. The final digestion of proteins takes place at the midgut cell microvillar membranes. It is interesting to note that *Z. subfasciatus* has digestive α-galactosidases that convert raffinose, which occurs in the seeds they ingest, to sucrose, later on hydrolyzed by α-glucosidases (Silva et al. 1999). The anterior midgut lacks a peritrophic membrane and its cells are covered by a vertebrate-like mucus. The posterior midgut has a peritrophic membrane, in agreement with visual observation. Transcriptomic data revealed that mucins are expressed along the whole midgut, whereas the proteins involved in the production of the peritrophic membrane, mainly in the posterior midgut. Despite the fact that the PM is found only in the posterior midgut, it is functional to explain the small excretory rate of chitinase and hence the role of PM in the digestive enzyme recycling (see Chap. 7 for technical details). These studies support the proposal that mucus is important for tissue protection and PM in enhancing digestive efficiency (Silva et al. 2022).

The superfamily Curculionoidea comprises several families of which the most important is Curculionidae like *Sphenophorus levis* and *Sitophilus oryzae*. The midgut of *Sphenophorus levis* is enlarged at the anterior midgut and decreases in diameter along the midgut (Fig. 2.2m). The anterior and middle midgut are acidic (5.5 to 6.5), lack a PM, and show a vertebrate-like mucus at the surface of midgut cells. The posterior midgut has a PM and the pH of its contents is 7.6. Amylase, maltase, and CAL (major endopeptidase) occur at luminal contents in decreasing amounts along the midgut, whereas aminopeptidase predominates at the microvillar membranes of posterior midgut cells. These results mean that all carbohydrate digestion occurs at the anterior midgut, where it starts protein digestion that ends at the surface of posterior midgut cells (Soares-Costa et al. 2011).

The midgut of *Migdolus fryanus* is similar to that of *S. levis* (Fig. 2.2m) and the overall pattern of digestion is also alike to that of *S. levis,* except that its main endopeptidase is trypsin (Dias, Terra, unpublished data, cited in Terra and Ferreira 2012). This is in accordance with preliminary work (Murdock et al. 1987), according to which Cerambycid larvae returned to a trypsin-based strategy. However, the transcriptome of another cerambycid (*Anoplophora glabripennis*) showed that their genes coding for CALs had undergone an expansion, some of which seem to be coding for digestive enzymes (Silva et al. 2022). Recently, a comprehensive transcriptome of *M. fryanus* has been published, and it is supposed to improve the knowledge of digestion and clarify the occurrence of digestive CALS and CABs in this insect (Nakayama et al. 2017).

13.10 Hymenoptera

Hymenoptera comprises several clades, once grouped under Symphyta, which is paraphyletic, and Apocrita (see Chap. 1). The basal lineages of Hymenoptera include the superfamily Tenthredinoidea (sawflies) that resembles caterpillars, like the sawfly *Themos malaise* (Argidae) and the wood wasps. The wood wasps have a chamber that stores fungi that are discharged in the wood together with the oviposition. The partly digested wood is then ingested by the wasps. Among the lineages of wood wasps that bore wood evolved Parasitoidea, which are small endoparasitic insects like *Bracon hebetor*. Aculeata (sting wasps) evolved from lineages close to Parasitoidea and originated ants, like *Camponotus rufipes,* and bees exemplified by *Apis mellifera* and *Scaptotrigona bipunctata* (other details in Chap. 1).

The midgut of larval *T. malaisei* is cylindrical with a ring of ceca in U form at the ventral side of the anterior midgut, from which protrude the largest ceca (Fig. 2.2n). In the anterior two-thirds of the midgut, the luminal pH of midgut contents is above 9.5 that is supposed to extract hemicelluloses from the plant cell walls ingested by the larvae, which in sequence are digested by a series of digestive enzymes together with plant cell contents. Amylase and trypsin (the major endopeptidase) decrease in activity along the midgut contents, suggesting they are recycled. Maltase is a soluble enzyme entrapped in the glycocalyx of the anterior midgut cells and aminopeptidase is bound to the microvillar membranes of posterior midgut cells (Dias, Ribeiro, Terra, unpublished data, cited in Terra and Ferreira 2005).

The endoparasitic *Bracon hebetor* attacks several pyralid moths and develops inside their bodies, ingesting hemolymph that is stored in the midgut, which occupies most of the body cavity. Digestion starts during the feeding stage, but peaks only after the initiation of cocoon formation, nearly finishing after an additional 24h, when the blind midgut connects to the hindgut. The midgut pH is 6.8, trypsin is the major endopeptidase with about 10% of its activity bound to the midgut cells, whereas aminopeptidase is completely cell bound (Baker and Fabrick 2000).

Amylase and maltase are soluble enzymes and together with trypsin and aminopeptidase are homogeneously distributed along the midgut. The pH optima of amylase (5.1) and maltase (6.1) are closer to the pH of midgut contents than that of trypsin (8.0) and aminopeptidase (8.0) (Dias, Ribeiro, Terra, unpublished data).

The larval bees of *Scaptotrigona bipuncata* feed on pollen and nectar that are digested under the action of only midgut digestive enzymes, as negligible digestive enzyme activities are recovered from salivary gland homogenates (Schumaker et al. 1993). This agrees with morphological and cytochemical data of Cruz-Landim and Mello (1981), according to which salivary glands are only involved in lubricating the food in larvae and in secreting a silken cocoon at the end of larval stage. Amylase and maltase predominate at the anterior midgut contents, aminopeptidase occurs similarly along the midgut, and trypsin is more active at the posterior midgut. This heterogeneous distribution of digestive enzymes is not expected, because due to peristalsis all enzymes should be translocated to the end of posterior midgut. It is then probable that the enzymes are recycled by countercurrent fluxes of fluid (Schumaker et al. 1993). This circulation should enhance digestive efficiency (see details in Chap. 7), but has no effect in avoiding enzyme excretion, as the midgut is closed at its posterior end and remains unconnected to the hindgut until pupation (Cruz-Landim and Mello 1981). The midgut contents are buffered around pH 5.8 along the whole midgut. There is an amylase, which is not activated by chloride, a major maltase that prefers as substrate sucrose, instead of maltose, and two minor ones preferring maltose. Trypsin is the major endopeptidase. Based on the fact that all digestive enzymes pass through the peritrophic membrane and on their molecular weights, the peritrophic membrane pores sizes are at least 8 nm in diameter (Schumaker et al. 1993).

Adult bees (*Apis mellifera*) ingest nectar and pollen. The nectar rich in sucrose is stored in the crop, known as the honey stomach (Fig. 2.2o), and the sucrose is hydrolyzed by the action of a sucrase and may be regurgitated to feed larvae or passed into the midgut, which lacks ceca. The ingested pollen grains are passed directly to the midgut, where they are digested (Barker and Lehner 1972 for references and see also Chap. 10). The absorption sites for glucose and leucine were found in the anterior midgut (Crailsheim 1988a, b) that together with ultrastructural and cytochemical data led Jimenez and Gilliam (1990) to propose enzyme recycling in the midgut of adult *Apis mellifera*.

Worker ants feed on nectar, honey, and plant exudates, which are rich in sucrose, obtaining most amino acids for growth from bacterial symbionts (Cook and Davidson 2006). Worker ants also feed on partially digested food regurgitated from their larvae (Dussutour and Simpson 2009). Thus, they have been said to lack digestive enzymes or only have digestive enzymes involved in intermediate or final digestion. This is true for leaf-cutting ants (Erthal Jr et al. 2007), which appear to depend only on monosaccharides produced by fungi, which they maintain (Silva et al. 2003). Nevertheless, this is not widespread, as exemplified by adult *Camponotus rufipes* (Formicinae). The gut of *C. rufipes* has a long esophagus, a dilated crop, followed by a wide midgut ending with a slender ileum and a dilated rectum (Caetano 1988). The pH of the contents of the foregut is very acid (pH 3.5) and of the midgut weekly acid (pH 6.2). *C. rufipes* adults have enzymes for all digestive steps. They have soluble amylase, trypsin (their major endopeptidase), maltase, and aminopeptidase enclosed in the peritrophic membrane. Only 14% amylase and less than 7% of the other digestive enzymes are delivered in feces during each midgut emptying, suggesting that *C. rufipes* have a digestive enzyme recycling mechanism (Dias, Ribeiro, Terra, unpublished data, cited in Terra and Ferreira 2005). Ongoing research on transcriptomicas and proteomics are expected to end the controversy regarding the presence of polymer hydrolases in adult ants.

13.11 Diptera

Diptera comprises two major branches: one polyphyletic branch (Nematocera) of basal insects distributed into four infraorders and a monophyletic branch (Brachycera) with three infraorders (see details of dipteran phylogeny, habits, and habitats in Chap. 1). Digestive physiology is better known in *Rhychcosciara americana* and *Aedes aegypti* among the basal clades and *Musca domestica*, *Drosophila melanogaster*, *Anastrepha*, and *Stomoxys calcitrans*, in the case of derived Diptera (Brachycera).

Rhychosciara americana (Brachycera: Sciaridae) larvae are litter decomposers in forests and their midguts are cylindrical with a peritrophic membrane with paste-like contents and two large anterior ceca that contain most ectoperitrophic fluid and (Marques 1976) (see Fig 2.1). These aspects turned easy the dissection of the peritrophic membrane without significant loss of contents and the collection of uncontaminated fluid by puncturing the ceca with a capillary. Thus, it is not unexpected that digestive enzyme compartmentalization and recycling were first discovered in *R. americana* (Terra et al. 1979; Ferreira and Terra 1980; Terra and Ferreira 1981; Ferreira and Terra 1982; Terra and Ferreira 1983) These data and the technical details to arrive to those conclusions are summarized in Chap. 2. The sites of water secretion and absorption causing the countercurrent flux of fluid that powers enzyme recycling were found by the use of dyes (see also Chap. 2) and supported by midgut cell ultrastructure (Ferreira et al. 1981) (see Chaps. 2 and 5). The pH values of the midgut are alkaline (pH 8.8-8.9) along the whole midgut and are thought to facilitate the extraction of hemicelluloses and cellulose from litter.

R. americana after metamorphosis to adults displays a crop, but still has anterior ceca, which are absent from the other adult Diptera (Snodgrass 1935). Based on the ultrastructure of midgut cells, *R. americana* adults apparently lost their water-secreting regions, while conserving their water-absorbing regions. Enzyme assays showed that after metamorphosis *R. americana* midguts have a decrease in the activity of trypsin and aminopeptidase, maintenance of amylase activity, and a very high increase in disaccharidase activities. This is in agreement with the nectar feeding behavior of *R. americana* adults, with occasional use of decaying products of animal or vegetal origin (Ferreira et al. 1993).

Like sciarids, the bibionid *Pentheria holocerica* larvae have an alkaline and oxidizing midgut without active fermentation microbiota (Sustr et al. 2014). The digestive process of larval mosquitoes is similar to those of sciarid and bibionid larvae, according to a microarray analysis of their midgut transcripts (Oviedo et al. 2008).

Digestion in adult mosquitoes differ from sciarids and bibionids. Males and females ingest nectar that is stored in the crops and digested and absorbed at their anterior midguts. Blood that is ingested only by females is directed to the posterior midgut, where digestion and absorption occur (Billingsley 1990). The anterior midgut of adult mosquitoes is maintained slightly acidic by the action of a H^+-ATPase to favor the action of α-glucosidase on nectar sucrose. As soon as the posterior midgut is filled with blood, its contents become alkaline to favor the action of the serine endopeptidases. The alkalization results from the amino acid-H^+-symporter (PAT) (Nepomuceno et al. 2017). Mosquitoes, like other blood-feeding arthropods, use hemoglobin as a major nutrient, which results in a large amount of heme. Heme is toxic because of the formation of oxygen-reactive species that affect nucleic acids, among other deleterious effects. Hence, counteracting the deleterious effects of heme constitutes the most important selective pressure in the evolution of blood-feeding insects. Heme toxicity counteracting processes include the formation of heme aggregates that may bound to the peritrophic membrane, heme degradation, antioxidant enzymes, and radical scavengers (Pascoa et al. 2002; Graça-Souza et al. 2006).

Among the Brachycera, Cyclorrhapha is the best-known grouping, including the superorders Muscoidea (e.g., *Musca domestica* and *Stomoxys calcitrans*) and Acalyptrata (Tephritidae and Drosophilidae).

The larvae of *Musca domestica* (Cyclorrhapha, Muscoidea, Muscidae) live in natural conditions in the highly infected dung (Brues 1946) and grow in to fertile adults having bacteria as their sole food (Levinson 1960). Then, *M. domestica* larvae are adapted to kill and digest bacteria. The midgut of *M. domestica* is differentiated into three regions with different luminal pH values: an anterior midgut (pH 6.1), a middle midgut (pH 2.3), and a posterior midgut (6.8). The luminal pH values may somewhat change depending on the food. The middle midgut has an embryological origin different from the cells of the anterior and posterior midgut (Poulson and Waterhouse 1960). The acidification of the middle midgut is caused by special cells, the oxyntic cells (*see details in* Chap. 6). A combination of enzyme assays, midgut pH determinations, showed how *M. domestica* larvae deal with bacteria. At first, the carbohydrate content of the ingested food is decreased in the anterior midgut, leaving bacteria more susceptible to the action of low pH, a special lysozyme, and a digestive cathepsin D at the middle midgut. The killed bacteria then pass into the posterior midgut, where most digestion occur carried out by luminal endopeptidases, carboxypeptidases, and amylases, and cell microvillar membrane-bound aminopeptidases and disaccharidases. The distribution of enzymes along the posterior midgut, the excretory rate of digestive enzymes, dye experiments, and ultrastructural research support the assertion that there is digestive enzyme recycling in the posterior midgut (Espinoza-Fuentes and Terra 1987; Espinoza-Fuentes et al.

1987; Terra et al. 1988a, b). Bacteria killing in the middle midgut is caused mainly by special enzymes active in very low pH: a lysozyme (Lemos et al. 1993) and a cathepsin D (Padilha et al. 2009) recruited from lysosomes (see also Chaps. 10 and 11). The molecular mechanisms underlying midgut pH buffering (Barroso et al. 2019) and midgut water fluxes (Barroso et al. 2020) were studied with a combination of in vivo midgut luminal pH determinations, action of drugs, transcriptomics, and proteomics (see details in Chap. 6).

More details in the *M. domestica* larval digestion of starch, proteins, and lipids and the absorption of their products were obtained by a combination of enzyme assays and transcriptomics and proteomics data. Digestion of starch is carried out by a soluble and a GPI-anchored amylase, followed by the action of soluble glycocalyx-associated and membrane-bound maltases. The resulting glucose molecules were absorbed by sugar transporters, some of which are upregulated by starch. Along the midgut, starch is first digested at the anterior midgut by amylase and maltase and the formed glucose molecules are absorbed at the middle midgut. After the breakdown of the walls of bacteria and fungi cells at the middle midgut, the released glycogen and sugars are digested and absorbed along the posterior midgut (Pimentel et al. 2018). Among lipids, phosphatides are digested by phospholipases at the anterior midgut. The released lysophosphatides emulsify triacylglycerols in the middle and posterior midgut, which are then digested by acid triacylglycerol lipases. Most resulting fatty acids are absorbed in the posterior midgut, where they may form diacylglycerols, triacylglycerols, and phosphatides or are oxidized along the midgut (Barroso et al. 2021). Protein digestion starts in the anterior midgut carried out by trypsin and aminopeptidase. Proteins released from bacteria and fungi in the middle midgut are digested by cathepsins D, followed mainly in the posterior midgut by serine endopeptidases, carboxypeptidases, and aminopeptidases. Absorption of amino acids and peptides is found along the whole midgut, but mainly at the posterior midgut (Barroso, Canettieri, Ferreira, Terra, manuscript in preparation).

M. domestica adults salivate (or regurgitate their crop contents) onto the meal. After ingesting the dispersed material, the digestion of starch takes place at the crop under the action of salivary amylase. Digestion of starch and other nutrients ended along the midgut, as described for larvae (Terra and Ferreira 2005, and see also above).

The blood-feeding stable fly *Stomoxys calcitrans* (Cyclorrhapha, Muscoidea, Muscidae) has a midgut with three major regions. The anterior midgut (thoracic midgut) and the middle midgut (reservoir region) store and dehydrate the blood meal and the posterior midgut is responsible for most digestion. Between the middle and posterior midgut, there is a region known as opaque zone, which is thought to be responsible for most secretion of digestive enzymes, including trypsin (Billingsley 1990; Moffatt and Lehane 1990; Wood and Lehane 1991). An immunocytological investigation of trypsin in *S. calcitrans* showed that at least part of the trypsin is released by apocrine secretion (Jordão et al. 1996).

Anastrepha fraterculus (Cyclorrhapha, Acalyptrata, Tephritidae) larvae inhabit in fruits of living plants. Despite that, *A. fraterculus* maintain the capacity of other

Cyclorrhapha to digesting bacteria, as they have an acidic middle midgut and an active special lysozyme with an acidic pH optimum and a luminal cathepsin D (Lemos and Terra 1991). These findings agree with previous works according to which bacteria may provide adequate nutrition to some Tephritidae (Drew et al. 1983; Drew and Lloyd 1987).

The major focus of the studies of larval *Drosophila melanogaster* (Cyclorrhapha, Acalyptrata, Drosophilidae) is cell-type diversity, tissue homeostasis, stem cell regeneration, cell fate decision, and nervous and endocrine systems (Buchon et al. 2013; Lemaitre and Miguel-Aliaga 2013; Dutta et al. 2015; Hung et al. 2020). Despite that emphasis, those studies resulted in some interesting findings from the point of view of molecular physiology of digestion. Thus, they reported the existence of a predominance of the expression of glycoside hydrolases and lipases in the anterior midgut and serine endopeptidases in the posterior midgut. Furthermore, the studies revealed that glucose transporters occur in the anterior midgut, whereas amino acid transporters (NAAT1, EAAT1, see Chap. 7 for definitions) are found in both anterior and posterior midgut.

13.12 Lepidoptera

Most Lepidoptera pertain to the clade Ditrysia that includes the majority of superfamilies best known from the point of view of molecular physiology. They are: Bombycoidea, Noctuoidea, Pyraloidea, and Tineoidea.

The midgut of *Erinnyis ello* (Bombycoidea: Sphingidae) is cylindrical with no ceca and its pH contents are highly alkaline. Amylase and trypsin are found in midgut contents and recovered in the regurgitated material. Aminopeptidase is bound to the microvillar membranes and α-glucosidase, β-glucosidase, and trehalase are associated with the glycocalyx of midgut cells. None of the last-mentioned enzymes occur in the regurgitate material. Due to the low excretory rate of digestive enzymes, their midgut luminal gradients, and ultrastructural data, it was concluded that there is enzyme recycling powered by midgut countercurrent fluxes of water in *E. ello* larvae. In addition to that, dye experiments support the view that the anterior midgut absorbs water and its posterior midgut secretes water (Santos et al. 1983, 1984; Santos and Terra 1984). The midgut of lepidopterans has, in addition to columnar cells, the goblet cells, which have a long neck (Fig. 5.1d) in the anterior midgut and are stalked (Fig. 5.1c) in the posterior midgut (Cioffi 1979; Santos et al. 1984). The function of goblet cells is to excrete excess K^+ ions absorbed from leaves ingested by larvae (Cioffi and Harvey 1981) and may assist the anterior midgut columnar cells in water absorption and the posterior midgut columnar cells in water secretion (Santos et al. 1984). The high pH determined in lepidopteran midguts is also caused by goblet cells (Dow 1992). It is not necessary, however, that a midgut cell must be a goblet cell to alkalize midgut contents, as exemplified by the columnar cells of sciarid and mosquito larvae (see above). It is more probable that goblet cells evolved as a compromise between the need to secrete potassium ions and the anterior midgut to absorb water and the posterior midgut to secrete water, besides a role in alkalization of the midgut contents (Terra 1988).

The organization of digestion in the larvae of *Spodoptera frugiperda* (Noctuoidea: Noctuidae) (Ferreira et al. 1994) is similar to that of *E. ello*. However, proteomic studies detailed the distribution of proteins in midgut columnar cell microvillar membranes and the midgut compartmentalization of digestive enzymes (Fuzita et al. 2019, 2022). The paper clarified the molecular mechanisms underlying vesicle budding in microapocrine secretion (see details in Chap. 5) and confirmed the retention in the ectoperitrophic fluid (outside the peritrophic membrane) of most enzymes involved in intermediate digestion (see details Chap. 7).

Bombyx mori (Bombycoidea: Bombycidae) seems to have the same organization of digestion of the other lepidopterans, taken into account data on proteomics of ectoperitrophic proteins, which include proteins restricted to the ectoperitrophic space or on course to the endoperitrophic space (Liu et al. 2018).

Lepidopteran larvae that ingest unique diets have some adaptations. Thus, *Tineola bisselliela* (Tineoidea: Tineidae) feed on wool and because of that have highly reducing midguts to open disulfide bonds in keratin to permit the access of proteolytic enzymes of the otherwise insoluble keratin (Terra and Ferreira 1994). The larva of *Hofmannophila pseudopretella* (Gelechioidea: Oecophoridae) also digest wool similarly with *T. bisselliela* (Christeller 1996).

An interesting mathematical model was developed by Woods and Kingsolver (1999) that predicted that absorption (or post-absorptive processes) was the limiting step to growth in a caterpillar feeding on artificial diets, whereas no such limit is predicted for caterpillar eating leaves.

Adult moths usually have nectar as their sole food. Due to pumping constraints, moths and butterflies feed on nectars containing 15-25% sugars. Bees, which lap up nectar, instead of pumping it, may feed on nectars having more than 50% sugar (Kingsolver and Daniel 1979). As soon as nectar is ingested by adult lepidopterans, it is stored in the crop, from where it is periodically moved into the midgut for digestion by an α-glucosidase (or β-fructosidase) and absorbed. In accordance with that, the adult lepidopteran midgut is water-absorbing along its whole extension (Terra et al. 1990). However, some adult lepidopterans have a salivary amylase and others a complete set of digestive enzymes (Terra et al. 1990), which may explain, at least in enzymological grounds, the adaptation of some adult lepidopteran adults to new feeding habits, such as blood (Banziger 1970) or pollen (Gilbert 1972) feeding.

References

Andersen PC, Brodbeck BV, Mizell RF (1989) Metabolism of amino acids, organic compounds and sugars extracted from the xylem fluid of four host plants by adult *Homalodisca coagulata*. Ent Exp Appl 50:149–159

Anstee JH, Charnley AK (1977) Effects of frontal removal and starvation on activity and distribution of six gut enzymes in Locusta. J Insect Physiol 23:965–974

Ashford DA, Smith WA, Douglas AE (2000) Living on a high sugar diet: The fate of sucrose ingested by a phloem-feeding insect, the pea aphid *Acyrthosiphon pisum*. J Insect Physiol 46:335–341

Azambuja P, Guimarães JA, Garcia ES (1983) Haemolytic factor from the crop of Rhodnius prolixus: evidence and partial characterization. J Insect Physiol 29:833–837

Azevedo DO, Neves CA, Mallet JRS et al (2009) Notes on midgut ultrastruture of *Cimex hemipterus* (Hemiptera: Cimicidae). J Med Entomol 46:435–441

Azevedo DO, Fialho MCQ, Vargas NC et al (2013) Morphology of the digestive tract of *Cladomorphus phyllinus* (Phasmatodea: Phasmidae). Florida Entomol Soc 96:1417–1423

Baines DM (1979) Studies of weight changes and movements of dyes in caeca and midgut of fifth-instar *Locusta migratoria migratorioides* (R& F) in relation to feeding and food deprivation Acrida 8:95–108

Baker JE, Fabrick A (2000) Host hemolymph proteins and protein digestion in larval *Habrobracon hebetor* (Hymenoptera: Braconidae). Insect biochem Mol Biol 30:937–946

Banziger H (1970) The piercing mechanism of the fruit-piercing moth *Calpe (Calyptra) thalictri* Bkh (Noctuidae) with reference to the skin-piercing blood-sucking moth *C. eustrigata* Hmps. Acta Trop 27:54–87

Barker RJ, Lehner Y (1972) A look at honey bee gut functions. Am Bee J 112:336–338

Barret FM (1982) Absorption of from the anterior midgut of *Rhodnius*. J Insect Physiol 28:335–341

Barroso IG, Santos CS, Bertotti M et al (2019) Molecular mechanisms associated with acidification and alkalization along the larval midgut of *Musca domestica*. Comp Biochem Physiol A:237. https://doi.org/10.1016/j.cbpa.2019.110535

Barroso IG, Fuzita FJ, Ferreira C et al (2020) Midgut fluxes and digestive enzyme recycling in *Musca domestica:* a molecular approach. Comp.Biochem. Physiol. A 241. https://doi.org/10.1016/j.cbpa.2019.110627

Barroso IG, Cardoso C, Ferreira C et al (2021) Transcriptomic and proteomic analysis of the underlying mechanisms of digestion of triacylglycerols and phosphatides and absorption and fate of fatty acids along the midgut of *Musca domestica*. Comp Biochem Physiol D 39. https://doi.org/10.1016/j.cbd.2021.100826

Bayon C (1981) Ultrastructure de l'épithelium intestinal et flore pariétale chez la larve xylophage d'*Oryctes nasicornis* L. (Coleoptera: Scarabaeidae). Int J Insect Morphol Embryol 10:359–371

Beckman N, Hurd LE (2003) Pollen feeding and fitness in praying mantis: the vegetarian side of a tritrophic predator. Environ Entomol 32:881–885

Benemann JR (1973) Nitrogen fixation in termites. Science 181:164–165

Bernays EA (1981) A specialized region of the gastric caeca in the locust Schistocerca gregaria. Physiol Entomol 6:1–6

Beton D, Guzzo CR, Ribeiro AF et al (2012) The 3D structure and function of digestive cathepsin L-like proteinases of *Tenebrio molitor* larval midgut. Insect Biochem Mol Biol 42:655–664

Biagio FP, Tamaki FK, Terra WR (2009) Digestive morphophysiology of *Gryllodes sigillatus* (Orthoptera: Gryllidae). J Insect Physiol 55:1125–1133

Bifano TD, Alegria TGP, Terra WR (2010) Transporters involved in glucose absorption in the *Dysdercus peruvianus* (Hemiptera: Pyrrhocoridae) anterior midgut. Comp. Biochem. Physiol. B 157:1–9

Bignell DE (1981) Nutrition and digestion. In: Bell WJ, Adiyodi KG (eds) The American cockroach. Chapman and Hall, London, pp 57–86

Billingsley PF (1990) The midgut ultrastructure of hematophagous insect. Annu Rev Entomol 35:219–248

Bolognesi R, Terra WR, Ferreira C (2008) Peritrophic membrane role in enhancing digestive efficiency. Theoretical and experimental models. J Insect Physiol 54:1413–1422

Borovsky D, Schlein Y (1988) Quantitative determination of trypsin-like and chymotrypsin-like enzymes in insects. Archs Insect Biochem Physiol 8:249–260

Brues CT (1946) Insect dietary. Harvard University Press, Cambridge

Brune A (2014) Symbiotic digestion of lignocellulose in termite guts. Nat Rev Microbiol 12:168–180

Buchon N, Osman D, David FA et al (2013) Morphological and molecular characterization in Drosophila. Cell Reports 3:1725–1738

Buonocore V, Poerio E, Silano V et al (1976) Physical and catalytic properties of alpha-amylase from Tenebrio molitor L. larvae. Biochem J 153:621–625

Caetano FH (1988) Anatomia, histologia do sistema digestivo e excretor de operárias de formigas (Hymenoptera, Formicidae). Naturalia 13:129–174

Caldeira W, Dias AB, Terra WR et al (2007) Digestive enzyme compartmentalization and recycling and sites of absorption and secretion along the midgut of *Dermestes maculatus* (Coleoptera) larvae. Arch Insect Biochem Physiol. 64:1–18

Cantón PE, Bonning BC (2020) Transcription and activity of digestive enzymes of Nezara viridula maintained on different plant diets. Front Physiol 10:1553

Chapman RF (1988) The relationship between diet and the size of the midgut caeca in grasshoppers (Insecta: Orthoptera: Acrididoidea). Zool L Linn Soc 94:319–338

Cheeseman MT, Pritchard G (1984) Peritrophic membranes in adults of two species of Carabid beetles (Coleoptera: Carabidae). Int J Insect Morphol Embryol 13:171–173

Cheung WWK, Marshall AT (1973a) Studies on water and ion transport in homopteran insects: ultrastructure and cytochemistry of the cicadoid and cercopoid midgut. Tissue Cell 5:651–669

Cheung WWK, Marshall AT (1973b) Water and ion regulation in cicadas in relation to xylem feeding. J Insect Physiol 19:1801–1916

Chouvenc T, Sobotnik J, Engel MS et al (2021) Termite evolution: mutualistic associations, key innovations, and the rise of Termitidae. Cell Mol Life Sci 78:2749–2769

Christeller JT (1996) Degradation of wood by *Hofmannophila pseudospretella* (Lepidoptera: Oecophoridae) larval midgut extracts under conditions simulating the midgut environment. Arch Insect Biochem Physiol 33:99–119

Cioffi M (1979) The morphology and fine structure of the larval midgut of the moth (Manduca sexta) in relation to active ion transport. Tissue Cell 11:467–479

Cioffi M, Harvey WR (1981) Comparison of potassium transport in three structurally distinct regions of the insect midgut. J Exp Biol 91:103–116

Cleveland LR, Hall SR, Sanders EP, Coller J (1934) The wood-feeding roach, *Cryptocercus*, its protozoa, and the symbiosis between protozoa and roach. Mem Amer Acad Arts Sci 17:185–343

Cohen AC (1993) Organization of protein digestion and preliminary characterization odf salivary trypsin-like enzymes in a predaceous heteropteran, *Zelus renardii*. J Insect Physiol 39:823–829

Cohen AC (1995) Extra-oral digestion in predaceous terrestrial Arthropoda. Annu Rev Entomol 40:85–104

Colepicolo-Neto P, Bechara EJH, Ferreira C (1986) Evolutionary considerations of the spatial organization of digestion in the luminescent predaceous larva of *Pyrearinus termitilluminans* (Coleoptera:Elateridae). Insect Biochem 16:811–817

Cook SC, Davidson DW (2006) Nutritional and functional biology of exudate-feeding ants. Ent Exp Appl 118:1–10

Cooper PD, Vulcano R (1997) Regulation of pH in the digestive system of the cricket, *Teleogryllus commodus* Walker. J Insect Physiol 43:495–499

Costa-Leonardo AM, Silva IB (2022) Gut anatomy and ultrastructural features of the paunch epithelium in the Neotropical termite *Serritermes serrifer* (Blattaria, Isoptera, Serritermitidae). J Comp Zool 300:75–81

Crailsheim K (1988a) Intestinal transport of sugars in the honey bee (*Apis mellifera* L). J Insect Physiol 34:839–845

Crailsheim K (1988b) Transport of leucine in the alimentary canal of the honeybee (*Apis melifera* L) and its dependence on season. J Insect Physiol 34:1093–1100

Cristofoletti PT, Terra WR (1999) Specificity, anchoring, and subsites in the active center of a microvillar aminopeptidase purified from *Tenebrio molitor* (Coleoptera) midgut cells. Insect Biochem Mol Biol 29:807–819

Cristofoletti PT, Ribeiro AF, Deraison C et al (2003) Midgut adaptation and digestive enzyme distribution in a phloem feeding insect, the pea aphid *Acyrthosiphon pisum*. J Insect Physiol 49:11–24

Cristofoletti PT, Ribeiro AF, Terra WR (2005) The cathepsin L-like proteinases from the midgut of *Tenebrio molitor* larvae: sequence, properties, immunocytochemical localization and function. Insect Biochem Mol Biol. 35:883–901

Cristofoletti PT, Mendonça-de-Sousa FA, Rahbé Y et al (2006) Characterization of a membrane-bound aminopeptidase purified from *Acyrthosiphon pisum* midgut cells. A major binding site for toxic mannose lectins. FEBS J 273:5574–5588

Cruz-Landim C, Mello RA (1981) Desenvolvimento e envelhecimento de larvas e adultos de *Scaptotrigona postica* Latreille (Hymenoptera: Apidae): aspectos histológicos e histoquímicos. Publ ACIESP São Paulo No 31:1–117

Damasceno TF, Dias RO, de Oliveira JR et al (2017) Active subsite properties, subsite residues and targeting to lysosomes or midgut lumen of cathepsins L from the beetle *Tenebrio molitor*. Insect Biochem Mol Biol 89:17–30

Del Bene G, Dallai R, Marchini D (1991) Ultrastructure of the midgut and the adhering tubular salivary glands of *Frankliniella occidentalis* (Pergande) (Thysanoptera: Thripidae). Int J Insect Morphol Embryol 29:15–24

Denecke S, Ioannidis P, Buer B et al (2020) A transcriptomic and proteomic atlas of expression in the *Nezara viridula* (Heteroptera: Pentatomidae) midgut suggests the compartmentalization of xenobiotic metabolism and nutrient digestion. BMC Genom 21:129

Deraison C, Darboux I, Duportets L et al (2004) Cloning and characterization of a gut-specific cathepsin L from the aphid *Aphis gossypii*. Insect Mol Biol 13:165–177

Dias RO, Cardoso C, Pimentel AC et al (2018) The roles of mucus-forming mucins, peritrophins and peritrophins with mucin domains in the insect midgut. Insect Mol Biol 27:46–60

Dias RO, Cardoso C, Leal CS et al (2019) Domain structure and expression along the midgut and carcass of peritrophins and cuticle proteins analogous to peritrophins in insects with and without peritrophic membrane. J Insect Physiol 114:1–9

Douglas AE (2006) Phloem-sap feeding by animals: problems and solutions. J ExpBot 57:747–754

Dow JAT (1981a) Localization and characterization of water uptake from the midgut of locust, Schistocerca gregaria. J Exp Biol 93:269–281

Dow JAT (1981b) Countercurrent flows, water movements and nutrient absorption in the locust midgut. J Insect Physiol 27:579–585

Dow JAT (1992) pH gradients in lepidopteran midgut. J Exp. Biol. 172:355–375

Drew RA, Lloyd AC (1987) Relationship of fruit flies (Diptera: Tephritidae) and their bacteria to host plants. Ann Ent Soc Am 80:629–636

Drew RA, Courtice AC, Teakle DS (1983) Bacteria as a natural source of food for adult fruit flies (Diptera: Tephritidae). Oecologia 60:279–284

Droste HJ, Zebe E (1974) Carbohydrasen und Kohhenhydratverdauung im Darmtrakt von *Locusta migratoria*. J Insect Physiol 20:1639–1657

Dussutour A, Simpson SJ (2009) Communal nutrition in ants. Curr Biol 19:740–744

Dutta D, Dobson AJ, Houtz PL et al (2015) Regional cell-specific transcritome mapping reveals regulatory complexity in the adult Drosophila midgut. Cell Reports 12:346–358

Elpidina EN, Tsybina TA, Dunaevsky YE et al (2005) A chymotrypsin-like proteinase from the midgut of Tenebrio molitor larvae. Biochimie 87:771–779

Engelmann F, Geraerts WPM (1980) The proteases and the protease inhibitor in the midgut of Leucophea maderae. J Insect Physiol 26:703–710

Erthal M Jr, Silva CP, Samuels RI (2007) Digestive enzymes in larvae of the leaf cutting ant, *Acromyrmex subterraneous* (Hymenoptera: Formicidae: Attinae). J Insect Physiol 53:1101–1111

Espinoza-Fuentes FP, Terra WR (1987) Physiological adaptations for digesting bacteria. Water fluxes and distribution of digestive enzymes in *Musca domestica* larval midgut. Insect Biochem 17:809–817

Espinoza-Fuentes FP, Ribeiro AF, Terra WR (1987) Microvillar and secreted digestive enzymes from *Musca domestica* larvae. Subcellular frationation of midgut cells with eléctron microscopy monitoring. Insect Biochem 17:819–827

Evans G (1977) The life of beetles. Allen & Unwin, London

Ferreira C, Terra WR (1980) Intracellular distribution of hydrolases in midgut caeca cells from an insect with emphasis on plasma membrane-bound enzymes. Comp Biochem Physiol B 66:467–473

Ferreira C, Terra WR (1982) Function of midgut caeca and ventriculus: microvilli bound enzymes from the cells of different midgut regions of straving and feeding Rhynchosciara americana larvae. J Insect Physiol 12:257–262

Ferreira C, Terra WR (1989) Spatial organization of digestion, secretory mechanisms and digestive enzyme properties in *Pheropsophos aequinoctialis* (Coleoptera:Carabidae). Insect Biochem 19:383–391

Ferreira C, Ribeiro AF, Terra WR (1981) Fine structure of the larval midgut of the fly *Rhynchosciara* and its physiological implications. J Insect Physiol 27:559–570

Ferreira C, Ribeiro AF, Garcia ES et al (1988) Digestive enzymes trapped between and associated with the double membranes of *Rhodnius prolixus* posterior midgut cells. Insect Biochem 18:521–530

Ferreira C, Oliveira MC, Terra WR (1990) Compartmentalization of the digestive process in *Abracris flavolineata* (Orthoptera: Acrididae). Insect Biochem 20:267–274

Ferreira C, Ribeiro AF, Terra WR (1993) Ultrastructural and biochemical aspects of digestion in the imagoes of the fly *Rhynchosciara americana*. Entomol Exp Appl 66:135–143

Ferreira C, Capella AN, Sitnik RE et al (1994) Digestive enzymes in midgut cells, endo- and ectoperitrophic contents, and peritrophic membranes of Spodoptera frugiperda (Lepidoptera) larvae. Archs Insect Biochem Physiol 26:299–313

Ferreira AHP, Marana SR, Terra WR et al (2001) Purification, molecular cloning, and properties of a β-glycosidase isolated from midgut lumen of *Tenebrio molitor* (Coleoptera) larvae. Insect Biochem Mol Biol 31:1065–1076

Ferreira AHP, Terra WR, Ferreira C (2003) Characterization of a β-glycosidase highly active on disaccharides and of a β-galactosidase from *Tenebrio molitor* midgut lumen. Insect Biochem Mol Biol 33:253–265

Ferreira C, Rebola KGO, Cardoso C et al (2015) Insect midgut carboxypeptidases with emphasis on S10 hemipteran and M14 lepidopteran carboxypeptidases. Insect Molec Biol 24:222–239

Fialho MCQ, Moreira NR, Zanuncio JC et al (2012) Prey digestion in the midgut of the predatory bug *Podisus nigrispinus* (Hemiptera: Pentatomidae). J Insect Physiol 58:850–856

Fialho MCQ, Terra WR, Moreira NR et al (2013) Ultrastructure ans immunocytolocalization of digestive enzymes in the midgut of *Podisus nigrispinus* (Heteroptera: Pentatomidae). Arthr Struct Devel 43:277–285

Fisher DB, Wright JP, Mittler TE (1984) Osmoregulation by the aphid *Myzus persicae*: a physiological role for honeydew oligosaccharides. J Insect Physiol 30:387–393

Fujita A, Abe T (2002) Amino acid concentration and distribution of lysozyme and protease activities in the guts of higher termites. Physiol Entomol 27:76–78

Fujita A, Hojo M, Aoyagi T et al (2010) Details of the digestive system in the midgut of *Coptotermes formosanus* Shiraki. J Wood Sci 56:222–226

Fuzita FJ, Pimenta DC, Palmisano G et al (2019) Detergent-resistant domains in *Spodoptera frugiperda* midgut microvillar membranes and their relation to microapocrine secretion. Comp Biochem Physiol B 325:8–18

Fuzita FJ, Palmisano G, Pimenta DC et al (2022) A proteomic approach to identify digestive enzymes, their exocytic and microapocrine secretory routes and their compartmentalization in the midgut of *Spodoptera frugiperda*. Comp Biochem Physiol B 257:110670

Garcia ES, Garcia MLM (1977) Control of protease action in the intestine of fifth instar larvae of *Rhodnius prolixus*. J Insect Physiol 23:247–251

Gelb SM, Filley TR, Hatcher PG et al (2008) Lignin degradation in wood-feeding insects. Proc Natl Acad Sci USA 105:12932–12937
Genta FA, Terra WR, Ferreira C (2003) Action pattern, specificity, lytic activities, and physiological role of five digestive β-glucanases isolated from *Periplaneta americana*. Insect Biochem Mol Biol 33:1085–1097
Genta FA, Blanes L, Cristofoletti PT et al (2006) Purification, characterization and molecular cloning of the major chitinase from *Tenebrio molitor* larval midgut. Insect Biochem Mol Biol 36:789–800
Genta FA, Bragatto I, Terra WR et al (2009) Purification, characterization and sequecing of the major beta-1,3-glucanase from the midgut of *Tenebrio molitor* larvae. Insect Biochem Mol Biol 39:861–874
Gilbert LE (1972) Pollen feeding and reproductive biology of *Heliconius* butterflies. Proc Natl Acad Sci Wash 69:1403–1407
Gomez A, Cardoso C, Genta FA et al (2013) Active site characterization and molecular cloning of *Tenebrio molitor* midgut trehalase and comments on their insect homologs. Insect Biochem Mol Biol 43:768–780
Gonçalves EK (2002) Estudo morfofuncional do Sistema digestivo de Haematopinus suis (Linnaeus, 1758) (Phthiraptera, Anoplura. PhD thesis. University of São Paulo
Goodchild AJP (1966) Evolution of the alimentary chanal in the Hemiptera. Biol Rev 41:97–140
Goptar IA, Shagin DA, Shagina IA et al (2013) A digestive prolyl carboxypeptidase in *Tenebrio molitor* larvae. Insect Biochem Mol Biol 43:501–509
Graça-Souza AV, Maya-Monteiro C, Paiva-Silva GO et al (2006) Adaptations against heme toxicity in blood-feeding arthropods. Insect Biochem Mol Biol 36:322–335
Grimaldi D, Engel MS (2005) Evolution of Insects. Cambridge University Press, New York
Hatle JD, Spring JH, Dow JAT (2002) Ion and water transport in the orthopteran alimentar canal: a comparison of Mantidae and Acrididae. J Orthop Res 11:19–24
Heinrich D, Zebe E (1973) Zur Feinstruktur der Mittleldarmzellen von in verschiedenen Phasen der Verdauung. Cytobiologie 7:315–326
Herbert JF, Thomas D, Cavalier A et al (1989) Structural and biochemical observations on specialized membranes of the filter chamber, a water shunting complex of sap-sucking homopteran insects. Biol Cell 66:155–163
Hill P, Campbell JA, Petrie IA (1976) *Rhodnius prolixus* and its symbiotic actinomycete: amicrobilogical, physiological and behavioral study. Proc Royal Soc Lond Ser B 194:501–525
Hodge C (1936) The anatomy and histology of the alimentary tract of the grasshopper, *Melanoplus differentialis* Thomas. J Morphol 59:423–439
Hodge C (1939) The anatomy and histology of the alimentary tract of Locusta migrratoria L. (Orthoptera: Acrididae). J Morphol 64:375–399
Hogan ME, Slaytor M, O'brien RW (1985) Transport of volatile fatty acids across the hindgut of the cockroach Panesthia cribrata, Sausurre and the temite, Mastotermes darwinensis Frogatt. J insect Physiol 31:587–591
Hogan M, Veivers PC, Slaytor M et al (1988) The site of cellulose breakdown in higher termites (*Nasutitermes walkeri* and *Nasutitermes exitiosus*). J Insect Physiol 34:891–899
Hongoh Y, Sharma VK, Prakash T et al (2008) Genome of an endosymbiont coupling N_2 fixation to cellulolysis within protist cells in termite gut. Science 322:1108–1109
Huang SW, Zhang HY, Marshall S et al (2010) The scarab gut: a potential bioreactor for bio-fuel production. Insect Sci 17:175–183
Hughes J, Vogler AP (2006) Gene expression in the gut of keratin-feeding clothes moths (Tineola) and keratin beetles (Trox) revealed by subtracted cDNA libraries. Insect Biochem Mol Biol 36:584–592
Hung R-J, Hu Y, Kirchner R, Liu Y et al (2020) A cell atlas of the adult Drosophila midgut. Proc Natl Acad Sci USA (Was.) 277:1514–1523
Jimenez DR, Gilliam M (1990) Ultrastructure of ventriculus of the honeybee *Apis mellifera* (L): cytochemical localization of the acid phosphatase, alkaline phosphatase and non-specific esterase. Cell Tissue Res 261:431–443

Jordão BP, Lehane MJ, Terra WR et al (1996) An immunocytochemical investigation of trypsin secretion in the midgut of the stablefly, *Stomoxys calcitrans*. Insect Biochem Mol Biol 26:445–453

Kingsolver JG, Daniel TL (1979) On the mechanics and energetics of nectar feeding in butterflies. J Theor Biol 76:167–179

Kitajima WF (1975) A peculiar type of glycocalyx on the microvilli of the midgut epithelial cells of the thrips *Frankliniella sp* (Thysanoptera: Thripidae). Cytobiologie 2:269–303

Köhler T, Dietrich C, Scheffrahn RH et al (2012) High-resolution analysis of gut environment and bacterial microbiota reveals functional compartmentation of the gut in wood-feeding higher termites (*Nasutitermes spp*.). Applied Environ Microbiol 78:4691–4701

Kollien AH, Waniek PJ, Pröls F et al (2004) Cloning and characterization of a trypsin-encoding cDNA of the human body louse *Pediculus humanus*. Insect Mol Biol 13:9–18

Kollien AH, Waniek PJ, Schaub GA (2007). Blutverdauung der Kleiderlaus *Pediculus humanus*. Berichte des naturwissenchaftlich-medizinschen vereins Innsbruk. (Suppl 17):109

Kumara RP, Saitoh S, Aoyama H et al (2016) Metabolic pathways in the mixed segment of the wood-feeding termite *Nasutitermmes takasagoensis* (Blattodea (Isoptera): Termitidae). Appl Entomol Zool 51:429–440

Le Caherec F, Guillam M-T, Beuron F et al (1997) Aquaporin-related proteins in the filter chamber of homopteran insects. Cell Tissue Res 290:143–151

Leal CS (2018) Comparative study of the Organization of the Digestive System of Representative Species of the Orders Hemiptera and Thysanoptera. MsC Dissertation. University of São Paulo

Lemaitre B, Miguel-Aliaga I (2013) The digestive tract of Drosophila melanogaster. Annu Rev Genet 47:377–404

Lemke T, Stingl U, Egert M et al (2003) Physicochemical conditions and microbial activities in the highly alkaline gut of the humus- feeding larva of *Pachnoda ephippiata* (Coleoptera:Scarabaeidae). Appl Environ Microbiol 69:6650–6658

Lemos FJA, Terra WR (1991) Digestion of bacteria and the role of midgut lysozyme in some insect larvae. Comp Biochem Physiol B 100:265–268

Lemos FJA, Campos FAP, Silva CP et al (1990) Proteinases and amylases of larval midgut of *Zabrotes subfasciatus* reared on cowpea (*Vigna unguiculata*) seeds. Entomol Exp Appl 56:219–227

Lemos FJA, Ribeiro AF, Terra WR (1993) A bacteria-digesting midgut lysozyme from *Musca domestica* (Diptera) larvae: purification, properties and secretory mechanism. Insect Biochem Mol Biol 23:533–541

Levinsky H, Birk Y, Applebaum SW (1997) Isolation and characterization of a new trypsin-like enzyme from *Tenebrio molitor* larvae. Int J Pep Prot Res 10:252–264

Levinson ZH (1960) Food of Housefly Larvae. Nature 188:427–428

Li X, Brune A (2005) Digestion of microbial mass, structural polysaccharides, and protein by the humivo- rous larva of *Pachnoda ephippiata* (Coleoptera: scarabaeidae). Soil Biol Biochem 37:107–116

Lindsay KL, Marshall AT (1981) The osmoregulatory role of the filter chamber in relation to phloem feeding in *Eurymela distincta* (Cicadeloidea: Homoptera). Physiol Entomol 6:413–419

Liu L, Qu M, Yang J et al (2018) The physiological differentiation along the midgut of Bombyx mori- inspirations from proteomics and gene expression patterns of the secreted proteins in the ectoperitrophic space. Insect Mol Biol 27:247–259

Lopes AR, Juliano MA, Marana SR et al (2006) Substrate specificity of insect trypsins and the role of their subsites in catalysis. Insect Biochem Molec Biol 36:130–140

Ma H, Liu Z-G, Bao Y et al (2009) Morphology and three-dimensional reconstruction of the digestive system of *Periplaneta americana*. J Med Entomol 46:165–168

Marana SR, Ribeiro AF, Terra WR et al (1997) Ultrastructure and secretory activity of *Abracris flavolineata* (Orthoptera: Acrididae) midguts. J Insect Physiol 43:465–473

Marques MD (1976) Aspectos anatomicos e histológicos dos estágios imaturos de *Rhynchosciara americana* (Wiedemann, 1821) (Diptera, Sciaridae). PhD thesis, University of São Paulo

Marshal AT, Cheung WK (1970) Ultrastruture and cytochemistry of an extensive plexiform surface coat on the midgut cells of a Fulgorid insect. J Ultratr Res 33:161–172

Moffatt MR, Lehane MJ (1990) Trypsin stored as an inactive zymogen in the midgut of *Stomoxys calcitrans*. Insect Biochem 20:719–723

Monteiro EC, Tamaki FK, Terra WR et al (2014) The digestive system of the stick bug *Cladomorphus phyllinus* (Phasmida: Phasmatidae): a morphological, physiological and biochemical analysis. Arthr Struct Develop 43:123–134

Moreira NR, Cardoso C, Ribeiro AF et al (2015) Insect midgut α-mannosidases from family 38 and 47 with emphasis on those of *Tenebrio molitor*. Insect Biochem Mol Biol 67:94–104

Moreira NR, Cardoso C, Dias RO et al (2017) A physiologically-oriented transcriptomic analysis of the midgut of *Tenebrio molitor*. J Insect Physiol 99:58–66

Muraleedharan D, Prabhu VKK (1979) Role of median neurosecretory cells in secretion of protease and invertase in the red cotton bug, *Dysdercus cingulatus*. J Insect Physiol 25:237–240

Murdock LL, Brookhart G, Dunn PE et al (1987) Cysteine digestive proteinases in Coleoptera. Comp Biochem Physiol 87B:783–787

Nakayama DG, Santos-Junior CD, Kishi LT et al (2017) A transcriptomic survey of *Migdolus fryanus* (sugarcane rhizome borer) larvae. PLoS One. https://doi.org/10.1371/journal.pone.0173059

Nepomuceno DB, Santos VC, Araujo RN et al (2017) pH control in the midgut of Aedes aegypti under different nutritional conditions. J Exp Biol 220:3355–3362

Noirot C, Noirot-Timothe C (1969) The digestive system. In: Krishna K, Weesner F (eds) Biology of termites. Academic, London, pp 107–119

O'Loughlin GT, Chambres TC (1972) Extracellular microtubules in the aphid gut. J Cell Biol 53:575–578

Ochanda JO, Oduor EAC, Galun R et al (2000) Partial purification of the aminopeptidase from the midgut of the human body louse, *Pediculus humanus humanus*. Physiol Entomol 25:242–246

Oviedo MN, VanEkeris L, Corena-Mcleod MDP et al (2008) A microarray-based analysis of transcriptional compartmentalization in the alimentary canal of *Anopheles gambiae* (Diptera: Culicidae) larvae. Insect Mol Biol 17:61–72

Padilha MHP, Pimentel AC, Ribeiro AF et al (2009) Sequence and function of lysosomal and digestive cathepsin D-Like proteinases of *Musca domestica* midgut. Insect Biochem Mol Biol 39:782–791

Panizzi AR, Lucine T, Mitchell PL (2021) Electronic monitoring of feeding behavior of phytophagous true bugs (Heteroptera). Springer Nature, Cham

Parenti P, Sachi FV, Hanozet GM et al (1986) Na-dependent uptake of phenylalanine in the midgut of a cockroach (*Blabera gigantea*). J Comp Physiol B 156:549–556

Pascoa V, Oliveira PL, Dansa-Petretski M et al (2002) *Aedes aegypti* peritrophic matrix and its interaction with heme during blood digestion. Insect Biochem Mol Biol 32:517–523

Pascual-Ruiz S, Carrilllo L, Álvarez-Alfageme F et al (2009) The effcts of different prey regimes in the proteolytic digestion of nymphs of the spined soldier bug, *Podisus maculiventris*, (Hemiptera: Pentatomidae). Bull Entomol Res 99:487–491

Peters W (1992) Peritrophic membranes. Springer-Verlag, Berlin

Pimentel AC, Fuzita FJ, Palmisano G et al (2017) Role of cathepsins D in the midgut of *Dysdercus peruvianus*. Comp Biochem Physiol B 204:45–52

Pimentel AC, Barroso IG, Ferreira JMJ et al (2018) Molecular machinery of starch digestion and glucose absorption along the midgut of *Musca domestica*. J Insect Physiol 109:11–20

Pimentel AC, Dias RO, Bifano TD et al (2020) Cathepsins L and B in *Dysdercus peruvianus*, *Rhodnius prolixus*, and *Mahanarva fimbriolata*. Looking for enzyme adaptations to digestion. Insect Biochem Mol Biol 127:103488

Pompon J, Quiring D, Goyer C et al (2011) A phloem-sap feeder mixes phloem and xylem sap to regulate osmotic potential. J Insect Physiol 57:1317–1322

Ponsen MB (1987) Alimentary tract. In: Mincks AK, Harrewijn P (eds) Aphids: their biology, natural enemies and control, vol 4. Elsevier, Amsterdam, pp 79–97

Ponsen MB (1991) Structure of the digestive system of aphids, in particular *Hyalopterus* and *Coloradoa*, and its bearing on the evolution of filterchambers in the Aphidoidea. Wageningen Agricultural University Papers 91–95, pp 3–61

Poulson DF, Waterhouse DF (1960) Experimental studies of pole cells and midgut differentiation in Diptera. Aust J Biol Sci 13:541–567

Ribeiro JMC (1987) Role of saliva in blood-feeding by arthropods. Ann Rev Entomol 32:463–478

Ribeiro JMC, Pereira MEA (1984) Midgut glycosidases of *Rhodnius prolixus*. Insect Biochem 14:103–108

Ribeiro JMC, Genta FA, Sorgine MHF et al (2014) An insight into the transcriptome of the digestive tract of the bloodsucking bug, Rhodnius prolixus. PLoS Negl Trop Dis 8:e2594

Salvucci ME (2003) Distinct sucrose isomerases catalyze trehalulose synthesis in whiteflies, *Bemisia argentifolii*, and *Erwinia rhapontici*. Comp Biochem Physiol B 135:385–395

Salvucci ME, Rosell RC, Brown JK (1998) Uptake and metabolism of leaf proteins by the silverleaf whitefly. Arch Insect Biochem Physiol 39:155–165

Santos CD, Terra WR (1984) Plasma membrane-associated amylase and trypsin. Intracellular distribution of digestive enzymes in the midgut of the cassava hornworm, *Erinnyis ello*. Insect Biochem 14:587–594

Santos CD, Ferreira C, Terra WR (1983) Consumption of food and spatial organization of digestion in the cassava hornworm, *Erinnyis ello*. J Insect Physiol 29:707–714

Santos CD, Ribeiro AF, Ferreira C et al (1984) The larval midgut of the cassava hornworm (Erinnyis ello). Ultrastructure, fluid fluxes and the secretory activity in relation to the organization of digestion. Cell Tissue Res 237:565–574

Sato PM, Lopes AR, Juliano L et al (2008) Subsite specificity of midgut insect chumotrypsins. Insect biochem Mol Biol 38:628–633

Saxena KN (1963) Mode of ingestion in a heteropterous insect *Dysdercus koeniggii* Fabr (Hemiptera: Pyrrhocoridae). J Insect Physiol 9:47–71

Schumaker TTS, Cristofoletti PT, Terra WR (1993) Properties and compartmentalization of digestive carbohydrases and proteases in *Scaptotrigona bipunctata* (Apidae: Meliponinae) larvae. Apidologie 24:3–17

Shelomi M, Kimsey LS (2014) Vital staining of the stick insect digestive system identifies appendices of the midgut as novel system of excretion. J Morphol 275:623–633

Shelomi M, Watanabe H, Arakawa G (2014a) Endogenous cellulase enzymes in the stick insect (Phasmatodea) gut. J Insect Physiol 60:25–30

Shelomi M, Jasper WC, Atallah J et al (2014b) Differential expression of endogenous plant cell wall degrading enzyme genes in the stick insect (Phasmatodea) midgut. BMC Genom 15:917

Shelomi M, Sitepu IR, Boundy-Mills KL et al (2015) Review of the gross anatomy and microbiology of the Phasmatodea digestive tract. J Orthoptera Res 24:1–12

Shemoli M, Wipfler B, Zhou X et al (2020) Multifunctional cellulase enzymes are ancestral in Polyneoptera. Insect Molec Biol 29:124–135

Shigenobu S, Watanabe H, Hattori M et al (2000) Genome sequence of the endocellular bacterial symbiont of aphids *Buchnera sp*. APS. Nature 407:81–86

Silva CP, Terra WR (1994) Digestive and absorptive sites along the midgut of the cotton seed sucker bug *Dysdercus peruvianus* (Hemiptera: Pyrrhocoridae). Insect Biochem Mol Biol 24:493–505

Silva CP, Xavier-Filho J (1991) Comparison between the levels of aspartic and cysteine proteinase of the larval mid- guts of *Callosobruchus maculatus* (F.) and *Zabrotes subfascia- tus* (Boh.) (Coleoptera: Bruchidae). Comp Biochem Physiol B 99:529–533

Silva CP, Ribeiro AF, Gulbenkian S et al (1995) Organization, origin and function of the outer microvillar (perimicrovillar) membranes of *Dysdercus peruvianus* (Hemiptera) midgut cells. J Insect Physiol 41:1093–1103

Silva CP, Ribeiro AF, Terra WR (1996) Enzyme markers and isolation of the microvillar and perimicrovillar membranes of dysdercus peruvianus (Hemiptera: Pyrrhocoridae) midgut cells. Insect Biochem Mol Biol 26:1011–1018

Silva CP, Terra WR, Xavier-Filho J et al (1999) Digestion in larvae of *Callosobruchus maculatus* and *Zabrotes subfasciatus*(Coleoptera: Bruchidae) with emphasis on α-amylases and oligosaccharidases. Insect Biochem Molec Biol 29:355–366

Silva LM, Lopes AR, Terra WR (2000). Preliminary characterization of digestive enzymes from collard aphid, *Brevicoryne brassicae*. In Procedings of the XXIX annual meeting of the Brazilian Society of Biochemistry and Molecular Biology, Caxambu, 2000, p 17

Silva A, Bacci M Jr, Siqueira CQ et al (2003) Survival of *Atta sexdens* workers on different food sources. J Insect Physiol 49:307–313

Silva CP, Silva JR, Vasconcelos FF et al (2004) Occurrence of midgut perimicrovillar membranes in paraneopteran insect orders with comments on their function and evolutionary significance. Arthropod Struct Dev 33:139–148

Silva CP, RO Dias, V Bernardes et al (2022) Recruitment of lysosomal cathepsins B,L and D as digestive enzymes in Coleoptera. Insect Mol Biol 31:225-240

Snodgrass RE (1935) Principles of Insect Morphology. McGraw Hill, New York

Soares-Costa A, Dias AB, Dellamano M et al (2011) Digestive physiology and characterization of a digestive cathepsin L-like proteinase from the sugar cane weevil *Sphenophorus levis*. J Insect Physiol 57:462–468

Song H, Amédegnato C, Cigliano MM et al (2015) 300 million years of diversification: ellucidating the patterns of orthoptera evolution based on comprehensive taxon and gene sampling. Cladistics 31:621–651

Souza G, Santos VC, Gontijo NF et al (2017) Morphophysiological study of digestive system litter-feeding termite *Cornitermes cumulans* (Kollar, 1832). Cell Tisue Res 368:579–590

Stafford-Banks CA, Rotenberg D, Johnson BR et al. (2014) Analysis of the salivary gland transcriptome of *Frankliniella occidentalis*. PLoS One:9 https://doi.org/10.1371/jourmalpone.009444

Strobl S, Maskos K, Betz M et al (1998) Crystal structure of yellow meal worm alpha-amylase at 1.64 A resolution. J Mol Biol 278:617–628

Sustr V, Stingl U, Brune A (2014) Microprofiles of oxygen, redox potential, and pH, and microbial fermentation products in the highly alkaline gut of the saprophagous larva of *Penthetria holoseicea* (Diptera: Bibionidae). J. Insect Physiol 67:64–69

Tamaki FK, Pimentel AC, Dias AB et al (2014) Physiology of digestion and the molecular characterization of the major digestive enzymes from *Periplaneta americana*. J Insect Physiol 70:22–35

Teo LH, Woodring JP (1985) Digestive enzymes in the house cricket Acheta domesticus with special reference to amylase. Comp Biochem Physiol A 82:871–877

Tereshchenkova VF, Goptar IA, Kulemzina IA et al (2016) Dipeptidyl peptidase 4 - An important digestive peptidase in *Tenebrio molitor* larvae. Insect Biochem Mol Biol 76:38–48

Tereshchenkova VF, Goptar IA, Zhuzhikov DP et al (2017) Prolidase is a critical enzyme for complete gliadin digestion in Tenebrio molitor larvae. Arch Insect Biochem Physiol:95. https://doi.org/10.1002/arch.21395

Terra WR (1988) Physiology and biochemistry of insect digestion: an evolutionary perspective. Braz J Med Biol Res 21:657–734

Terra WR (2001) The origin and functions of the insect peritrophic membrane and peritrophic gel. Arch Insect Biochem Physiol 47:47–61

Terra WR, Ferreira C (1981) The physiological role of the peritrophic membrane and trehalase: digestive enzymes in the midgut and excreta of starved larvae of Rhynchosciara. J Insect Physiol 27:325–331

Terra WR, Ferreira C (1983) Further evidence that enzymes involved in the final stages of digestion by R*hynchosciara* do not enter the endoperitrophic space. Insect Biochem 13:143–150

Terra WR, Ferreira C (1994) Insect digestive enzymes: properties, compartmentalization and function. Comp Biochem Physiol B 109:1–62

Terra WR, Ferreira C (2005) Biochemistry of digestion. In: Gilbert LI, Iatrov K, Gill S (eds) Comprehensive molecular insect science, vol 4. Elsevier, Oxford, pp 171–224

Terra WR, Ferreira C (2012) Biochemistry and molecular biology of digestion. In: Gilbert LI (ed) Insect molecular biology and biochemistry. Academic/Elsevier, London, pp 365–418

Terra WR, Ferreira C, De Bianchi AG (1979) Distribution of digestive enzymes among the endo- and ectoperitrophic spaces and midgut cells of *Rhynchosciara* and its physiological significance. J Insect Physiol 25:487–494

Terra WR, Ferreira C, Bastos F (1985) Phylogenetic considerations of insect digestion: Disaccharidases and the spatial organization of digestion in the *Tenebrio molitor* larvae. Insect Biochem 15:443–449

Terra WR, Espinoza-Fuentes FP, Ribeiro AF et al (1988a) The larval midgut of the housefly (Musca domestica): Ultrastructure, fluid fluxes and ion secretion in relation to the organization of digestion. J Insect Physiol 34:463–472

Terra WR, Ferreira C, Garcia ES (1988b) Origin, distribution, properties and functions of the major *Rhodnius prolixus* hydrolases. Insect Biochem 18:423–434

Terra WR, Santos CD, Ribeiro AF (1990) Ultrastructural and biochemical basis of the digestion of nectar and other nutrients by the moth *Erinnyis ello*. Entomol Exp Appl 56:277–286

Terra WR, Dias RO, Ferreira C (2019) Recruited lysosomal enzymes as major digestive enzymes in insects. Biochem Soc Trans 46:615–623

Thomas KK, Nation JL (1984) Absorption of glucose, glycine and palmitic acid by isolated midgut and hindgut from crickets. Comp Biochem Physiol A 70:289–295

Tokuda G, Watanabe H (2007) Hidden cellulases in termites: a revision of an old hypothesis. Biol Lett 3:336–339

Tokuda G, Watanabe H, Matsumoto T et al (1997) Cellulose digestion in the wood-eating termite, *Nasutitermes takasagoensis* (Shiraki): distribution of celluloses and properties of endo-β-1,4-glucanase. Zool Sci 14:83–93

Treherne JE (1957) Glucose absortption in the cocokroach. J Exp Biol 34:478–485

Treherne JE (1958a) The absorption of glucose from the alimentary canal of the locust *Schistocerca gregaria* (Forsk.). J Exp Biol 35:297–306

Treherne JE (1958b) The digestion and absorption of tripalmitin in the cockroach, *Periplaneta americana*. J Exp Biol 35:611–625

Treherne JE (1959) Amino acid absorption Forsk. J Exp Biol 36:533–545

Treherne JE (1967) Gut absorption. A Rev Ent 12:43–58

Tsai JH, Perrier JL (1966) Morphology of digestive and reproductive system of *Dalbulus maidis* and *Graminella nigrifons* (Homoptera: Cicadellidae). Florida Entomol 79:563–578

Ursic-Bedoya RJ, Nazzari H, Cooper D et al (2008) Identification characterization of two novel form Rhodnius prolixus, a vector of Chagas disease. J Insect Physiol 54:593–603

Utiyama AH, Terra WR, Ribeiro AF (2016) The digestive system of the leafhopper *Bucephalogonia xanthopis* (Hemiptera: Cicadellidae). J Ent Res 40:339–346

Vinokurov K, Taranushenko Y, Krishnan N et al (2007) Proteinase, amylase, and proteinase-inhibitor activities in the gut of six cockroach species. J. Insect Physiol 53:794–802

Waniek PJ (2009) The digestive system of human lice: current advances and potential applications. Physiol Entomol 34:203–210

Waniek PJ, Kollien AH, Schaub GA (2002) Digestive enzymes of the human body louse (*Pediculus humanus*). Joint annual meeting of the german and dutch societies for parasitology (DGP and NVP), Lübeck, p 78

Waniek PJ, Hendgen-Cotta UB, Stock P et al (2005) Serine proteinases of the human body louse (*Pediculus humanus*): sequence characterization and expression patterns. Parasitol Res 97:486–500

Watanabe H, Tokuda G (2010) Cellulotic systems in insects. Annu Rev Entomol 55:609–632

Wigglesworth VB (1972) The principles of insect physiology. Methuen, London

Wood AR, Lehane MJ (1991) Relative contributions of apocrine and eccrine secretion of digestive enzyme release from midgut cells of *Stomoxys calcitrans* (Insecta: Diptera). J Insect Physiol 37:161–166

Woodring J, Lorenz MW (2007) Feeding, nutrient flow, and functional gut morphology in the cricket Gryllus maculatus. J Morphol 268:815–825

Woods HA, Kingsolver JG (1999) Feeding rate and the structure of protein digestion and absorption in Lepidoptera midguts. Arch Insect Biochem Physiol 42:74–87

Zhang F, Zhang C, Dai W et al (2012) Morphology and histology of the digestive system of the vector leafhopper Psammotettix striatus (L.) (Hemiptera: Cicaddellidae). Micron. 43:725–738

Zhang J, Zhang Y, Li J et al (2016) Midgut transcriptome of the cockroach *Periplaneta americana* and its microbiota: digestion, detoxification and oxidative stress response. PLoS One 11:e0155254

Zhong H, Zhang Y, Wei C (2013) Anatomy and the fine structure of the alimentary canal of the spittlebug *Lepyronia coleoterata* (L.) (Memiptera: Cercopoidea). Arhr Struc Devel 42:521–530

Chapter 14
General Trends in the Evolution of Digestive Systems

Abstract The findings regarding insect digestive features and phylogeny led to the proposal that the overall pattern of digestion of the insects derived from basic plans corresponding to ancestors. In the Neoptera ancestor, midgut digestive enzymes are translocated to the crop, where digestion starts, ending in the midgut. Enzymes active in initial, intermediate, and final digestion move freely among gut compartments. The excretion of digestive enzymes is decreased by midgut countercurrent fluxes caused by the secretion of water by the Malpighian tubules and its absorption back into the ceca. Condylognatha ancestors lost ceca, water-secreting regions, and acquired microvilli-associated membranes to facilitate nutrient absorption. The Holometabola insects have countercurrent fluxes of water caused by the secretion of fluid in the posterior midgut and its absorption into the ceca or, in their absence, by the anterior midgut. The Hymenoptera-Panorpoid ancestor has no enzymes in crop, only enzymes of initial digestion pass through the peritrophic membrane, but have midgut countercurrent fluxes like the Holometabola ancestors. The Cyclorrhapha ancestor acquired a highly acid middle midgut rich in lysozyme and cathepsin D to digest bacteria, and its countercurrent flux of water is absorbed in the middle posterior midgut. The Lepidoptera ancestor differs from the Hymenopteran-Panorpoid ancestor in having a highly alkaline midgut with goblet cells to excrete K^+.

It is customary to organize insect species and their digestive enzymes according to gut morphologies and/or the types of diets they consume (Wigglesworth 1972; Dow 1986). However, both viewpoints are oversimplifications. For example, there is an obvious similarity among the morphology of the guts of cockroaches (generalists), grasshoppers (solid/plant feeder), and mantids (solid/animal feeder) (see Chap. 13). On physiological grounds, there is a striking difference in how blood is digested by hematophagous hemipterans and mosquitoes (see Fig. 2.2a, c, e and see Chap. 13). The remarkable difference observed cannot result from diet but from adaptations of different ancestors: mosquitoes from ancestors able to digest polymers in adult stage and hematophagous hemipterans from sap-sucking hemipterans. Finally, on enzymological grounds, the chemical composition of the diet does not necessarily

W. R. Terra et al., *Molecular Physiology and Evolution of Insect Digestive Systems*, Entomology in Focus 7, https://doi.org/10.1007/978-3-031-39233-7_14

determine the type of digestive enzymes. This is exemplified by the nectar-feeding adult blackfly *Cnephia docotens*is (Diptera: Simuliidae) that has a high trypsin activity in their midguts, whereas caterpillars (Santos et al. 1983) and grasshoppers (Morgan 1976) that ingest leaves do not digest cellulose. The data obtained from several sources suggest that all insects have the full complement of ordinary enzymes, the relative activities of each changing in response to diet composition. This change may take place during the feeding of one individual or may result from the adaptation of the ancestor of an insect taxon to a specific diet, causing a permanent increase of some enzymes in relation to others (Terra and Ferreira 1994). In addition to the ordinary enzymes, the adaptation of ancestors to specific diets may result from the recruitment of enzymes from lysosomes (Chap. 9) or acquisition by horizontal transference from fungi or bacteria (Chap. 10).

As a consequence of the above discussion, adaptive features found in insects of the same taxon, but with different habits, looked as evolved from identical basic digestive patterns. On the contrary, insects with similar feeding habits, but pertaining to different taxa, have different organization of digestion and absorption, in spite of numerous convergent features. Hence, the features of the digestive process have a stronger correlation with the phylogeny of the insect groups than to their diet as seen in Chap. 13. This conclusion was also supported by a cladogram prepared with midgut morphological and physiological characteristics of 23 species pertaining to 8 different orders. The obtained cladogram is closer to the phylogenetic trees described in Chap. 1, than to a cladogram prepared, taking into account features of the insect diets (Terra and Ferreira 2005; Dias 2004).

The findings regarding the correlation of insect digestive features and phylogeny led Terra (1988, 1990) to propose that the overall pattern of digestion of insects pertaining to different orders follow the basic plans of ancestral forms. The hypothetical basic plans of Terra (1988, 1990) were improved over the years (Terra and Ferreira 2009, 2020). The hypothetical plans of digestion are outlined in Fig. 14.1.

The basic plan of the digestive physiology for Neoptera (Neoptera ancestor, Fig. 14.1a) was hypothesized from studies of insects pertaining to the Polyneopteran orders (Dow 1981a, 1981b; Ferreira et al. 1990; Marana et al. 1997; Biagio et al. 2009; Woodring and Lorenz 2007; Monteiro et al. 2014) (reviewed in Chap. 13). In this ancestor, most digestion occurs in its large crop by digestive enzymes moved forward from the midgut by antiperistalsis. Each time the insect initiates a new meal, the ceca contents are discharged into the crop. The function of saliva is, as a rule, restricted to lubrication and carbohydrate digestion. After a while, following ingestion, the crop contracts, transferring digestive enzymes and incompletely digested food into the midgut. The anterior midgut is acid with high activity of glycoside hydrolases and the posterior midgut is alkaline with high endopeptidases activity. The food bolus is propelled along the gut by peristalsis. Polymeric food, once digested up to small molecules able to traverse the peritrophic membrane, diffuses into the space outside the peritrophic membrane (ectoperitrophic space) (Fig. 14.1a). Then, the enzymes and nutrients are moved toward the ceca by a countercurrent flux of water caused by water secretion by Malpighian tubules and its absorption by ceca cells. Due to the countercurrent flux of water, the activity of

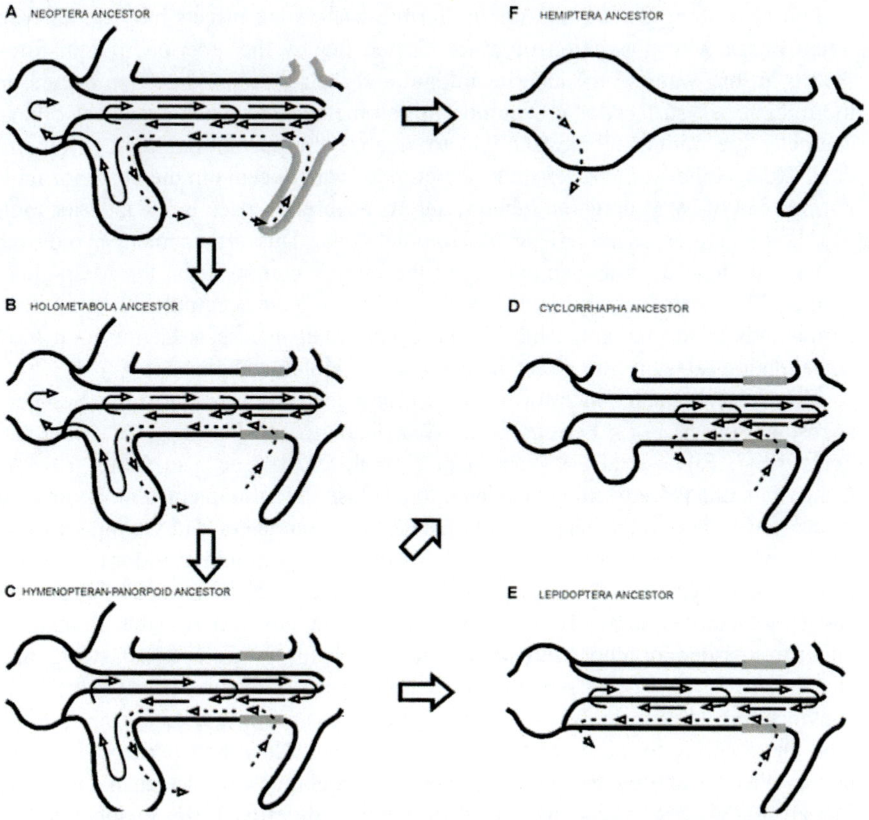

Fig. 14.1 Diagrammatic representation of water fluxes (dotted arrows) and of the circulation of digestive enzymes (solid arrows) in putative insect ancestors that correspond to the major basic gut plans. In Neoptera ancestors (**a**), midgut digestive enzymes pass into the crop. Countercurrent fluxes depend on the secretion of fluid by the Malpighian tubules and its absorption by the ceca. Enzymes involved in initial, intermediate, and final digestion circulate freely among gut compartments. Holometabola ancestors (**b**) are similar except that secretion of fluid occurs in the posterior ventriculus. The ancestors of hymenopteran and panorpoid (Lepidoptera and Diptera assemblage) insects (**c**) display countercurrent fluxes like Holometabola ancestors, midgut enzymes are not found in the crop, and only the enzymes involved in initial digestion pass through the peritrophic membrane. Enzymes involved in intermediate digestion are restricted to the ectoperitrophic space and those responsible for terminal digestion are immobilized at the surface of midgut cells. Cyclorrhapha ancestors (**d**) have a reduction in ceca, an acidic middle midgut, absorption of water in the middle posterior midgut, and the anterior midgut playing a storage role. Lepidoptera ancestors (**e**) are similar to Hymenoptera-Panorpoid ancestors, except that the anterior midgut replaces the ceca in fluid absorption and the midgut has goblet cells. Hemiptera ancestors (**f**) have lost crop, ceca, and fluid-secreting regions. Fluid is absorbed in the anterior midgut. (Reprinted with permission from Terra and Ferreira (2009) © Elsevier)

digestive enzymes decreases along the midgut and they are excreted with a lower rate. Divergences from the Neoptera ancestor plan are observed among Polyneoptera orders. Among Dictyoptera, cockroaches and termites have an enlargement of hindgut structures associated mainly with wood-feeding. Some cockroaches have the midgut acid along the whole midgut. Termites and stick insects have no anterior midgut ceca with absorption of water carried out by the anterior midgut. Stick insects absorb water in the anterior midgut and their terminal digestion occurs on the midgut cell surface, as is common in holometabolans. Stick insects also have characteristic midgut tubules thought to alkalize the posterior midgut (Monteiro et al. 2014). Finally, in crickets, the secretion of water occurs in the posterior midgut, instead of by Malpighian tubules, and its absorption back in the anterior midgut, both features characteristic of Holometabola. This arrangement avoids the accumulation of noxious compounds in the midgut coming from the Malpighian tubules. The loss of ceca is another way of decreasing the accumulation of noxious compounds in the midgut, which is more important in insects having high food consumption rates that are more characteristic of Holometabola.

The Condylognatha ancestor changed dramatically in relation to the Neoptera ancestor (Ponsen 1987; Ferreira et al. 1988; Ribeiro and Pereira 1984; Silva and Terra 1994; Silva et al. 1995; Pimentel et al. 2020) (see also Chap. 13). As Condylognatha is frequently considered to be close to Phthiraptera, a more detailed discussion is necessary here. Condylognatha (Thysanoptera and Hemiptera) and Psocodea (Psocoptera and Phthiraptera) do not have a common ancestor (see Misof et al. 2014 and Chap. 1). Condylognatha diverged from a clade formed by Psocodea and Holometabola, arguably by adapting to a plant sap-feeding habit. Plant saps either from xylem or phloem are poor in proteins (with few exceptions), starch, and amino acids, but may be rich in sucrose. Thus, a sap-feeding insect must have enzymes to hydrolyze only sucrose and be able to absorb nutrients from a dilute diet. On adapting to these conditions, Condylognatha lost their peritrophic membranes (PMs) (unnecessary in the absence of luminal digestion, see details in Chap. 2), part of PM peritrophins, midgut chitin synthase, digestive serine endopeptidases (Dias et al. 2019), and water-secreting regions. At the same time, Condylognatha acquired microvilli-associated membranes, supposed to aid in nutrient absorption from a dilute medium (see Chap. 7) and structures to eliminate excess of ingested water (Goodchild 1966 and see below). Under the evolutionary pressure caused by defense proteins and other proteins available on ingesting sap, Condylognatha recruited lysosomal cysteine and aspartic endopeptidases as digestive enzymes (Pimentel et al. 2020, for Hemiptera, and unpublished work of the authors for Thysanoptera; see below).

The evolutionary path of Psocodea was different from that of Condylognatha. Based only on Phthiraptera, for which there are sufficient data (see Chap. 13), Psocodea retained their serine endopeptidases, did not acquire microvilli-associated membranes (Silva et al. 2004), retained their midgut chitin synthase and some PM peritrophins, but lost their peritrophic membranes (Dias et al. 2019).

Evolving from the Neoptera ancestor, the Holometabola ancestor (Fig. 14.1b) secretes water from the posterior midgut, instead from the Malpighian tubules. However, like the Neoptera ancestor, all steps of digestion occur inside the peritrophic membrane (endoperitrophic space). This proposal was based on studies performed with Coleoptera (Baker 1981; Terra et al. 1985) and those used in the proposition of the Hymenoptera-Panorpoid (assemblage of Diptera and Lepidoptera) ancestor (Hobson 1931; Greenberg and Paretsky 1955; Terra et al. 1979; Espinoza-Fuentes and Terra 1987; Santos et al. 1983; Jimenez and Gilliam 1990; Schumaker et al. 1993) (see also Chap. 13).

Coleoptera ancestors were like the Holometabola ancestors, but despite having complete metamorphosis, their young forms (larvae) and adults have the same feeding habits. Larvae and adults may both be plant feeders, in which case adults generally feed on aerial parts and larvae, on the roots of the same plant, or are both predatory (Crowson 1981). This condition is reminiscent of that of the Polyneoptera orders. However, there is an evolutionary trend toward a decrease in the importance of crop digestion. Thus, except for adult Adephaga, which may store pieces of the prey in their crops, most predatory beetles have pre-oral digestion (Crowson 1981). Other evolutionary trends in Coleoptera are the enlargement of hindgut structures as fermentations chambers (Chap. 13) and the recruitment of lysosomal proteins as digestive enzymes (Chap. 9) and the reception by horizontal transfer of microbial genes coding for cell wall-degrading enzymes (Chap. 10). Those enzymes turned out possible beetle stem and wood feeding and leaf and seed mining (Chap. 13).

The Hymenoptera-Panorpoid ancestor (Fig 14.1d) has midgut countercurrent fluxes of water like the Holometabola ancestor, but differs from it in the lack of digestive enzymes in crop, presence inside the membrane peritrophic only of enzymes involved in initial digestion, retention outside the peritrophic membrane of the enzymes that carried out intermediate digestion, immobilization of the enzymes responsible for terminal digestion at the surface of midgut cells and, finally, in the differentiation in midgut luminal pH. The selective pressures leading from the Holometabola ancestor to the Hymenoptera-Panorpoid ancestor is an increase in the digestive efficiency to support faster life cycles. Faster life cycles make possible the development of more generations in a fixed time, thus permitting the survival of more individuals, even if the mortality is high in each generation, and hence, favoring them to occupy habitats exposed to predation or ephemeral ecological niches. In accordance with this, the life span of a beetle is about 12 months, whereas that of a fly or butterfly is only 6 weeks (Sehnal 1985). Associated with a decrease in life span, there is an increase in relative food consumption and growth rates. Indeed, the relative food consumption rate (dry weights of food ingested per biomass) average (range) are: 0.6 (0.02–1.4) for Coleoptera, 1.8 (0.27–6.9) for Lepidoptera, and 2.3 (0.9–3.6) for Hymenoptera, whereas the relative growth rate (increase in biomass per initial mass per day) average (range) are: 0.07 (0.01–0.16) for Coleoptera; 0.3 (0.03–1.5) for Lepidoptera, and 0.21 (0.08–0.4) for Hymenoptera (Slansky and Scriber 1985).

The Hymenopteran ancestor evolved from the Hymenoptera-Panorpoid ancestor by acquiring surprising features. The hymenopteran basal lineages, for example, Tenthredinoidea, have larvae that look like caterpillars and have advanced digestive systems similar to those of Lepidoptera: alkaline midgut contents, maltase associated with the glycocalyx, and aminopeptidase membrane-bound to the microvilli of midgut cells. More derived hymenopterans (Aculeata: bees and ants) have less advanced digestive systems: most digestive enzymes are soluble, and midgut pH luminal is nearly neutral, characteristics similar to those of Coleoptera ancestors.

This condition is thought to be the result of the evolution of Aculeata from endoparasitic Apocrita that have simplified digestive systems on adapting to endoparasitism (see review in Chap. 13).

Based on studies carried out, the basal (Terra et al. 1979; Lemos and Terra 1991) and higher lineages (Hobson 1931; Greenberg and Paretsky 1955; Espinoza-Fuentes and Terra 1987; Regel et al. 1998) of Diptera ancestors are not much different from the Hymenoptera-Panorpoid ancestors. However, Cyclorrhapha ancestors (Fig. 14.1d) differ significantly from the Hymenoptera-Panorpoid ancestors, as a consequence of their adaptation to feed on bacteria and fungi infecting plant or animal remains. For this, they acquired a new midgut region, the middle midgut, that has an embryological origin different from the cells of the anterior and posterior midguts (Poulson and Waterhouse 1960). The middle midgut has a low luminal pH, high activity of a digestive lysozyme, and a cathepsin D (Espinoza-Fuentes and Terra 1987, see also Chap. 13), conditions that favor killing and opening bacterial and fungi cells. The anterior midgut of Cyclorrhapha is mostly devoted to storage, but carries some digestion. The posterior midgut is where the majority of the compartmentalized digestion occurs and countercurrent fluxes of water take place, propelled by water secretion at the end of the posterior midgut and its absorption back in the middle posterior midgut (Chap. 13). The Cyclorrhapha ancestor has a reduction of ceca in relation to the Hymenoptera-Panorpoid ancestor, and the enzymes involved in terminal digestion are bound to the microvillar membrane of midgut cells.

The basic gut plan of Lepidoptera (Fig. 14.1e) was initially proposed from studies carried out by Santos et al. (1983), Ferreira et al. (1994), and Liu et al. (2018). According to them, the Lepidoptera ancestors differ from the Hymenoptera-Panorpoid ancestor in having the anterior midgut replacing the missing ceca in water absorption and in having midgut goblet cells, responsible for K^+ excretion and in raising the midgut luminal pH. The immobilization of enzymes involved in terminal digestion is caused by their entrapping in the cell glycocalyx (except for aminopeptidase).

References

Baker JE (1981) Localization of proteolytic enzymes in the midguts of larvae of the black carpet beetle. J Georgia Entomol Soc 16:495–500

Biagio FP, Tamaki FK, Terra WR et al (2009) Digestive morphophysiology of *Gryllodes sigillatus* (Orthoptera: Gryllidae). J Insect Physiol 55:1125–1133

Crowson RA (1981) The biology of the Coleoptera. Academic, London

Dias Jr AB (2004) Evolução do Sistema digestivo dos insetos. PhD thesis, University of São Paulo

Dias RO, Cardoso C, Leal CS et al (2019) Domain structure and expression along the midgut and carcass of peritrophins and cuticle proteins analogous to peritrophins in insects with and without peritrophic membrane. J Insect Physiol 114:1–9

Dow JAT (1981a) Localization and characterization of water uptake from the midgut of the locust, *Schistocerca gregaria*. J Exp Biol 93:269–281

Dow JAT (1981b) Countercurrent flow, water movements and nutrient absorption in the locust midgut. J Insect Physiol 27:579–585
Dow JAT (1986) Insect midgut function. Adv Insect Physiol 19:187–328
Espinoza-Fuentes FP, Terra WR (1987) Physiological adaptations for digesting bacteria. Water fluxes and distribution of digestive enzymes in *Musca domestica* larval midgut. Insect Biochem 17:809–817
Ferreira C, Ribeiro AF, Garcia ES et al (1988) Digestive enzymes trapped between and associated with the double membranes of *Rhodnius prolixus* posterior midgut cells. Insect Biochem 18:521–530
Ferreira C, Oliveira MC, Terra WR (1990) Compartmentalization of the digestive process in *Abracris flavolineata* (Orthoptera: Acrididae) adults. Insect Biochem 20:267–274
Ferreira C, Capella AN, Sitnik R et al (1994) Digestive enzymes in midgut cells, endo-and ectoperitrophic contents and peritrophic membranes of *Spodoptera frugiperda* (Lepidoptera) larvae. Arch Insect Biochem Physiol 26:299–313
Goodchild AJP (1966) Evolution of the alimentary canal in the Hemiptera. Biol Rev 41:97–140
Greenberg B, Paretsky D (1955) Proteolytic enzymes in the housefly, *Musca domestica* (L.). Ann Entomol Soc America 48:46–50
Hobson RP (1931) Studies on the nutrition of blow-fly larvae. I. Structure and function of the alimentary canal. J Exp Biol 8:109–123
Jimenez DR, Gilliam M (1990) Ultrastructure of ventriculus of the honeybee *Apis mellifera* (L): cytochemical localization of the acid phosphatase, alkaline phosphatase and non-specific esterase. Cell Tissue Res 261:431–443
Lemos FJA, Terra WR (1991) Digestion of bacteria and the role midgut lysozyme in some insect larvae. Comp Biochem Physiol B 100:265–268
Liu L, Qu M, Yang J et al (2018) The physiological differentiation along the midgut of Bombyx mori- inspirations from proteomics and gene expression patterns of the secreted proteins in the ectoperitrophic space. Insect Mol Biol 27:247–259
Marana SR, Ribeiro AF, Terra WR et al (1997) Ultrastructure and secretory activity of *Abracris flavolineata* (Orthoptera: Acrididae) midguts. J Insect Physiol 43:465–473
Misof B, Liu S, Meusemann K et al (2014) Phylogenomics resolves the timing and pattern of insect evolution. Science 346:763–767
Monteiro EC, Tamaki FK, Terra WR et al (2014) The digestive system of the "stick bug" *Cladomorphus phyllinus* (Phasmida, Phasmatidae): a morphological, physiological and biochemical analysis. Arthr Struct Devel 43:123–134
Morgan MRJ (1976) Gut carbohydrases in locusts and grasshoppers. Acrida 5:45–58
Pimentel AC, Dias RO, Bifano TD et al (2020) Cathepsins L and B in *Dysdercus peruvianus*, *Rhodnius prolixus*, and *Mahanarva fimbriolata*. Looking for enzyme adaptations to digestion. Insect Biochem Mol Biol 127:103488
Ponsen MB (1987) Alimentary tract. In: Mincks AK, Harrewijn P (eds) Aphids: their biology, natural enemies and control, vol 4. Elsevier, Amsterdam, pp 79–97
Poulson DF, Waterhouse DF (1960) Experimental studies of pole cells and midgut differentiation in Diptera. Aust J Biol Sci 13:541–567
Regel R, Matioli SR, Terra WR (1998) Molecular adaptation of *Drosophila melanogaster* lysozymes to a digestive function. Insect Biochem Mol Biol 28:309–319
Ribeiro JMC, Pereira MEA (1984) Midgut glycosidases of *Rhodnius prolixus*. Insect Biochem 14:103–108
Santos CD, Ferreira C, Terra WR (1983) Consumption of food and spatial organization of digestion in the cassava hornworm, *Erinnyis ello*. J Insect Physiol 29:707–714
Schumaker TTS, Cristofoletti PT, Terra WR (1993) Properties and compartmentalization of digestive carbohydrases and proteases in *Scaptotrigona bipunctata* (Apidae: Meliponinae) larvae. Apidologia 24:3–17
Sehnal F (1985) Growth and life cycles. In: Kerkut A, Gilbert LI (eds) Comprehensive insect physiology, biochemistry and pharmacology, vol 4. Pergamon, Oxford, pp 163–187

Silva CP, Terra WR (1994) Digestive and absorptive sites along the midgut of the cotton seed sucker bug *Dysdercus peruvianus* (Hemiptera: Pyrrhocoridae). Insect Biochem Mol Biol 24:493–505

Silva CP, Ribeiro AF, Gulbenkian S et al (1995) Organization, origin and function of the outer microvillar (perimicrovillar) membranes of *Dysdercus peruvianus* (Hemiptera) midgut cells. J Insect Physiol 41:1093–1103

Silva CP, Silva JR, Vasconcelos FF et al (2004) Occurrence of midgut perimicrovillar membranes in paraneopteran insect orders with comments on their function and evolutionary significance. Arthropod Struct Dev 33:139–148

Slansky F, Scriber JM (1985) Food consumption and utilization. In: Kerkut A, Gilbert LI (eds) Comprehensive insect physiology, biochemistry and pharmacology, vol 2. Pergamon, Oxford, pp 1–86

Terra WR (1988) Physiology and biochemistry of insect digestion: an evolutionary perspective. Braz J Med Biol Res 21:675–734

Terra WR (1990) Evolution of digestive systems of insects. Annu Rev Entomol 35:181–200

Terra WR, Ferreira C (1994) Insect digestive enzymes: properties, compartmentalization and function. Comp Biochem Physiol B 109:1–62

Terra WR, Ferreira C (2005) Biochemistry of digestion. In: Gilbert LI, Iatrov K, Gill S (eds) Comprehensive molecular insect science, vol v. 4. Elsevier, Oxford, pp 171–224

Terra WR, Ferreira C (2009) Digestive system. In: Resh VH, Cardé RT (eds) Encyclopedia of insects, 2nd edn. Academic, San Diego, pp 273–281

Terra WR, Ferreira C (2020) Evolutionary trends of digestion and absorption in the major insect orders. Arthrop Struct Develop 56:100931

Terra WR, Ferreira C, De Bianchi AG (1979) Distribution of digestive enzymes among the endo- and ectoperitrophic spaces and midgut cells of *Rhynchosciara* and its physiological significance. J Insect Physiol 25:487–494

Terra WR, Ferreira C, Bastos F (1985) Phylogenetic considerations of insect digestion. Disaccharidases and the spatial organization of digestion in the *Tenebrio molitor* larvae. Insect Biochem 15:443–449

Wigglesworth VB (1972) The principles of insect physiology, 7th edn. Methuen, London

Woodring J, Lorenz MW (2007) Feeding, nutrient flow, and functional gut morphology in the cricket, *Gryllus maculatus*. J Morphol 268:815–825

Chapter 15
New Technologies of Insect Control That Act Through the Gut

Abstract In this chapter, we cover the new insect control techniques that target the intestinal tract and are not based on chemical insecticides. Crystal (Cry) and Cytolitic (Cyt) protein families from *Bacillus thuringiensis* (Bt) and related toxins are used worldwide for insect control. Their primary action is to lyse midgut epithelial cells via binding to specific brush border receptors and elicit the formation of pores or trigger a necrotic pathway, which leads to the destruction of the midgut. We discuss in some detail recent findings on their mode of action. The use of proteinaceous inhibitors expressed in transgenic plants directed to enzyme targets, such as endopeptidases, α-amylases, and polygalacturonases, is reviewed. The success of transgenic crop plants expressing Cry toxin genes has paved the way for the development of genetically modified plants expressing double-stranded RNA (dsRNA) and CRISPR/Cas that can suppress the expression of genes coding for selected vital proteins. We discuss the recent developments in dsRNA and CRISPR/Cas technologies for insect pest management focusing on their actions in insect midgut.

15.1 Initial Considerations

In the previous chapters of this book, the readers could realize that the intestinal tract tissues are very important interfaces between the internal environment of the insect and the ingested dietary material. Of the three subdivisions of the insect intestinal tract, the midgut tissue is of particular importance because it is not lined by cuticular deposition, as are the foregut and the hindgut. In other words, the midgut epithelial cells are more exposed to ingested material or living organism and virus that reaches the midgut lumen. The midgut cells are also responsible for the release of digestive enzymes, absorption of nutrients, maintenance of the luminal pH milieu, establishment of fluid flows, and division of the lumen into sub-compartments because of the peritrophic membrane formation. It is not a surprise that the insect midgut is the target for several insecticidal compounds. Many of these insecticidal compounds are proteins produced by different organisms such as bacteria, fungi, and plants.

W. R. Terra et al., *Molecular Physiology and Evolution of Insect Digestive Systems*, Entomology in Focus 7, https://doi.org/10.1007/978-3-031-39233-7_15

Membrane pore-forming insecticidal proteins from the bacterium *Bacillus thuringiensis* (Bt) represent an important arsenal used in crop protection methods and are today the most successful alternative to synthetic insecticides. The progressive knowledge about the molecular structure of these bacterial insecticidal proteins, their modes of action, and the identification of genes encoding them enabled the appearance of a very profitable biotechnological industry based on the supply of sprayable formulations or breeding of transgenic plants resistant to some major insect pests. Since 1996, the adoption of genetically modified plants expressing Bt insecticidal proteins to confer insect resistance has revolutionized crop pest management. In addition to the very successful use of Bt insecticidal proteins, other insecticidal proteins of plant or fungus origin such as α-amylase inhibitors, endopeptidase inhibitors, and polygalacturonase inhibitors can compromise insect development when ingested from either dietary formulations or from transgenic plants.

The increasing access to whole genomes and transcriptomes from model and non-model insects and from plants has provided valuable information about the gene encoding the insect digestive enzymes and their proteinaceous inhibitors. This makes possible the design of stable and active structures under field conditions, as well as the expression of mature active inhibitors in genetically modified plants. This chapter focuses on insecticidal proteins that target different structures in the midgut or that interfere with the digestive process of insect pests. Although the greatest advances in this area are related to integrated pest management, the knowledge bases on how these gene products act on insect digestion can help both in the management of vector insects and improve the breeding of beneficial insects.

15.2 Bacterial Insecticidal Proteins Targeting Insect Midgut Tissue

15.2.1 Membrane Pore-Forming Bt Toxins

The most successful insecticidal proteins effective under field conditions, whether applied in sprayable formulations or expressed in transgenic plants, belong to a diverse group of toxins produced by different strains of the bacterium *Bacillus thuringiensis*, known as Bt toxins or δ-endotoxins. A large number of Bt toxins have been identified so far. They are classified according to their primary structures and structural similarities. The largest group belongs to the crystal-derived three-domain protein family known as 3D Cry toxins. Another group is formed by the single domain cytolytic Bt toxins (Cyt toxins) and the third group comprises the Vegetative Insecticidal Proteins (VIP toxins). Cry and Cyt proteins are synthesized during the sporulation phase and are accumulated in parasporal crystals that remain inside the bacterial cells. To date, more than 270 3D Cry toxins are known, many of them active against Lepidoptera and Coleoptera, while Cyt toxins are active against Diptera. VIPs are produced during the vegetative growth phase and secreted to the growth medium as soluble protoxins. There are reports of more than 100 VIPs most of which are active to a wide range of lepidopteran species (Syed et al. 2020; Estruch et al. 1996). Like Cry toxins, VIPs also have high selectivity at the species level. VIPs from the Vip3 family are of particular importance as they are widely used in spray formulations for application in the field or expressed in transgenic plants. Ingestion of Cry, Cyt, and VIP variants by susceptible insect species leads to pore formation and extensive damage to the midgut cells (Silva-Filha et al. 2021; Gomis-Cebolla et al. 2017; Lee et al. 2003).

Despite more than 390 different Bt toxins known to date, a small number of them are commercialized in agriculture or in the combat against vector mosquitoes (Crickmore et al. 2021; Jurat-Fuentes et al. 2021; Silva-Filha et al. 2021). To date, only eight natural Cry toxins, three engineered Cry proteins, and two natural Vip3 proteins are currently produced by commercialized Bt maize, soybean, and cotton crops. In transgenic plants, they can be expressed as either protoxins or bioactive toxins. One of the advantages of using Bt toxins in agriculture and in public health is the high degree of specificity of these toxins. In general, Bt toxins act on closely related insect species and are also specific to the developmental stages of insect pests belonging to the orders Coleoptera, Diptera, and Lepidoptera. These proteins do not present toxicity against mammals, and the advance in the use of Bt toxins has led to a significant decrease in the use of chemical insecticides, which have a much greater impact on the environment. Among the different types of Bt toxins, those belonging to the 3D family of Cry toxins are the most used in agriculture, having been commercially introduced since 1996 and becoming the dominant biopesticides in many parts of the world since 2010 (Acheuk et al. 2022; Jurat-Fuentes et al. 2021). The successful use of Bt toxins is also reflected in the extension of areas planted with this technology and in the growing number of countries that approved the use of genetically modified crops with these insecticidal proteins (Acheuk et al. 2022). Data from the International Service for the Acquisition of Agri-Biotech Applications for 2019 showed that 108.7 million hectares of transgenic Bt crops were grown worldwide, representing more than 53% of the global cultivated area of transgenic plants (ISAAA 2019). Among the crops where Bt toxins are employed, maize, cotton, and soybean stand out (Acheuk et al. 2022). Transgenic plants that express more than one Bt toxin gene at the same time are also used, a strategy known as defense protein pyramiding (Jurat-Fuentes et al. 2021; Chen et al. 2017).

15.2.2 Mode of Action of Bt Toxins

15.2.2.1 Mode of Action of 3D Cry Toxins

Mechanistic details of the action of Bt toxins are still the subject of intense research and great advances in recent years, especially when more and more structures of these proteins are unraveled (Byrne et al. 2021; Heckel 2021). A common aspect of the mode of action is the formation of pores in midgut cell microvillar membranes of susceptible insects. Another aspect which is common to different Bt toxins is the activation step through the activity of trypsin-like peptidases in the lumen of the midgut after the ingestion of the protoxins. Following partial proteolysis and crossing through the peritrophic membrane, the interactions of active toxins with their midgut cell-specific receptors elicit pore formation. Known receptors for Cry toxins include ATP-binding cassette transporter proteins (ABC proteins), cadherins, GPI-anchored alkaline phosphatases (ALPs), and GPI-anchored aminopeptidases-N (APNs) (Adang et al. 2014). The nucleation mediated by the association of Bt toxins subunits with their receptors leads to the formation of pores in the apical plasma membrane of midgut cells in different parts of the midgut epithelium, leading to cell death and collapse of the intestinal epithelium. After the destruction of the midgut epithelium, septicemia may eventually occur due to the invasion of Bt and the midgut-resident microbiota to the internal environment of the insect, leading to its death (Jurat-Fuentes et al. 2021; Adang et al. 2014).

The heterologous expression of ABC proteins and other receptors of Bt toxins in membranes of insect cell lines represents an important tool in studies of the mode of action of Cry toxins. The expression of a mutant ABC protein from *Bombyx mori* in Sf9 cells demonstrated the participation of these proteins and the synergism with another type of receptor, the cadherins (Tanaka et al. 2013). The highest level of synergism of ABC proteins and cadherins was observed when both proteins were expressed in the same cells. The synergism of ABC proteins and cadherins has also been shown to occur in the lepidopterans *Heliothis virescens* and *Helicoverpa armigera* (Zhang et al. 2021a; Bretschneider et al. 2016; Wang et al. 2016). It has been proposed that the binding of 3D Cry toxins to cadherins spatially guides the toxins to interact with ABC proteins which leads to subsequent pore formation in the microvillar membrane. However, the molecular details involved in this synergism between cadherins, APNs and ALPs, and the ABC proteins are still not fully understood. Increasing experimental evidence suggest that the interaction of active Cry toxins with ABC proteins is critical for high insecticidal activity, while interaction with the other receptors may increase toxicity due to their synergistic roles (Sun et al. 2022; Heckel 2021). The interaction of Cry toxins with their receptors is often species-specific and complex, involving multiple steps until oligomerization and insertion of the toxin into the cell membrane is triggered. The three domains of 3D Cry toxins participate in different events in the mechanism that leads to pore formation and cell death. Domain 1 is proposed to play a role in membrane perturbation, domain 2 has a role in receptor binding, and domain 3 has been implicated in target specificity (Ma et al. 2022; Heckel 2021).

According to Heckel (2021), there are three non-mutually exclusive models that can explain the cadherin-ABC protein synergism: One model called trans-acting mechanisms predicts that the synergism involves actions of cadherins and the ABC proteins with these proteins separated from each other. Another possibility is that synergism depends on a close physical interaction, called a cis-acting mechanism. There are two models of cis-acting mechanisms described by Heckel (2021). In one model, it is predicted that cadherin binds to Cry toxin and brings it to ABC protein, and this interaction facilitates pore formation. Another cis-acting mechanism predicts that cadherin pulls the active toxin away from the ABC transporter, allowing it to enter the membrane to form the pore and releasing the ABC transporter for the next interaction (Heckel 2021; Bretschneider et al. 2016). As pointed out by Heckel (2021), a more advanced understanding of the structures of different ABC proteins and their interactions with other receptors will certainly facilitate the understanding of pore formation by Cry toxins at the molecular level.

15.2.2.2 Mode of Action of VIPs

The knowledge about the mode of action of VIPs has advanced driven by new gene editing and omic techniques in addition to the flood of new resolved molecular structures of these toxins, as, for example, the first in silico modeling of a VIP3 family member, Vip3Af1 (Banyuls et al. 2018), the Vip3B variant Vip3B2160 (PDB 6V1V), Vip3Aa (Núñez-Ramírez et al. 2020), and Vip3Bc1 (Byrne et al. 2021). Interestingly, the identity of VIP receptors in the microvillar membranes of insect midgut cells is still controversial (Jin et al. 2022; Byrne et al. 2021; Jiang et al. 2018; Singh et al. 2010). An important candidate for VIP protoxin receptor is the scavenger receptor class C-like protein (Sf-SR-C) from the midgut cells of *S. frugiperda*, which seems to mediate the endocytosis of Vip3Aa and its insecticidal activity (Jiang et al. 2018). VIPs organize themselves as tetramers to exert their pore-forming activity in the microvillar membrane and trigger the midgut cell apoptosis (Jin et al. 2022; Byrne et al. 2021). Like Cry toxins, VIPs must be activated by partial proteolysis catalyzed by trypsin-like enzymes giving rise to active forms of the toxin. Data from the Vip3Bc1 variant showed that VIP monomers have five well-defined domains. Domains 1 and 2 form the N-terminal portion of the protein, which is rich in α-helices, while domains 3–5 make up the C-terminal portion, with a greater prevalence of β-strands. The central portion of the protein is mainly formed by disordered structures (Byrne et al. 2021; Banyuls et al. 2018). Although there is no great sequence identity between the VIPs and the Cry toxins, there is structural homology between domains found in VIPs with the three domains present in the 3D Cry toxins (Byrne et al. 2021).

The overall structure of domain 2 from Vip3Bc1 is like domain 1 of 3D Cry toxins, which is thought to be directly involved in membrane perturbation associated with pore formation. Evidence that domains 1 and 2 of VIPs are directly involved in pore formation came from alanine scanning mutagenesis experiments of the Vip3Af1 protein showing that mutations of amino acid residues in domain 2 compromised the insecticidal activity against *Spodoptera frugiperda* and *Agrotis segetum* (Banyuls et al. 2018). The domain 3 of Vip3Bc1 resembles domain 2 of 3D Cry toxins, which is involved in receptor binding. Domains 4 and 5 of Vip3Bc1 have similar structures to each other and their conformations bear structural similarity with domain 3 of 3D Cry toxins which has been implicated in target insect specificity (Byrne et al. 2021).

Recently, Byrne and colleagues using techniques for inserting the mutant Vip3Bc1 into liposome membranes associated with images obtained by cryo-electron tomography (cryo-EM) very elegantly proposed a model of how this toxin interacts with the plasma membrane to form a pore (Byrne et al. 2021). The interfaces between domains 1 and 2 of each of the monomers generate a conical shape to the activated tetramer, with the other domains having no direct role in the oligomerization. It is in domain 2 that lysine residues are located as trypsin cleavage sites. In the case of Vip3Bc1, K205 is the main cleavage site, but not the only one, suggesting that there is redundancy in this initial step of toxin activation. After this partial cleavage, the α-helices of domains 1 and 2 rearrange themselves so that the activated tetramer takes on an "open umbrella" shape, with an extended helical bundle stalk protruding with a tip ready to be inserted in the membrane environment. This stalk can be visualized penetrating the membrane of liposomes in a video posted as supplementary material of this article (Byrne et al. 2021). The insertion of the stalk in the membrane forms a ~ 5 Å diameter pore. The results combining heterologous expression, fluorescent dye release assays from unilamellar vesicles, modeling of the quaternary structure of Vip3Bc1, and use of cryo-EM to document the interaction between toxin and liposomes brought a great enrichment of information that reinforces that VIPs toxicity is related to the formation of pores in membranes.

Another recent study with a completely different experimental approach carried out by Jin et al. (2022) combined analysis of RNA-seq and proteomic data from the midgut tissues of larval *S. frugiperda* treated with sublethal doses of Vip3Aa. The results indicated that Vip3Aa could induce the caspase-mediated apoptosis pathway, with the involvement of the MAPK signaling and endocytosis pathways, while simultaneously the hormone biosynthesis pathways were significantly downregulated in the midgut cells. These data add further support to previous results that point out that receptor-mediated endocytosis followed by apoptosis via MAPK signaling of insect midgut epithelial cells is an important outcome following the interaction of VIPs with midgut cell membrane components (Jin et al. 2022; Jiang et al. 2018). For some authors, these cellular responses represent a way of developing tolerance to these toxins by the midgut cells (Guo et al. 2020; Jiang et al. 2018; Los et al. 2011). However, the high concentrations of VIPs used in spray formulations and in transgenic plants exacerbate these responses combined with the disruption of membranes by pore formation, leading to the high toxicity of these Bt toxins.

Despite convincing data that VIPs form pores in biological membranes and can trigger midgut cell apoptosis, the identity of receptors for these toxins is still a matter of intensive research (Jin et al. 2022; Byrne et al. 2021; Jurat-Fuentes et al. 2021; Jiang et al. 2018; Singh et al. 2010). Literature data indicate that Cry toxins and VIPs do not share the same receptors (Byrne et al. 2021; Chen et al. 2017; Singh et al. 2010). This is good news, as it reinforces the idea of pyramiding different types of Bt toxins to make it difficult to select cross-resistance among insect pests and vectors. The emergence of some resistant races of Lepidoptera and Coleoptera lit the yellow light on the need for diversification in the use of different Bt toxins, which should boost research on VIPs, as happened before with 3D Cry toxins.

15.2.3 Practical Resistance to Bt Toxins

The success of Bt crop cultivation witnessed in the last decades is now threatened by the field-evolved resistance to various types of Cry toxins (Heckel 2021; Jurat-Fuentes et al. 2021). To date, 21 resistant races of insect pests to Cry toxins are recognized in 6 species of Lepidoptera and 2 species of Coleoptera (Ma et al. 2022; Heckel 2021; Jurat-Fuentes et al. 2021; Tabashnik and Carrière 2019). All eight commercial strains of Cry toxins have at least one case of documented practical resistance. In most cases of Cry toxin resistance, impairment of receptor expression on membranes due to mutations is involved in some way. Early studies on field-evolved resistant insect strains and insect laboratory colonies helped in understanding the mechanisms of action of Bt toxins.

Efforts to understand the mechanisms of resistance have shown the central role that ABC proteins and cadherins play in the early stages of interaction with Cry toxins to trigger pore formation (Heckel 2021; Jurat-Fuentes et al. 2021; Zhang et al. 2021a; Guo et al. 2020). It has been observed that mutations in the genes encoding the ABC proteins and cadherins are often correlated with high-level resistance to Cry toxins (Zhang et al. 2021a), but mutations in the genes encoding cadherins, APNs, and ALPs are also correlated with resistance (Ma et al. 2022; Jurat-Fuentes et al. 2021; Tabashnik et al. 2013). For example, in the lepidopteran *Trichoplusia ni* (the cabbage looper), the resistance to Cry1Ac is due to a mutation in the ABC protein-encoding gene (*ABCC2*) and an additional mutation or more mutations related to the genes encoding two aminopeptidases (*APN1* and *APN6*) (Ma et al. 2022).

The resistance can also be correlated to altered expression of receptors in response to constitutive expression of mitogen-activated protein kinases (MAPKs). In the diamondback moth *P. xylostella*, the expression levels of ABCC2, ABCC3, and ALP were found to be under control of an MAPK signaling pathway (Guo et al. 2015). The resistance to Cry1Ac in this lepidopteran species was caused by the downregulation of these receptors in the microvillar membranes. The constitutive expression of the MAP4K4 resulted in low levels of ABCC2, ABCC3, and ALP in midgut cell membranes and practical resistance. As expected, gene silencing of MAPK4 using RNAi resulted in the restoration of the moth's susceptibility to the Cry1Ac toxin (Guo et al. 2015, 2020).

The pressure exerted by the emergence of resistant strains led to an intense search for modification in the structure of already engineered Bt toxins and an increase in attempts to pyramid different Bt toxins, such as the association of Cry and VIP toxins in the same transgenic plant (Eghrari et al. 2022; Jurat-Fuentes et al. 2021; Chen et al. 2017). In addition to these measures, precautions such as maintaining areas with unmodified plants to shelter susceptible insects and increasing doses applied or expressed in plants have been also tested (Jurat-Fuentes et al. 2021).

15.3 Digestive Enzyme Inhibitors

15.3.1 Proteinase, α-Amylase, and Polygalacturonase Inhibitors

As discussed in Chap. 2, digestion in insects can be divided into three phases: Initial, intermediate, and final digestion. The enzymes that act in the first stage of digestion, the so-called polymer hydrolases, are the preferred targets of plant proteins that act as inhibitors of these enzymes. In plants, it is much more common to find inhibitors of endopeptidases, α-amylases, and polygalacturonases than inhibitors of enzymes involved in the other phases of digestion. This observation illustrates a tendency in nature to inhibit important biological processes at an early stage. Thus, the strategy of inhibiting the initial phase of polysaccharide and protein digestion is widespread among plant defenses against herbivores.

The extensive use of proteinaceous inhibitors also allows plants to adapt to changes in target enzyme specificity by selecting for mutated forms of the inhibitor encoding gene. The result throughout evolution can be witnessed as a dynamic adaptation of insects to inhibitors found in plant tissues and, on the other hand, the expression of inhibitors with different specificities conferring an adaptive advantage to plants, allowing them to overcome the pressure of herbivory from insects and other animals. Among the proteins of plant origin with inhibitory activity, endopeptidase inhibitors, α-amylase inhibitors, and polygalacturonase inhibitors stand out. And among these inhibitors, serine endopeptidase inhibitors are the most common in plant tissues. This can be understood by the fact that serine endopeptidases, mainly trypsin and chymotrypsin, are major digestive endopeptidases of most insect species and herbivorous animals in general. The inhibition of protein digestion may represent an important anti-nutritional effect. Among the carbohydrates of plants, starch and hemicelluloses are the most used as nutrients by phytophagous insects. This explains the occurrence of α-amylase-inhibiting and polygalacturonase-inhibiting proteins as effective defense molecules against some insect species (Chiu et al. 2021; Zhang et al. 2021b; Chotechung et al. 2016).

15.3.1.1 Plant-Derived Endopeptidase Inhibitors (PIs)

The search for alternative biopesticides to Bt has concentrated on those derived from plants. Since the early 1970s, it has been shown that proteins capable of reducing the activity of endopeptidases were not only constitutively present in plant tissues, but the genes that express these proteins that inhibit proteolytic activity were overexpressed in response to mechanical damage or bites from herbivorous insects. The pioneering works of Dr. Clarence Ryan at Washington State University showed that tomato and potato plants accumulated large amounts of trypsin inhibitors when exposed to attack by the Colorado potato beetle, *Leptinotarsa decemlineata* (Coleoptera: Chrysomelidae) (Divekar et al. 2022b; Ryan 1990; Green and Ryan

1972). Since then, several studies have shown that chronic exposure of insects to plant-derived endopeptidases inhibitors, such as SKTI (Soybean Kunitz Trypsin Inhibitor) and SBBI (Soybean Bowman-Birk Inhibitor), affect the development of herbivorous insects (Divekar et al. 2022a; Singh et al. 2020; Dang and Van Damme 2015; Silva and Samuels 2011). Plant-defensive endopeptidase inhibitors (PIs) are diverse and can be classified according to phylogenetic relationships based on amino acid sequence, being divided into 82 families, and subdivided into 39 clans (Rawlings et al. 2018). Traditionally, these inhibitors can also be classified according to the type of proteinase inhibited into four categories: aspartic, cysteine, serine, and metalloendopeptidase inhibitors (Silva and Samuels 2011; Laskowski and Qasim 2000). The pioneering work of Hilder et al. (1987) showed that transformation of tobacco plants expressing the cowpea (*Vigna unguiculata*) trypsin inhibitor enhanced resistance to the lepidopteran *Heliothis virescens*. However, despite other examples of successful transformations and substantial progress in the understanding of the defensive potential of PIs (Haq et al. 2004; Hilder and Boulter 1999), transgenic plants expressing endopeptidase inhibitors have not established themselves as viable insect-resistant crops under field conditions and on a large scale (Divekar et al. 2022a; Silva and Samuels 2011).

The most common types of PIs are those that inhibit serine and cysteine endopeptidases, which are the most common proteolytic enzymes found in phytophagous insects (Divekar et al. 2022a, b; Sardoy et al. 2021; Singh et al. 2020). The competitive inhibition of endopeptidases responsible for initial digestion of dietary proteins and degradation of even more toxic plant proteins compromise the insect's digestive process affecting its growth and development (Elden 2000; Dang and Van Damme 2015; Ryan 1990). It has been documented that ingestion of PIs leads to overexpression of PI-sensitive and PI-insensitive endopeptidases, which can lead to extra energy expenditure and impaired development of the insect (Zhu-Salzman and Zeng 2015; War et al. 2012; Broadway 1997; Jongsma and Bolter 1997). The importance of PIs as defensive macromolecules can be illustrated by the following observations: a) PIs are found in some plant organs in high concentrations; b) PIs can be induced in response to insect attack; c) herbivorous insects can modify the expression of insensitive endopeptidase in response to ingestion of plant PIs; d) the higher abundance of serine and cysteine endopeptidase inhibitor parallels the major occurrence of these enzymes in insects; e) transgenic plants that express PIs become resistant to some insect species (Divekar et al. 2022a, 2022b; Silva and Samuels 2011; Hilder et al. 1987). These adverse effects were attributed to the inhibition of digestive enzymes responsible for digestion of dietary proteins, which supposedly leads to compromised supply of amino acids for insect metabolism. In fact, the reactive sites of PIs interact with the active site of endopeptidases, often forming stable complexes and greatly decreasing enzymatic activity, resulting in a deficiency of free amino acids and, consequently, delaying metamorphosis, reducing the survival of immature forms and the adult fertility. PIs tend to have effects against a narrow spectrum of pests, particularly those that inhibit serine or cysteine endopeptidases, such as cystatins. As seen in Chap. 3, plant tissues generally have a suboptimal protein content. Thus, nitrogen often becomes a limiting factor in the nutrition of many herbivorous insects. Therefore, efficient digestion of dietary proteins to obtain essential amino acids is crucial for insect pest survival, and its inhibition is a common defense strategy observed in plants. However, the effects of PIs can be more complex than just reducing protein digestion. It has been shown that mechanisms of tolerance and adaptation to PIs lead to the hyperproduction of endopeptidases to compensate for the activity of inhibited enzymes as mentioned above and inhibition of endopeptidases can decrease the inactivation of other toxic dietary proteins. A common feature of many PIs is that their effects tend to be chronic rather than acute, as seen with synthetic insecticides. Thus, the effects observed on a population of insect pests are much less slow than in the case of synthetic insecticides. PIs rarely produce 100% insect mortality, tending to increase mortality to a limited extent, but still being able to significantly delay insect pest development and population growth. It is important to note that, even without producing an acute effect, the impairment in the development of the insect can keep its population below the intervention threshold, facilitating integrated management due to a longer time window for the intervention of the farmer.

Due to the high ability of insects to adapt to PIs and the long period of exposure necessary to cause effective control rates, the use of these molecules in agriculture has not been effective so far. In this sense, even efforts to express exogenous PIs constitutively (e.g., as is done with Bt toxins) in plants have not been effective since the initial attempt in 1987. Despite the ability of insects to adapt to ingestion of PIs, these macromolecules are undoubtedly an efficient defense mechanism against insects and, therefore, a valuable target for the development of new low environmental impact insecticides. In fact, the mechanisms of resistance to PIs in herbivorous insects are not yet fully understood and, therefore, PIs with characteristics that may prevent the insect from overcoming the effect of these resistance and tolerance mechanisms can generate promising results. So far, much of the work testing the insecticidal effects of PIs has used purified molecules and genes cloned from plants. It is believed that the use of inhibitors from other sources, such as animals, may also be effective since herbivorous insects were not under selection pressure to adapt to such inhibitors. Another alternative would be the transformation of plants that express different insecticidal proteins directed to the midgut, such as the combination of the expression of PIs and Bt toxins (Mesén-Porras et al. 2020).

15.3.1.2 α-Amylase Inhibitors

Another important class of plant defense proteins that act on insect digestive enzymes are α-amylase inhibitors (αAIs). The effects caused by the ingestion of αAIs are attributed to the impairment of the digestion of starch granules by the insect (Franco et al. 2002; Morton et al. 2000; Valencia et al. 2000; Chrispeels et al. 1998). Since the supply of glucose as the main water-soluble fuel is diminished by inhibiting the initial digestion of starch, animal growth is reduced. It has been demonstrated that proteinaceous α-amylase inhibitors isolated or purified from cereals or different legume species affected various developmental parameters of plant-consuming insects, mainly beetles, such as lower female fecundity, higher larval mortality, and longer mean development period.

There are many types of αAIs purified from plant tissues, mainly from seeds, but αAIs produced by microorganisms are also known. They are classified as proteinaceous and non-proteinaceous. In the case of non-proteinaceous inhibitors, they are mainly low-molecular-weight organic structural analogs of carbohydrates or cyclic oligosaccharides that can act as competitive or non-competitive inhibitors of α-amylases. Within this class of inhibitors are compounds such as acarbose, acarviosine-glucose, isoacarbose, and cyclodextrins (Del Valle 2004; Franco et al. 2002). The proteinaceous α-amylase inhibitors can be classified according to their tertiary structure into different classes that include six families purified from plants and two identified in bacteria, such as helianthamide from *Stichodactyla helianthus* and tendamistat from *Streptomyces tendae* (Franco et al. 2002). The six plant families of αAIs include lectin-like, knottins, cereal bifunctional α-amylase/trypsin inhibitors (ATIs), Kunitz-like, γ-purothionine, and thaumatin (Franco et al. 2002; Richardson 1991). These classes of αAIs have structural differences and have different specificities in relation to the type of targeted α-amylase. Some of these inhibitors are capable of inhibiting insect and mammalian α-amylases, while others are active only against insect or vertebrate amylases. Interestingly, αAIs of certain classes can be bifunctional and act both on α-amylases of glycoside hydrolase family 13 (GH13) and serine endopeptidases (Drula et al. 2022).

The transgenic plant approach by using αAI genes provides another option in addition to Bt toxins and PIs as an eco-friendly alternative to synthetic insecticides and could be adopted for Integrated Pest Management (IPM). The most studied and promising αAIs for use in insect pest control or for IPM are the lectin-type inhibitors, purified from legume seeds, and the bifunctional inhibitors found in cereals. Of the lectin-type inhibitors, αAIs purified from common bean, *Phaseolus vulgaris* called αAI-1 and αAI-2, have already been used in the transformation of peas (*Pisum sativum*), including field trials (Ludvíková and Griga 2022; de Sousa-majer et al. 2007; Morton et al. 2000; Shade et al. 1994). These studies showed excellent results in relation to larval mortality and in relation to the delay in larval development. Interestingly, the results were more pronounced with pea transformed with the αAI-1 gene compared to transformation with αAI-2. These results can be attributed to differences in inhibitor specificity. It is known that αAI-2 is more effective against α-amylases from the bean-beetle *Zabrotes subfasciatus*, while αAI-1 is more effective at inhibiting amylases from Old World bruchids such as the pea-beetle *Bruchus pisorum*.

Despite the negative effects on the population of *B. pisorum* and other Old World bruchids such as *Callosobruchus maculatus* and *Callosobruchus chinensis*, the economic viability of transgenic plants expressing αAI-1 is still debatable. The bruchid-resistant transgenic pea expressing AI is the one closest to the commercialization stage (Ludvíková and Griga 2022). However, the range of resistance is limited, as some seed-beetles can adapt to the ingestion of these inhibitors. In a co-evolved bruchid species that normally develops in *P. vulgaris* grains, such as *Z. subfasciatus*, there is an overexpression of induced α-amylases that are insensitive to αAI-1 (Silva et al. 2001a, b). This phenomenon is like the adaptation of insects to endopeptidase inhibitors, where there is a modulation in the expression of enzymes involved in the initial phase of digestion, whether proteins or starch. As with the induction of endopeptidases, the molecular details of how the expression of α-amylases is modulated are still unknown and deserve further investigation.

15.3.1.3 Polygalacturonase-Inhibiting Proteins (PGIPs)

The enzymes involved in the initial phase of protein and starch digestion, respectively, endopeptidases and α-amylases, are major digestive enzymes and have been extensively studied in insects. However, in the last decades, genes that express enzymes active in plant cell wall degradation have been found in the genomes, intestinal transcriptomes, and intestinal proteomes of phytophagous insects belonging to the orders Coleoptera and Phasmatodea (Kirsch et al. 2014, 2016; Shelomi et al. 2016). These digestive enzymes are known as plant cell wall-degrading enzymes (PCWDEs) and are involved in the digestion of plant cell walls including carbohydrate esterases, glycoside hydrolases, and polysaccharide lyases (Kirsch et al. 2014; Calderón-Cortés et al. 2012; see also Chap. 10). These enzymes together are necessary for the complete or partial degradation of polysaccharides such as cellulose, hemicelluloses, and pectins (Calderón-Cortés et al. 2012). Among the PCWDEs, the polygalacturonases of the GH28 family stand out for the gene expansion observed in several beetle species belonging to the Polyphaga clade, which harbors the group of herbivorous beetles with the highest adaptive radiation, resulting in the largest clade of animals found on Earth (Kirsch et al. 2016). The action of this type of enzyme is crucial in the digestion of the plant cell wall by cleaving the bonds between galacturonic acid residues that make up pectin, an abundant polymer that coats cellulose and hemicellulose fibers (see Chap. 3).

The enzymatic properties and amino acid sequence comparisons of the GH28 family polygalacturonases found in Polyphaga revealed that these genes were incorporated into the genome of these insects because of Horizontal Gene Transfer (THG) events originating from fungal and/or bacterial-like microorganisms (Kirsch et al. 2014). After THG, there is a lot of evidence that these enzymes underwent gene duplication events followed by sub-functionalization and neofunctionalization (Tokuda 2019). In contrast, plants express genes encoding polygalacturonase inhibitor proteins, abbreviated here as PGIPs. PGIPs are highly conserved glycoproteins that belong to the LRR (leucine-rich repeat) protein superfamily and can selectively bind and inhibit the activity of polygalacturonases from bacteria, fungi, nematodes, and insects (Zhang et al. 2021b; Juge 2006). These PGIPs can inhibit the pectin degradation of the plant cell wall, inhibiting the growth of fungi, bacteria, phytophagous nematodes, and the digestion of herbivorous insects that express digestive endo-polygalacturonases. Although less studied, the inhibition of polygalacturonases can also be attributed to a decrease in the insect's ability to digest the plant cell wall or gain access to the interior of starch granules, which can result in less access to nutrients present in plant cells (Rathinam et al. 2020; Chotechung et al.

2016). The expression of PGIPs is common in higher plants, and it is common to find multigene families of these inhibitors (Zhang et al. 2021b; Kirsch et al. 2020).

There is no doubt that polygalacturonases are digestive enzymes secreted into the lumen of the midgut, just as there is evidence that PGIPs ingested in the diet can interact with them as they pass with the bolus through the digestive tract of insects. These observations reinforce the defensive role exerted by PGIPs and that these proteins can maintain their inhibitory activity under conditions prevailing in the midgut lumen. The polygacturonases expressed in Polyphaga beetles are part of multigenic families, and it is common to find numbers above 10 isoforms. Interestingly, part of the secreted polygalacturonases does not have enzymatic activity, many being products of pseudogene expression. These inactive forms are believed to have a higher affinity for some PGIPs, resulting in a sort of scavengers of these inhibitors and thus allowing active enzymes to be less inhibited resulting in the degradation of dietary pectin (Kirsch et al. 2019).

Evidences that PGIPs are part of the plant defense proteins arsenal are similar to those previously mentioned for PIs, such as (a) genes encoding these inhibitors are overexpressed when plants are attacked by fungi, bacteria, nematodes, or insects; (b) the incorporation of these proteins in the insect diet leads to antibiosis, and (c) gene loci of PGIPs that correlate with seed resistance to pathogens and seed-beetles attacks have already been found (Gamage et al. 2022; Zhang et al. 2021b; Kirsch et al. 2020; Kaewwongwal et al. 2017; Chotechung et al. 2016).

In contrast to plant pathogens such as bacteria and fungi, comparably few studies are available about the impact of PGIPs in crop protection against insects. Several studies report the expression of PGIPs in model and nonmodeled plants, such as *Arabidopsis thaliana*, tobacco, and cabbage, and how the transformation enhanced their resistance against bacterial and fungal infections (Chiu et al. 2021; Hwang et al. 2010; Manfredini et al. 2005). Nevertheless, different studies have shown that PGIPs are not only effective against bacterial and fungal attack, but they are also involved in plant defense against the herbivory exerted by Phytophaga beetles, which express digestive polygalacturonases, and some herbivorous Hemiptera that inject these enzymes in plant tissues (Haeger et al. 2020; Kirsch et al. 2020). Recent studies have demonstrated that a single PGIP can negatively compromise pectin degradation in herbivorous insects. For example, Haeger et al. (2021) have demonstrated that a single PGIP is sufficient to negatively affect pectin digestion and larval growth of the leaf beetle *Phaedon cochleariae* (Coleoptera: Chrysomelidae) feeding on transformed Chinese cabbage (*Brassica rapa* ssp. *pekinensis*) (Haeger et al. 2020).

The transgenic plant strategy by using PGIP genes could provide another option in addition to Bt toxins, PIs, and αAIs as an alternative to chemical insecticides and could be used in IPM programs (Chiu et al. 2021; Haeger et al. 2020). Although there has been progress in transgenesis of PGIP in plants, future work is required to determine if this strategy will be sustainable in field conditions against insects. Some challenges remain unsolved, as, for example, if the insertion of PGIP genes into suitable plants will be stable and what effects on the microbial community will be achieved for different PGIPs. In addition, more studies are required to test if there are advantages in pyramiding PGIP and other inhibitors of digestive enzymes, such as PIs and αAIs against insect pests.

15.4 Application of RNAi and CRISPR/Cas Technologies in Insect Control

With the discovery of mechanisms of gene silencing involving RNA interference (RNAi) or genome editing through CRISPR/Cas9 technologies combined with new achievements in the sequencing of genes that encode vital proteins for insects, researchers quickly realized that these mechanisms of interference in gene expression can be a very important tool in the control and integrated management of insect pests.

15.4.1 Application of RNAi Technology in Insect Pest Control

RNA interference (RNAi) is a biological mechanism in which the internalization of double-stranded RNA (dsRNA) into eukaryotic cells triggers, in a sequence-specific manner, the degradation of the complementary messenger RNA (mRNA) leading to gene "silencing," in which only the mRNA cognate to dsRNA is specifically cleaved. Silencing of certain genes could result in insect death, leading to reduced pest populations and lower damage to crops. Therefore, this technology has emerged as a potential strategy for the IPM in agriculture. There are three main routes of dsRNA delivery into insects: injections into the hemolymph or eggs, topical application (as a sprayable formulation, for example), and oral ingestion (in transgenic plants, for example). Among the different forms of dsRNA application, the oral ingestion of dsRNA has been shown to be the most suitable for insect control (Li et al. 2022a, b). Plant-incorporated dsRNA technique combines the advantages of RNAi and resistant transgenic crops.

In this chapter, we focus on RNAi silencing through insect feeding on transgenic plants. Still as a recent technique, the use of RNAi presents certain limitations in its effective use in field conditions, such as the stability of the dsRNA both in the topical application mode and during its passage through the gut, the unwanted effects on non-target insect species, etc. Although there are still some challenges of RNAi technology in field conditions, intense research is being carried out to address these issues. Overall, RNAi technology holds great promise for the future of a sustainable agriculture and environmentally friendly crop production systems. Table 15.1 illustrates some of the most successful cases involving transgenic plants expressing dsRNA or oral ingestion of dsRNA targeting midgut proteins. In some cases, dsRNA ingestion led to worse performance in insect development, while in other cases gene silencing was used to show proof of concept, as in cases where inhibition of trypsin or aminopeptidase N expression led to a decrease in the action of Bt toxins.

Table 15.1 Experiments using RNAi technology to silence genes related to digestion in insect pests

Target insect	Target gene product	Form of dsRNA delivery	References
Bemisia tabaci	α-Glucosidase	dsRNA-transformed *Nicotiana tabacum*	Raza et al. (2016)
Lipaphis erysimi	α-Glucosidase	dsRNA-transformed *Brassica juncea*	Dhatwalia et al. (2022)
Diaphorina citri	α-Glucosidase	Oral delivery	Santos-Ortega and Killiny (2018)
D. citri	Cathepsin D	Oral delivery	Galdeano et al. (2017)
Myzus persicae	Cathepsin L	dsRNA-transformed *N. tabacum*	Rauf et al. (2019)
Myzus persicae	Serine peptidase	dsRNA-transformed *Arabidopsis thaliana*	Bhatia et al. (2012)
Nilaparvata lugens	Trypsin Carboxipeptidase Hexose transporter	dsRNA-transformed *Oryza sativa*	Zha et al. (2011)
Helicoverpa armigera	Trypsin	Oral delivery	Jayachandran et al. (2013)
Spodoptera litura	Aminopeptidase N	Injection in hemolymph	Rajagopal et al. (2002)
Chilo suppressalis	Aminopeptidase N	dsRNA-transformed *O. sativa*	Qiu et al. (2017)
Coptotermes formosanus	Endoglucanase	Injection and oral delivery	Wu et al. (2019)
Cryptolestes ferrugineus	Chitin synthase 2	Injection and feeding of chitosan/dsRNA nanoparticles	Zhang et al. (2020)

15.4.2 Application of CRISPR/Caspase Technology in Insect Pest Control

The use of gene silencing technology with the use of RNAi has some limitations, like all pest control and management techniques. One of the limitations of using RNAi in crop plants is that some insects are naturally less susceptible to gene silencing with the ingestion of dsRNA, such as insects of the orders Lepidoptera, Diptera, and Hemiptera, and variability in success and efficiency has been reported (Li et al. 2022a, b). RNAi silencing is sometimes transient and insect species have been observed with RNAi resistance. CRISPR/Cas gene editing technology has some advantages over the use of RNAi. CRISPR/Cas is the acronym for Clustered Regularly Interspaced Short Palindromic Repeats, and Cas stands for CRISPR-associated system (consists of nucleases called caspases). It is a technology that targets a specific DNA section and makes a precise cut at the target site, silencing the gene or replacing one version of the gene with another.

Table 15.2 Engineered insects using CRISPR/Cas technology to silence genes expressed in midgut tissues

Target insect	Target gene product	Form of dsRNA delivery	References
Chilo suppressalis	Chitin synthase 2	Microinjection	Zeng et al. (2022)
Helicoverpa armigera	Cadherin	Microinjection in eggs	Wang et al. (2016)
H. armigera	ABC transporter	Microinjection in eggs	Wang et al. (2017)
H. armigera	Cholesterol transporter	Microinjection in eggs	Zheng et al. (2020)
Spodoptera frugiperda	ABC transporter	Microinjection in eggs	Jin et al. 2021 and Li et al. (Li et al. 2022a, 2022b)
Mythimna separata	Sterol transporter	Microinjection in eggs	Tang et al. (2022)

The CRISPR-Cas mechanism relies on the nuclease activity of caspases (usually one called caspase 9 or Cas9 or Cas12) and a guide sequence (gRNA). The gRNA guides the Cas9 nuclease to a target site to cleave the DNA. The key feature of gRNA is its complementarity with the target sequence. The primary difference between CRISPR/Cas and RNAi technologies is that CRISPR/Cas results in complete loss-of-function effects, while RNAi generally causes a partial loss of gene function. For some authors, CRISPR/Cas generates higher reproducibility and consistent phenotypes in comparison to RNAi (Moon et al. 2022; Tyagi et al. 2020). Crop plants can be modified using CRISPR/Cas technology so that they express or do not express a particular protein or a substance whose synthesis is dependent on a primary gene product. Alternatively, a modified crop plant can deter insect pests from encountering its exposed organs or can attract insect predators of the insect that is feeding on the plant (Moon et al. 2022). Table 15.2 illustrates some of the experiments involving gene edition targeting midgut proteins.

15.5 Conclusions and Prospects

Although there has been progress in cloning proteinaceous inhibitor genes and in obtaining successful plant transformation with some of them, future work is still required to determine how effective is the integration of these heterologous inhibitor genes into a plant genome to improve plant insect resistance in field conditions. In addition to the public concern in several countries about the use of transgenic crops, some challenges remain unsolved and further research is needed to develop heterologous inhibitors into practical use. As PGIPs are effective against bacteria and fungi, the impact of these inhibitors toward these organisms should be evaluated. For example, we do not know if the expression of heterologous PGIP will affect beneficial fungi, such as those that play an essential role in the mycorrhizal plant-fungi interactions. We do not know yet how the expression of plant digestive inhibitors will affect unwanted targets or if these heterologous inhibitors can negatively impact the environment. However, some undesired effects can still be observed, such as the high allergenic potential that some αAIs have, for example wheat inhibitors, which can make the use of these αAIs in plant resistance programs unfeasible.

Extensive research has still to be conducted into the inhibitors of insect digestive enzymes, with the specificity and effectiveness of various inhibitors being necessary to be studied in more depth. It has become necessary to build a database of inhibitors and their genes to manage the discovery of novel inhibitors that are effective against insects and that are stable under field conditions, such as the Bt toxins are. Moreover, with the development of sequence analysis and bioinformatic tools, the specificity of different classes of inhibitors has been solved and this can be of significant importance for the artificial screening and obtaining either transgenic plants or formulations for application to the aerial parts of plants.

In summary, engineering crop plants to express Bt toxins, inhibitors of insect digestion or gene editing approaches such as RNAi and CRISPR/Cas technologies, have immense potential for the integrated management of insect pests. The gene editing-based approaches are more specific and environmentally friendly than traditional chemical-based methods, making them an attractive alternative for sustainable pest control in agriculture. The development of transgenic plants for the management of insect pests based on the novel techniques discussed in this chapter represents an important area of innovation in agriculture targeting the development of sustainable and eco-friendly crop production systems.

References

Acheuk F, Basiouni S, Shehata AA et al (2022) Status and prospects of botanical biopesticides in Europe and Mediterranean countries. Biomol Ther 12:311. https://doi.org/10.3390/biom12020311

Adang MJ, Crickmore N, Jurat-Fuentes JL (2014) Diversity of *Bacillus thuringiensis* crystal toxins and mechanism of action. Adv Insect Physiol 47:39–87. https://doi.org/10.1016/B978-0-12-800197-4.00002-6

Banyuls N, Hernández-Rodríguez CS, Van Rie J, Ferré J (2018) Critical amino acids for the insecticidal activity of Vip3Af from *Bacillus thuringiensis*: inference on structural aspects. Sci Rep 8:1–4. https://doi.org/10.1038/s41598-018-25346-3

Bhatia V, Bhattacharya R, Uniyal PL et al (2012) Host generated siRNAs attenuate expression of serine protease gene in *Myzus persicae*. PLoS One 12:e46343. https://doi.org/10.1371/journal.pone.0046343

Bretschneider A, Heckel DG, Pauchet Y (2016) Three toxins, two receptors, one mechanism: mode of action of Cry1A toxins from *Bacillus thuringiensis* in *Heliothis virescens*. Insect Biochem Mol Biol 76:109–117. https://doi.org/10.1016/j.ibmb.2016.07.008

Broadway RM (1997) Dietary regulation of serine proteinases that are resistant to serine proteinase inhibitors. J Insect Physiol 43:855–874

Byrne MJ, Iadanza MG, Perez MA et al (2021) Cryo-EM structures of an insecticidal Bt toxin reveal its mechanism of action on the membrane. Nat Commun 12:2791. https://doi.org/10.1038/s41467-021-23146-4

Calderón-Cortés N, Quesada M, Watanabe H et al (2012) Endogenous plant cell wall digestion: a key mechanism in insect evolution. Annu Rev Ecol Evol Syst 43:45–71. https://doi.org/10.1146/annurev-ecolsys-110411-160312

Chen WB, Lu GQ, Cheng HM et al (2017) Transgenic cotton coexpressing Vip3A and Cry1Ac has a broad insecticidal spectrum against lepidopteran pests. J Invertebr Pathol 149:59–65. https://doi.org/10.1016/j.jip.2017.08.001

Chiu T, Behari A, Chartron JW, Putman A, Li Y (2021) Exploring the potential of engineering polygalacturonase-inhibiting protein as an ecological, friendly, and nontoxic pest control agent. Biotechnol Bioeng 118:3200–3214. https://doi.org/10.1002/bit.27845

Chotechung S, Somta P, Chen J et al (2016) A gene encoding a polygalacturonase-inhibiting protein (PGIP) is a candidate gene for bruchid (Coleoptera: Bruchidae) resistance in mungbean (*Vigna radiata*). Theor Appl Genet 129:1673–1683. https://doi.org/10.1007/s00122-016-2731-1

Chrispeels MJ, de Sa MF, Higgins TJ (1998) Genetic engineering with α-amylase inhibitors makes seeds resistant to bruchids. Seed Sci Res 8:257–264

Crickmore N, Berry C, Panneerselvam S et al (2021) A structure-based nomenclature for *Bacillus thuringiensis* and other bacteria-derived pesticidal proteins. J Invertebr Pathol 186:107438. https://doi.org/10.1016/j.jip.2020.107438

Dang LY, Van Damme EJM (2015) Toxic proteins in plants. Phytochemistry 117:51–64. https://doi.org/10.1016/j.phytochem.2015.05.020

de Sousa-majer M, Hardie DC, Turner NC, Higgins TJ (2007) Bean alpha amylase inhibitor in transgenic peas inhibit the development of pea weevil larvae. J Econ Entomol 100:1416–1422. https://doi.org/10.1093/jee/100.4.1416

Del Valle EMM (2004) Cyclodextrins and their uses: a review. Process Biochem 39:1033–1046. https://doi.org/10.1016/S0032-9592(03)00258-9

Dhatwalia D, Aminedi R, Kalia V et al (2022) Host-mediated attenuation of gut sucrase in mustard aphid *Lipaphis erysimi* impaired its parthenogenetic reproduction on Indian mustard *Brassica juncea*. Pest Manag Sci 78:803–811. https://doi.org/10.1002/ps.6694

Divekar P, Majumder S, Halder J et al (2022a) Chapter 8. Trilogy: plants-proteases-insects. In: Banik S, Halder J (eds) Plant protection: present developments and future strategies. Today & Tomorrow's Printers and Publishers, New Delhi

Divekar PA, Rani V, Majumder S et al (2022b) Protease inhibitors: an induced plant defense mechanism against herbivores. J Plant Growth Regul 28:1–7. https://doi.org/10.1007/s00344-022-10767-2

Drula E, Garron ML, Dogan S et al (2022) The carbohydrate-active enzyme database: functions and literature. Nucleic Acids Res 7:D571–D577. https://doi.org/10.1093/nar/gkab1045

Eghrari K, Oliveira SC, Nascimento AM et al (2022) The implications of homozygous vip3Aa20- and cry1Ab-maize on *Spodoptera frugiperda* control. J Pest Sci 95:115–127. https://doi.org/10.1007/s10340-021-01362-7

Elden TC (2000) Influence of a cysteine proteinase inhibitor on alfalfa weevil (Coleoptera: Curculionidae) growth and development over successive generations. J Entomol Sci 35:70–76. https://doi.org/10.18474/0749-8004-35.1.70

Estruch JJ, Warren GW, Mullins MA et al (1996) Vip3A, a novel *Bacillus thuringiensis* vegetative insecticidal protein with a wide spectrum of activities against lepidopteran insects. Proc Natl Acad Sci U S A 93:5389–5394

Franco OL, Rigden DJ, Melo FR, Grossi-de-Sá MF (2002) Plant α-amylase inhibitors and their interaction with insect α-amylases: structure, function and potential for crop protection. Eur J Biochem 269:397–412. https://doi.org/10.1046/j.0014-2956.2001.02656.x

Galdeano DM, Breton MC, Lopes JR et al (2017) Oral delivery of double-stranded RNAs induces mortality in nymphs and adults of the Asian citrus psyllid, *Diaphorina citri*. PLoS One 12:e0171847. https://doi.org/10.1371/journal.pone.0171847

Gamage SI, Kaewwongwal A, Laosatit K et al (2022) Tandemly duplicated genes encoding polygalacturonase inhibitors are associated with bruchid (*Callosobruchus chinensis*) resistance in moth bean (*Vigna aconitifolia*). Plant Sci 323:111402. https://doi.org/10.1016/j.plantsci.2022.111402

Gomis-Cebolla J, de Escudero IR, Vera-Velasco NM et al (2017) Insecticidal spectrum and mode of action of the *Bacillus thuringiensis* Vip3Ca insecticidal protein. J Invertebr Pathol 142:60–67. https://doi.org/10.1016/j.jip.2016.10.001

Green TR, Ryan CA (1972) Wound-induced proteinase inhibitors in plant leaves: a possible defense mechanism against insects. Science 175:776–777

Guo Z, Kang S, Chen D et al (2015) MAPK signaling pathway alters expression of midgut ALP and ABCC genes and causes resistance to *Bacillus thuringiensis* Cry1Ac toxin in diamondback moth. PLoS Genet 11:e1005124. https://doi.org/10.1371/journal.pgen.1005124

Guo Z, Kang S, Sun D et al (2020) MAPK-dependent hormonal signaling plasticity contributes to overcoming *Bacillus thuringiensis* toxin action in an insect host. Nat Commun 11:1–4. https://doi.org/10.1038/s41467-020-16608-8

Haeger W, Henning J, Heckel DG et al (2020) Direct evidence for a new mode of plant defense against insects via a novel polygalacturonase-inhibiting protein expression strategy. J Biol Chem 295:11833–11844. https://doi.org/10.1074/jbc.RA120.014027

Heckel DG (2021) The essential and enigmatic role of ABC transporters in Bt resistance of noctuids and other insect pests of agriculture. Insects 12:389. https://doi.org/10.3390/insects12050389

Hilder VA, Boulter D (1999) Genetic engineering of crop plants for insect resistance - a critical review. Crop Prot 18:177–191

Hilder VA, Gatehouse AMR, Sheerman SE et al (1987) A novel mechanism of insect resistance engineered into tobacco. Nature 330:160–163

Hwang BH, Bae H, Lim HS et al (2010) Overexpression of polygalacturonase-inhibiting protein 2 (PGIP2) of Chinese cabbage (*Brassica rapa* ssp. *pekinensis*) increased resistance to the bacterial pathogen *Pectobacterium carotovorum* ssp. *carotovorum*. Plant Cell Tissue Organ Cult 103:293–305. https://doi.org/10.1007/s11240-010-9779-4

ISAAA (2019) Global status of commercialized of biotech/GM crops in 2019: biotech crops drive socio-economic development and sustainable environment in the new frontier; International Service for the Acquisition of Agri-biotech applications: ISAAA brief 55-2019, Ithaca, NY, USA

Jayachandran B, Hussain M, Asgari S (2013) An insect trypsin-like serine protease as a target of microRNA: utilization of microRNA mimics and inhibitors by oral feeding. Insect Biochem Mol Biol 43:398–406. https://doi.org/10.1016/j.ibmb.2012.10.004

Jiang K, Hou XY, Tan TT et al (2018) Scavenger receptor-C acts as a receptor for *Bacillus thuringiensis* vegetative insecticidal protein Vip3Aa and mediates the internalization of Vip3Aa via endocytosis. PLoS Pathog 14:e1007347. https://doi.org/10.1371/journal.ppat.1007347

Jin MH, Tao JH, Qi LI et al (2021) Genome editing of the SfABCC2 gene confers resistance to Cry1F toxin from *Bacillus thuringiensis* in *Spodoptera frugiperda*. J Integr Agr 20:815–820. https://doi.org/10.1016/S2095-3119(19)62772-3

Jin M, Shan Y, Peng Y et al (2022) An integrative analysis of transcriptomics and proteomics reveals novel insights into the response in the midgut of *Spodoptera frugiperda* larvae to Vip3Aa. Toxins 14:55. https://doi.org/10.3390/toxins14010055

Jongsma MA, Bolter C (1997) The adaptation of insects to plant protease inhibitors. J Insect Physiol 43:885–895

Juge N (2006) Plant protein inhibitors of cell wall degrading enzymes. Trends Plant Sci 11:359–367. https://doi.org/10.1016/j.tplants.2006.05.006

Jurat-Fuentes JL, Heckel DG, Ferré J (2021) Mechanisms of resistance to insecticidal proteins from *Bacillus thuringiensis*. Annu Rev Entomol 66:121–140. https://doi.org/10.1146/annurev-ento-052620-073348

Kaewwongwal A, Chen J, Somta P et al (2017) Novel alleles of two tightly linked genes encoding polygalacturonase-inhibiting proteins (VrPGIP1 and VrPGIP2) associated with the Br locus that confer bruchid (*Callosobruchus* spp.) resistance to mungbean (*Vigna radiata*) accession V2709. Front Plant Sci 8:1692. https://doi.org/10.3389/fpls.2017.01692

Kirsch R, Gramzow L, Theißen G et al (2014) Horizontal gene transfer and functional diversification of plant cell wall degrading PGs: key events in the evolution of herbivory in beetles. Insect Biochem Mol Biol 52:33–50. https://doi.org/10.1016/j.ibmb.2014.06.008

Kirsch R, Heckel DG, Pauchet Y (2016) How the rice weevil breaks down the pectin network: enzymatic synergism and sub-functionalization. Insect Biochem Mol Biol 71:72–82. https://doi.org/10.1016/j.ibmb.2016.02.007

Kirsch R, Kunert G, Vogel H, Pauchet Y (2019) Pectin digestion in herbivorous beetles: impact of pseudoenzymes exceeds that of their active counterparts. Front Physiol 10:685. https://doi.org/10.3389/fphys.2019.00685

Kirsch R, Vurmaz E, Schaefer C et al (2020) Plants use identical inhibitors to protect their cell wall pectin against microbes and insects. Ecol Evol 10:3814–3824. https://doi.org/10.1002/ece3.6180

Laskowski M, Qasim MA (2000) What can the structures of enzyme-inhibitor complexes tell us about the structures of enzyme substrate complexes? Biochim Biophys Acta 1477:324–337. https://doi.org/10.1016/S0167-4838(99)00284-8

Lee MK, Walters FS, Hart H et al (2003) The mode of action of the *Bacillus thuringiensis* vegetative insecticidal protein Vip3A differs from that of Cry1Ab δ-endotoxin. Appl Environ Microbiol 69:4648–4657. https://doi.org/10.1128/AEM.69.8.4648-4657.2003

Li Q, Jin M, Yu S et al (2022a) Knockout of the ABCB1 gene increases susceptibility to emamectin benzoate, beta-cypermethrin and chlorantraniliprole in *Spodoptera frugiperda*. Insects 13:137. https://doi.org/10.3390/insects13020137

Li X, Liu X, Lu W et al (2022b) Application progress of plant-mediated RNAi in pest control. Front Bioeng Biotechnol 8:1364. https://doi.org/10.3389/fbioe.2022.963026

Los FC, Kao CY, Smitham J et al (2011) RAB-5-and RAB-11-dependent vesicle-trafficking pathways are required for plasma membrane repair after attack by bacterial pore-forming toxin. Cell Host Microbe 9:147–157. https://doi.org/10.1016/j.chom.2011.01.005

Ludvíková M, Griga M (2022) Pea transformation: history, current status and challenges. Czech J Genet Plant Breed 58:127–161. https://doi.org/10.17221/24/2022-CJGPB

Ma X, Shao E, Chen W et al (2022) Bt Cry1Ac resistance in *Trichoplusia ni* is conferred by multi-gene mutations. Insect Biochem Mol Biol 140:103678. https://doi.org/10.1016/j.ibmb.2021.103678

Manfredini C, Sicilia F, Ferrari S et al (2005) Polygalacturonase-inhibiting protein 2 of *Phaseolus vulgaris* inhibits BcPG1, a polygalacturonase of *Botrytis cinerea* important for pathogenicity, and protects transgenic plants from infection. Physiol Mol Plant Pathol 67:108–115. https://doi.org/10.1016/j.pmpp.2005.10.002

Mesén-Porras EA, Dahdouh-Cabia S, Jimenez-Quiros C et al (2020) Soybean protease inhibitors increase *Bacillus thuringiensis* subs. *Israelensis* toxicity against *Hypothenemus hampei*. Agron Mesoam 1:461–478. https://doi.org/10.15517/am.v31i2.36573

Moon TT, Maliha IJ, Khan AA et al (2022) CRISPR-Cas genome editing for insect pest stress management in crop plants. Stress 2:493–514. https://doi.org/10.3390/stresses2040034

Morton RL, Schroeder HE, Bateman KS et al (2000) Bean α-amylase inhibitor 1 in transgenic peas (*Pisum sativum*) provides complete protection from pea weevil (*Bruchus pisorum*) under field conditions. Proc Natl Acad Sci U S A 97:3820–3825. https://doi.org/10.1073/pnas.070054597

Núñez-Ramírez R, Huesa J, Bel Y et al (2020) Molecular architecture and activation of the insecticidal protein Vip3Aa from *Bacillus thuringiensis*. Nat Commun 11:1–9. https://doi.org/10.1038/s41467-020-17758-5

Qiu L, Fan J, Zhang B et al (2017) RNA interference knockdown of aminopeptidase N genes decrease the susceptibility of *Chilo suppressalis* larvae to Cry1Ab/Cry1Ac and Cry1Ca-expressing transgenic rice. J Invertebr Pathol 145:9–12. https://doi.org/10.1016/j.jip.2017.03.001

Rajagopal R, Sivakumar S, Agrawal N et al (2002) Silencing of midgut aminopeptidase N of *Spodoptera litura* by double-stranded RNA establishes its role as *Bacillus thuringiensis* toxin receptor. J Biol Chem 277:46849–46851. https://doi.org/10.1074/jbc.C200523200

Rathinam M, Rao U, Sreevathsa R (2020) Novel biotechnological strategies to combat biotic stresses: Polygalacturonase inhibitor (PGIP) proteins as a promising comprehensive option. Appl Microbiol Biotechnol 104:2333–2342. https://doi.org/10.1007/s00253-020-10396-3

Rauf I, Asif M, Amin I et al (2019) Silencing cathepsin L expression reduces *Myzus persicae* protein content and the nutritional value as prey for *Coccinella septempunctata*. Insect Mol Biol 28:785–797. https://doi.org/10.1111/imb.12589

Rawlings ND, Barrett AJ, Thomas PD et al (2018) The MEROPS database of proteolytic enzymes, their substrates and inhibitors in 2017 and a comparison with peptidases in the PANTHER database. Nucleic Acids Res 46:D624–D632. https://doi.org/10.1093/nar/gkx1134

Raza A, Malik HJ, Shafiq M et al (2016) RNA interference based approach to down regulate osmoregulators of whitefly (*Bemisia tabaci*): Potential technology for the control of whitefly. PLoS One 11:e0153883. https://doi.org/10.1371/journal.pone.0153883

Richardson M (1991) Seed storage proteins: the enzyme inhibitors. In: Rogers LJ (ed) Methods in plant biochemistry, Amino Acids, vol 5. Proteins and Nucleic Acids. Academic Press, London, pp 259–305

Ryan CA (1990) Protease inhibitors in plants: genes for improving defenses against insects and pathogens. Phytopathology 28:425–449

Santos-Ortega Y, Killiny N (2018) Silencing of sucrose hydrolase causes nymph mortality and disturbs adult osmotic homeostasis in *Diaphorina citri* (Hemiptera: Liviidae). Insect Biochem Mol Biol 101:131–143. https://doi.org/10.1016/j.ibmb.2018.09.003

Sardoy P, Ilina N, Borniego L et al (2021) Proteases inhibitors-insensitive cysteine proteases allow *Nezara viridula* to feed on growing seeds of field-grown soybean. J Insect Physiol 132:104250. https://doi.org/10.1016/j.jinsphys.2021.104250

Shade RE, Schroeder HE, Pueyo JJ et al (1994) Transgenic pea seeds expressing the α-amylase inhibitor of the common bean are resistant to bruchid beetles. Bio/Technology 12:793–796

Shelomi M, Danchin EG, Heckel D et al (2016) Horizontal gene transfer of pectinases from bacteria preceded the diversification of stick and leaf insects. Sci Rep 23:1–9. https://doi.org/10.1038/srep26388

Silva CP, Samuels RI (2011) Plant defense proteins that inhibit insect peptidases. In: Ecological aspects of nitrogen metabolism in plants. Wiley, West Sussex, pp 280–307

Silva CP, Terra WR, De Sá MG et al (2001a) Induction of digestive α-amylases in larvae of *Zabrotes subfasciatus* (Coleoptera: Bruchidae) in response to ingestion of common bean α-amylase inhibitor 1. J Insect Physiol 47:1283–1290. https://doi.org/10.1016/S0022-1910(01)00115-9

Silva CP, Terra WR, Xavier-Filho J et al (2001b) Digestion of legume starch granules by larvae of *Zabrotes subfasciatus* (Coleoptera: Bruchidae) and the induction of α–amylases in response to different diets. Insect Biochem Mol Biol 31:41–50. https://doi.org/10.1016/S0965-1748(00)00103-X

Silva-Filha MH, Romão TP, Rezende TM et al (2021) Bacterial toxins active against mosquitoes: mode of action and resistance. Toxins 13:523. https://doi.org/10.3390/toxins13080523

Singh G, Sachdev B, Sharma N et al (2010) Interaction of *Bacillus thuringiensis* vegetative insecticidal protein with ribosomal S2 protein triggers larvicidal activity in *Spodoptera frugiperda*. Appl Environ Microbiol 76:7202–7209. https://doi.org/10.1128/AEM.01552-10

Singh S, Singh A, Kumar S et al (2020) Protease inhibitors: recent advancement in its usage as a potential biocontrol agent for insect pest management. Insect Sci 27:186–201. https://doi.org/10.1111/1744-7917.12641

Sun D, Zhu L, Guo L et al (2022) A versatile contribution of both aminopeptidases N and ABC transporters to Bt Cry1Ac toxicity in the diamondback moth. BMC Biol 20:1–6. https://doi.org/10.1186/s12915-022-01226-1

Syed T, Askari M, Meng Z et al (2020) Current insights on vegetative insecticidal proteins (Vip) as next generation pest killers. Toxins 12:522. https://doi.org/10.3390/toxins12080522

Tabashnik BE, Carrière Y (2019) Global patterns of resistance to Bt crops highlighting pink bollworm in the United States, China, and India. J Econ Entomol 112:2513–2523. https://doi.org/10.1093/jee/toz173

Tabashnik BE, Brévault T, Carrière Y (2013) Insect resistance to Bt crops: lessons from the first billion acres. Nat Biotechnol 31:510–521. https://doi.org/10.1038/nbt.2597

Tanaka S, Miyamoto K, Noda H et al (2013) The ATP-binding cassette transporter subfamily C member 2 in *Bombyx mori* larvae is a functional receptor for cry toxins from *Bacillus thuringiensis*. FEBS J 280:1782–1794. https://doi.org/10.1111/febs.12200

Tang R, Li S, Liang J et al (2022) Optimization of the application of the CRISPR/Cas9 system in *Mythimna separata*. Entomol Exp Appl 170:593–602. https://doi.org/10.1111/eea.13184

Tokuda G (2019) Plant cell wall degradation in insects: recent progress on endogenous enzymes revealed by multi-omics technologies. Adv Insect Physiol 57:97–136. https://doi.org/10.1016/bs.aiip.2019.08.001

Tyagi S, Kesiraju K, Saakre M et al (2020) Genome editing for resistance to insect pests: an emerging tool for crop improvement. ACS Omega 5:20674–20683. https://doi.org/10.1021/acsomega.0c01435

Valencia A, Bustillo AE, Ossa GE, Chrispeels MJ (2000) α-Amylases of the coffee berry borer (*Hypothenemus hampei*) and their inhibition by two plant amylase inhibitors. Insect Biochem Mol Biol 30:207–213. https://doi.org/10.1016/S0965-1748(99)00115-0

Wang J, Zhang H, Wang H et al (2016) Functional validation of cadherin as a receptor of Bt toxin Cry1Ac in *Helicoverpa armigera* utilizing the CRISPR/Cas9 system. Insect Biochem Mol Biol 76:11–17. https://doi.org/10.1016/j.ibmb.2016.06.008

Wang J, Wang H, Liu S et al (2017) CRISPR/Cas9 mediated genome editing of *Helicoverpa armigera* with mutations of an ABC transporter gene HaABCA2 confers resistance to *Bacillus thuringiensis* Cry2A toxins. Insect Biochem Mol Biol 87:147–153. https://doi.org/10.1016/j.ibmb.2017.07.002

War AR, Paulraj MG, Ahmad T et al (2012) Mechanisms of plant defense against insect herbivores. Plant Signal Behav 7:1306–1320. https://doi.org/10.4161/psb.21663

Wu W, Gu D, Yan S, Li Z (2019) RNA interference of endoglucanases in the formosan subterranean termite *Coptotermes formosanus* shiraki (Blattodea: Rhinotermitidae) by dsRNA injection or ingestion. J Insect Physiol 112:15–22. https://doi.org/10.1016/j.jinsphys.2018.11.007

Zeng B, Chen FR, Sun H, Liu Y, Wu SF, Bass C et al (2022) Molecular and functional analysis of chitin synthase genes in *Chilo suppressalis* (Lepidoptera: Crambidae). Insect Sci 30:1–16. https://doi.org/10.1111/1744-7917.13134.3

Zha W, Peng X, Chen R et al (2011) Knockdown of midgut genes by dsRNA-transgenic plant-mediated RNA interference in the hemipteran insect *Nilaparvata lugens*. PLoS One 6:e20504. https://doi.org/10.1371/journal.pone.0020504

Zhang M, Du MY, Wang GX et al (2020) Identification, mRNA expression, and functional analysis of chitin synthase 2 gene in the rusty grain beetle, *Cryptolestes ferrugineus*. J Stored Prod Res 87:101622. https://doi.org/10.1016/j.jspr.2020.101622

Zhang D, Jin M, Yang Y et al (2021a) Synergistic resistance of *Helicoverpa armigera* to Bt toxins linked to cadherin and ABC transporters mutations. Insect Biochem Mol Biol 137:103635. https://doi.org/10.1016/j.ibmb.2021.103635

Zhang Q, Yan Q, Yuan X et al (2021b) Two polygalacturonase-inhibiting proteins (VrPGIP) of *Vigna radiata* confer resistance to bruchids (*Callosobruchus* spp.). J Plant Physiol 258:153376. https://doi.org/10.1016/j.jplph.2021.153376

Zheng JC, Yue XR, Kuang WQ et al (2020) NPC1b as a novel target in controlling the cotton bollworm, *Helicoverpa armigera*. Pest Manag Sci 76:2233–2242. https://doi.org/10.1002/ps.5761

Zhu-Salzman K, Zeng R (2015) Insect response to plant defensive protease inhibitors. Annu Rev Entomol 60:233–252. https://doi.org/10.1146/annurev-ento-010814-020816